THE HYDRAULICS OF STEPPED CHUTES AND SPILLWAYS

To my family.

To Ya-Hui for her support and patience,

To Bernard for his love,

To Nicky with tenderness.

THE HYDRAULICS OF STEPPED CHUTES AND SPILLWAYS

HUBERT CHANSON

Reader in Environmental Fluid Mechanics and Water Engineering, Department of Civil Engineering, The University of Queensland, Brisbane, Australia.

A.A. BALKEMA PUBLISHERS / LISSE / ABINGDON / EXTON (PA) / TOKYO

Library of Congress Cataloging-in-Publication Data

Chanson, Hubert.
The hydraulics of stepped chutes and spillways / Hubert Chanson.
p. cm.
Includes bibliographical references and index.
ISBN 9058093522
1. Spillways. 2. Hydraulics. I. Title.

TC555 .C4623 2001
627'.883--dc21

2001037881

Cover design: Studio Jan de Boer, Amsterdam, The Netherlands.
Typesetting: Palm Produkties, Nieuwerkerk a/d IJssel, The Netherlands.
Printed by: Grafisch Produktiebedrijf Gorter, Steenwijk, The Netherlands.

ISBN 90 5809 352 2 (hardback)

Contents

Preface

Stepped channel design has been used for more than 3,500 years. A significant number of dams were built with overflow stepped spillways during the 19th century and early 20th century, before the design technique became outdated with the progresses in hydraulic jump stilling basin design. Recent advances in technology (e.g. RCC, polymer-coated gabion wire) have triggered a renewal of interest in the stepped design, although much expertise has been lost in the past 80 years. The steps increase significantly the rate of energy dissipation taking place along the chute and reduce the size of the required downstream energy dissipation basin. Stepped cascades are used also in water treatment plants to enhance the air-water transfer of atmospheric gases (e.g. oxygen, nitrogen) and of volatile organic components (VOC).

Research on stepped spillway hydraulics has been active between 1980 and 2000. For the period 1985-2000, the international database Science Citation Index® lists over sixteen journal papers and twenty-six discussions and closures on stepped chute hydraulics, all but two published between 1990 and 2000. A 1985 paper (Sorensen, *Jl Hyd Engrg*) was cited seventeen times during the period; two papers published in 1994 (Chanson, *Jl of Hyd Res*, No. 2 and 3) were cited fourteen and eleven times respectively[1]. The international database Global Books in Print® lists one book (Chanson, 1995). In 1998, Professor Ohtsu and Dr Yasuda organised a workshop on the hydraulic characteristics of stepped channel flows in Tokyo, attended by over seventy participants. Two years later, Professors Minor and Hager organised an international workshop on hydraulics of stepped spillways in Zürich. The workshop attracted over forty participants from Europe, North America, Iran, and Australia; the sponsorship of the American Society of Civil Engineers (ASCE), International Association for Hydraulic Engineering and Research (IAHR) and Swiss National Committee on Large Dams demonstrated the importance of the event.

This book presents the state of the art in stepped chute hydraulics. It is based upon the research expertise of the writer, his professional experience as an expert-consultant, and his experience in teaching stepped spillway hydraulics to undergraduate students, postgraduate research students and professionals since 1982 (Fig. P1.1). Results from more than forty five laboratory studies and four prototype investigations were re-analysed and compared. The book provides a new understanding of stepped channel hydraulics, and it is aimed at both the research and professional community.

In the introduction (Chapter 1), the basic concepts of stepped channels and stepped chute flows are described. A clear distinction is made between the

[1]Ref.: Science Citation Index®.

main flow regimes; nappe flow, transition and skimming flow. The historical progress of stepped channels and spillways from Antiquity up to the present is charted in Chapter 2. The monograph then reviews the hydraulic characteristics of stepped channel flows. Three different flow regimes may take place depending upon the flow conditions and chute geometry; nappe flow regime for small discharges, transition flow and skimming flow regime. The hydraulics of each flow regime are described in Chapters 3, 4 and 5. The effects of flow aeration and air bubble entrainment are also discussed, and the gas transfer processes taking place above the stepped chute (aeration, re-oxygenation, stripping and de-nitrification) are described in Chapter 6. Later practical examples of hydraulic design are presented in Chapter 7; stepped fountains, stepped weirs, gabion stepped spillways, earth dam spillways with precast concrete blocks, roller compacted concrete (RCC) weirs, and debris dams. In Chapter 8 a critical review of accidents and failures with stepped channels is given, highlighting that the hydrodynamic forces on the step faces are much larger than on smooth chute inverts. Wave phenomena and instabilities are reviewed and discussed in a separate section within Chapter 9. In the next and final chapter, the key issues are summarised and questions answered.

At the beginning of the book, the reader will find the Table of Contents, a List of Symbols and a Glossary of technical terms and names. A detailed list of References and an Index can be found at the end of the book. Chapter 10 is followed by a list of physical and chemical properties of fluids expressed in SI Units (App. 1). A table of unit conversions is given in Appendix 2. Several appendices detail particular calculations: nappe trajectory at a drop structure (App. 3), bubble rise velocity calculations (App. 4), modelling form drag and flow resistance in skimming flows (App. 5), void fraction distributions in chute flows (App. 6), chute calculations in skimming flow (App. 7), modelling air-water gas transfer in skimming flows (App. 8).

Appendix 9 contains a correction form, where readers who find an error in the book are welcome to record the error on the page and to send a copy to the author. Corrections and updates will be posted on the Internet at :

{http://www.uq.edu.au/~e2hchans/reprints/book4.htm}

Finally, a list of relevant internet resources are listed below:

Stepped spillway resources	
Photographs of stepped spillways	{http://www.uq.edu.au/~e2hchans/photo.html#Step_spillways}
Hydraulic design of stepped cascades	{http://www.uq.edu.au/~e2hchans/dpri/topic_2.html}
Embankment overflow stepped spillways. Earth dam spillways with precast concrete blocks	{http://www.uq.edu.au/~e2hchans/over_st.html}
The formal water garden. Stepped cascades, fountains and water staircases	{http://www.uq.edu.au/~e2hchans/wat_gard.html}
Timber Crib Weirs in Queensland, Australia	{http://www.uq.edu.au/~e2hchans/tim_weir.html
Free-surface aeration	{http://www.uq.edu.au/~e2hchans/self_aer.html}
Chicago Calumet Waterway plant	{http://www.mwrdgc.dst.il.us/plants/sepa.htm}
Corrections and updates	{http://www.uq.edu.au/~e2hchans/reprints/book4.htm}

General resources	
Gallery of photographs	{http://www.uq.edu.au/~e2hchans/photo.html}
Reprints of research papers	{http://www.uq.edu.au/~e2hchans/reprints.html}
NASA Earth observatory	{http://earthobservatory.nasa.gov/}
NASA rain, wind and air-sea gas exchange research	{http://bliven2.wff.nasa.gov/index.htm}

Figure P1.1. Field trip to the Gold Creek Dam spillway (Brisbane, Australia) with undergraduate students on 9 Aug. 2000.

Acknowledgements

I wish to thank Professor Colin Apelt (University of Queensland, Australia) for his continuous and generous support to this work. The manuscript was reviewed by Professor Iwao Ohtsu and Dr. Youichi Yasuda (Nihon University, Japan), and Dr. Jorge Matos (IST, Portugal), and I am most grateful for their valuable input. The assistance of Dr. L. Toombes (Australia) for proof-reading parts of the work, and helpful comments from Dr. Sergio Montes (University of Tasmania, Australia) are also acknowledged.
The following students and colleagues are also thanked for their assistance: Mr. M. Eastman, M. Condon, L. Rowlands, N. Van Schagen, Dr. L. Toombes, Mr. G. Illidge, Professor C.J. Apelt.

I would also like to express gratitude to the following persons who made available various photographs of interest:
Dr. R. Baker, University of Salford, United Kingdom;
Mr. Hans Bandler, Turramurra NSW, Australia;
Mr. Errol Beitz, Queensland Department of Natural Resources, Australia;
Mr. F. Botton, Lyon, France;
Mr. M. Burridge and Mrs L. Burridge, Brisbane, Australia;
Mr. L. Stuart Davies, Welsh Water, United Kingdom;
Mr. K.D. Gardiner, North West Water Engineering, United Kingdom;
Dr. M.R. Gourlay, Brisbane, Australia;
Mr. D. Patrick James, Neutral Bay NSW, Australia;
Mr. D. Jeffery, Goulburn-Murray Water, Australia;
Pr. J. Knauss, Münich University of Technology, Germany;
Mr. Michel Lefebvre, Montréal, Québec, Canada;
Ms. M. Mengel, USA;
Mr. G. Michell, Coliban Region Water Authority, Australia;
Mr. J. Mitchell, Brisbane, Australia;
Mr. Dennis Murphy, Houston, USA;
Musée de l'Histoire de Rueil-Malmaison, France;
Officine Maccaferri, Italy, and Maccaferri Gabions South Pacific, Australia;
Professor Y. Pravdivets, Moscow Institute of Civil Engineers, Russia;
Queensland Department of Natural Resources, Dam Safety Section;
Mr. Marq Redeker, Ruhrverband, Germany;
Mr. P. Royet, Cemagref, France;
Services Techniques, Ville de Firminy, France;
Mr. Rod Smyth, Coliban Water, Australia;
The Texas Historical Society, Texas, Australia;
Mr. Ross Tomkins, Gold Coast QLD, Australia;

Dr. L. Toombes, Brisbane, Australia;
US Bureau of Reclamation;
Victoria Rural Water Corporation, Australia;
Mr. Gordon Williams, Geolab Group, Australia;
Dr. Youichi Yasuda, Nihon University, Japan;
Mr. A.R. Yazdany, Iran.

The permission of the American Society of Civil Engineers for the reprint of a number of photographs is also acknowledged.

I would also like to thank the following people:
Mr. E.A. Battison, Windsor, Vermont, USA; Mr. E. Beitz, Queensland Department of Natural Resources, Australia; Mr. F. Botton, Lyon, France; Madame C.M. Chanson, Paris, France; Ms. Chou Y.H., Brisbane, Australia; Mr. L. Stuart Davies, Welsh Water, United Kingdom; Direction Départementale de l'Equipement, Côte d'Or; Mr. B.A. Forbes, GHD, Australia; Madame Gaumont, Montsauche-les-Settons, France; Madame D. Helot-Lécroart, Société Historique de Rueil-Malmaison, France; Mr. D. Jeffery, Goulburn-Murray Water, Australia; Mrs Kalenitchenko, Musée de l'Histoire de Rueil-Malmaison, France; Pr. J. Knauss, Münich University of Technology, Germany; Mr. D. Ledingham, LHO Group, Australia; Pr. A. Lejeune, University of Liège, Belgium; Ms. C. Litchfield, Brisbane, Australia; Officine Maccaferri, Italy; Dr. J. Matos, IST, Lisbon, Portugal; Ms. Melissa Mengel, USA; Mr. J. Mitchell, Brisbane, Australia; Mr. Dennis Murphy, Houston, USA; Professor C. O'Connor, Brisbane, Australia; Professor Iwao Ohtsu, Nihon University, Japan; Dr. A.N. Pinheiro, IST, Lisbon, Portugal; Mr. F. Portet, Fraisses, France; Mr. P. Pulicani, Direction Départementale de l'Equipement Côte d'Or, France; Ms. Qiao G.L., China University of Mining and Technology, P.R. China; Mr. O. De Quinsonas, Château du Touvet, France; Dr. P. Reichert, EAWAG, Switzerland; Mr. A. Serre, Inglewood, Australia; Dr. Shien C.L., National Cheng Kung University, Taiwan R.O.C.; Mr. Rod Smyth, Coliban Water, Australia; Mr. Brit Storey, US Bureau of Reclamation, USA; The Texas Historical Society, Texas QLD, Australia (Mrs E. Craig and Mrs C. Glasser); Dr. M.J. Tozzi, Universidade Federal do Paranã, Brazil; Mr. Richard Tumman, Brookfield, Australia; La Ville de Saint-Etienne (M. Bonnefoy, Direction Générale), France; Mr. R. Wheeler, Queensland Department of Natural Resources, Australia; Professor R.L. Whitmore, Brisbane, Australia; Mr. R. Wingate, US Bureau of Reclamation, USA; Dr. Youichi Yasuda, Nihon University, Japan; Mr. A.R. Yazdany, Iran.

I am particularly indebted to Dr. R. Baker, University of Salford (UK), for providing valuable information on embankment overflow protection systems in Russia and UK.

This book was developed during two Special Study Programs at National Cheng Kung University (Taiwan) and Toyohashi University of Technology (Japan). I would therefore like to thank both Professor Lee Jaw Fang, Department of Hydraulics and Ocean Engineering, National Cheng Kung University and Dr. Shin-ichi Aoki, Department of Architecture and Civil Engineering, Toyohashi University of Technology for their support.

List of symbols

A	gas-liquid interface area (m^2);
A_w	cross-section area (m^2) of the water flow;
A_1, A_2	constants;
A_3, A_4	constants;
a	specific interface area (m^{-1}) defined as the air-water surface area per unit volume of air and water;
a_{mean}	mean interface area (m^{-1}) in a cross-section of fluid normal to the flow direction: $a_{mean} = \frac{1}{Y_{90}} * \int_0^{Y_{90}} a * dy$
B_1, B_2, B_3	constants;
C	air concentration defined as the volume of air per unit volume; (Note : it is also called void fraction);
C_D	dimensionless discharge coefficient;
C_{DS}	downstream dissolved gas concentration (kg/m^3);
C_{US}	upstream dissolved gas concentration (kg/m^3);
C_e	equilibrium depth averaged air concentration for uniform flow;
C_{gas}	concentration of dissolved gas in water (kg/m^3);
C_{mean}	depth averaged air concentration defined as: $(1 - Y_{90}) * C_{mean} = d$;
C_s	gas saturation concentration in water (kg/m^3);
C_*	mean air concentration at a reference location ($s = s_*$);
Chl	chlorinity;
D_{gas}	molecular diffusivity of gas in water (m^2/s);
D_H	hydraulic diameter (m) defined as: $D_H = 4 * \frac{A_w}{P_w}$; $\left(D_H = \frac{4 * d * W}{W + 2 * d} \text{ for a rectangular channel}\right)$
$(D_H)_i$	hydraulic diameter (m) of a jet impacting a pool of water;
D_t	turbulent diffusivity (m^2/s) of air bubble in air-water flows;
D'	dimensionless air bubble diffusivity (defined by Chanson, 1997a);
d	1- flow depth measured normal to the channel slope at the edge of a step; 2- characteristic depth (m) defined as: $\int_0^{Y_{90}} (1 - C) * dy$;; 3-nappe thickness (m);
d_I	flow depth at the inception point (m);
d_M	sidewall standing wave height (m) (Chapter 9);
d_{ab}	air bubble diameter (m);

d_b	flow depth at the brink of a step (m);
d_c	critical flow depth (m); for a rectangular channel : $d_c = \sqrt[3]{q_w^2/g}$;;
d_i	jet thickness (m) at the impact of the nappe with the receiving pool in nappe flow regime;
d_o	1- dam break characteristic depth (m), equal to the initial flow depth in an ideal dam break (Chapter 9); 2- inflow depth (m);
d_p	flow depth in the pool beneath the nappe (m);
d_t	1- tailwater depth (m); 2- pool height (m);
d_1	flow depth (m) upstream of a hydraulic jump;
d_2	flow depth (m) downstream of a hydraulic jump;
d_{50}	median boulder diameter (m);
d_*	flow depth (m) at a reference location ($s = s_*$);
E	1- aeration efficiency defined as: $E = (C_{DS} - C_{US})/(C_S - C_{US})$ (Chapter 6); 2- specific energy (m);
E_C	kinetic energy correction coefficient (Coriolis Coefficient);
E_i	aeration efficiency of the i-th step;
F	nappe oscillation frequency (Hz);
F_{ab}	bubble count rate (Hz): i.e., number of bubbles impacting a probe per seconds;
$(F_{ab})_{max}$	maximum bubble count rate (Hz), often observed for C = 50%;
Fr	Froude Number;
Fr_b	Froude Number defined in terms of the brink depth: $Fr_b = q_w/\sqrt{g * d_b^3}$
Fr_1	Froude Number upstream of a hydraulic jump: $Fr_1 = q_w/\sqrt{g * d_1^3}$
F_*	Froude Number defined in terms of the roughness height : $q_w/\sqrt{g * \sin\alpha * (k_s')^3}$ for smooth chute: $q_w/\sqrt{g * \sin\alpha * k_s^3}$ for skimming flow;
f	1- Darcy Friction Factor; 2- Darcy Friction Factor for clear-water (non-aerated) flow;
f_d	(equivalent) Darcy Friction Factor estimate of the form drag;
f_e	Darcy Friction Factor for air-water flow;
f_i	Darcy Friction Factor estimate neglecting the effects of air entrainment;
g	gravity constant (m/s^2) or acceleration of gravity; numerical values of the gravity constant are given for several countries in Appendix 1;
H	total head (m);

H_{dam}	dam height (m): i.e. dam crest head above downstream toe;
H_{gas}	Henry's Law Constant
H_{max}	maximum head available (m) using the spillway toe as datum;
H_o	free-surface elevation (m) above the spillway crest;
H_{res}	residual head (m) at the downstream end of the spillway, using the spillway toe as datum: $H_{res} = H_{max} - \Delta H$;
H_1	total head (m) upstream of a hydraulic jump;
h	height of steps (m) (measured vertically);
I	1- integer;
	2- number of wavelengths in a free-falling nappe;
i	integer;
κ	Von Karman Constant;
K	inverse of the spreading rate of a turbulent shear layer;
K_L	liquid film coefficient (m/s);
K_M	coefficient of mass transfer (m/s);
K_o	dimensionless term (Chapter 5);
K'	integration constant (defined by Chanson, 1997a);
k	dimensionless parameter;
k'	constant (defined by Avery & Novak, 1978);
k_{AN}	constant (defined by Avery & Novak, 1978);
k_{N1}, k_{N2}	constants (defined by Nakasone, 1987);
k_{N3}	constant (defined by Nakasone, 1987);
k_o	dimensionless term (Chapter 5);
k_s	1 roughness height (m) measured perpendicular to the channel bed;
	2- step dimension (m) measured normal to the flow direction: $k_s = h * \cos\alpha$;
k_s'	surface (skin) roughness height (m);
	Note: for a smooth-invert channel: $k_s' = k_s$;
L	spillway chute length (m);
L_d	length (m) of the drop measured from the vertical face of the step;
L_I	distance from the start of growth of boundary layer to the inception point of air entrainment;
L_r	roller length (m) of a fully-developed hydraulic jump;
L_s	roughness spacing (m);
l	horizontal length of steps (m) (measured perpendicular to the vertical direction);
l_{cav}	cavity length (m): $l_{cav} = L_s - l_s$;
l_s	roughness length (m);
Ln	Neperian logarithm;
Log_{10}	decimal logarithm: $Log_{10}(X) = Ln(X)/Ln(10)$;
M_{gas}	mass of dissolved gas (kg);
N	exponent of the velocity power law;
N_{step}	number of steps;
n_{gabion}	gabion porosity (or void fraction);

$n_{Manning}$ Gauckler-Manning Coefficient ($s/m^{1/3}$);
P pressure (Pa);
P_N pressure gradient number: $P_N = \Delta P/(\rho_w * g * d)$;
P_{Nb} pressure gradient number defined in terms of the brink depth: $P_{Nb} = \Delta P/(\rho_w * g * d_b)$;
P_{atm} atmospheric pressure (Pa);
P_{gas} partial pressure of gas in air (Pa);
P_{hyd} hydrostatic pressure (Pa): $P_{hyd} = \rho_w * g * d * \cos\alpha$;
P_s mean stagnation pressure (Pa) at the impact of a falling nappe (Chapter 8);
Pstd standard pressure: Pstd = 1 atm = 1.01325+5 Pa;
P_v vapour pressure (Pa);
P_w wetted perimeter (m);
Q discharge (m^3/s);
Q_{des} design flow rate (m^3/s);
Q_w' incoming discharge (m^3/s) into a side channel;
q discharge per unit width (m^2/s);
q_w' incoming discharge per unit width (m^2/s) into a side channel;
$(q_w)_c$ characteristic discharge per unit width (m^2/s) for which the growing boundary layer reaches the free-surface at the spillway end;
r deficit ratio: $(C_S - C_{US})/(C_S - C_{DS})$;
r_T deficit ratio at temperature T;
r_{15} deficit ratio at 15 Celsius;
r_{20} deficit ratio at 20 Celsius;
Sal salinity;
S_f friction slope: $S_f = - \partial H/\partial s$;
S_o bed slope: $S_o = \sin\theta$;
s curvilinear coordinate (m): i.e., distance measured along the channel bottom;
s_* reference location coordinate (m) (Chapter 5);
s' dimensionless coordinate: $s' = (s-s_*)/d_*$ (Chapter 5);
T temperature (K);
TC temperature expressed in Celsius: TC = TK - 273.16;
Teta coefficient;
TK temperature expressed in Kelvin: TK = TC - 273.16;
T_o reference temperature (K);
Tu turbulence intensity defined as: Tu = u'/V;
t time (s);
t', t'' time (s);
U_w clear-water flow velocity (m/s): $U_w = q_w/d = q_w/\int_0^{Y_{90}} (1 - C) * dy$;
u' root mean square of longitudinal component of turbulent velocity (m/s);

u_r	bubble rise velocity (m/s);
$(u_r)_{Hyd}$	bubble rise velocity (m/s) in a hydrostatic pressure gradient;
V	velocity (m/s);
V_b	flow velocity (m/s) at the brink of a step;
V_c	critical velocity (m/s); for a rectangular channel: $V_c = \sqrt[3]{g * q_w}$;
V_{ent}	entrainment velocity (m/s) (Chapter 5);
V_i	velocity (m/s) of the falling nappe at the intersection of the nappe and the receiving pool in nappe flow regime;
V_{max}	1- maximum velocity (m/s) in a shear layer; 2- ideal fluid flow velocity (m/s)
V_o	free-stream velocity (m/s);
V_x	velocity component m/s) in the x-direction;
V_y	velocity component m/s) in the y-direction;
V_*	shear velocity (m/s);
V_1	mean flow velocity (m/s) upstream of a hydraulic jump: $V_1 = q_w/d_1$;
V_{90}	characteristic velocity (m/s) where the air concentration is 90%;
v	local velocity (m/s);
v'	root mean square of lateral component of turbulent velocity (m/s);
W	channel width (m);
W'	width (m) (Chapter 7);
W_1	width (m) of side channel (Chapter 7);
x	longitudinal distance (m);
x_s	flood wave front coordinate (m);
x'	dimensionless distance;
Y_{50}	location where C = 50%;
Y_{90}	characteristic depth (m) where the air concentration is 90%;
y	1- distance (m) from the bottom measured perpendicular to the spillway surface; 2- distance (m) from the pseudo-bottom (formed by the step edges) measured perpendicular to the flow direction;
y'	dimensionless distance: 1- $y' = y/Y_{90}$; 2- $y' = y_1/d_b$ (App. 3);
y_1	vertical distance (App. 3);
y_{50}	distance (m) normal to the jet centreline where $V = V_o/2$;
Z	Morton Number defined as : $\frac{g * \mu_w^4}{\rho_w * \sigma^3}$;;
z	bed elevation (or altitude) positive upwards (m);

Greek symbols

α channel slope;

α_b angle between the invert and horizontal at the first step brink (Chapter 5, App. 3);

α' spillway slope;

β sidewall deflection angle;

ΔH 1- total head loss (m) : $\Delta H = H_{max} - H_{res}$;
2- fall height (m);

ΔP subpressure (Pa) in the cavity beneath the nappe: i.e., atmospheric pressurse minus the pressure in the cavity;

ΔT average period (s) between fluid ejection events (Chapter 5);

Δt fluid ejection event duration (s) (Chapter 5);

ΔV difference between the upstream and downstream velocities (m/s);

δ step slope: i.e. angle between the step and the horizontal:
$\delta > 0$ inclined upward step;
$\delta < 0$ inclined downward step;

δ_{BL} boundary layer thickness (m);

Γ_y vertical acceleration (m/s^2) of a free-falling nappe;

μ dynamic viscosity ($N.s/m^2$);

μ_T dynamic viscosity ($N.s/m^2$) at temperature T;

ν kinematic viscosity (m^2/s);

ν_T kinematic viscosity (m^2/s) at temperature T;

ν_t turbulent kinematic viscosity (m^2/s);

π p = 3.141592653589793238462643...;

θ angle of the impinging jet and the horizontal step;

θ_b initial angle of the streamlines at the brink of a step;

θ_s shock wave angle with the main flow direction;

ρ density (kg/m^3);

ρ_{gas} gas density (kg/m^3);

ρ_s density of rockfill (kg/m^3);

ρ_T density (kg/m^3) at temperature T;

σ surface tension between air and water (N/m);

σ_c compressive strength (Pa);

τ shear stress (Pa);

τ_{max} maximum shear stress (Pa);

τ_o average bottom shear stress (Pa);

$\varnothing$ diameter (m);

$\varnothing_s$ circular strip roughness diameter (m);

Subscript

air air flow;

b flow conditions at the brink of a step;

c critical flow conditions;

I	inception point of free-surface aeration;
w	water flow;
1	upstream flow conditions;
15, 20, 25	at 15 Celsius, 20 Celsius, 25 Celsius;

Superscript

HJ	hydraulic jump;
jet	plunging jet;
lower	lower nappe;
upper	upper nappe;

Abbreviations

D/S	downstream;
HJ	hydraulic jump;
NA	nappe flow;
NA1	nappe flow with fully developed hydraulic jump;
NA2	nappe flow with partially developed hydraulic jump;
NA3	nappe flow without hydraulic jump;
RCC	roller compacted concrete;
SK	skimming flow regime;
SK1	skimming flow on flat chute (regime 1);
SK2	skimming flow regime 2;
SK3	skimming flow on steep chute (regime 3);
TRA	transition flow regime;
U/S	upstream.

Glossary

ABUTMENT: Part of the valley side against which the dam is constructed. Artificial abutments are sometimes constructed to take the thrust of an arch where there is no suitable natural abutment.
ADVECTION: Movement of a mass of fluid which causes change in temperature or in other physical or chemical properties of fluid.
AIR: Mixture of gases comprising the atmosphere of the Earth. The principal constituents are Nitrogen (78.08%) and Oxygen (20.95%). The remaining gases in the atmosphere include Argon, Carbon Dioxide, water vapour, Hydrogen, Ozone, Methane, Carbon Monoxide, Helium, and Krypton.
AIR CONCENTRATION: Concentration of un-dissolved air defined as the volume of air per unit volume of air and water.
APELT: C.J. Apelt is an Emeritus Professor in Civil Engineering at the University of Queensland (Australia).
AQUEDUCT: A conduit for conveying a large quantity of flowing waters. The conduit may include canals, siphons, pipelines.
ARCH DAM: Dam curved in plan dependent upon arch action for its stability.
ARCHED DAM: Gravity dam which is curved in plan. Alternatives include "curved-gravity dam" and "arch-gravity dam".
ASSYRIA: A land to the North of Babylon comprising, in its greatest extent, a territory between the Euphrates and the mountain slopes east of the Tigris. The Assyrian Kingdom lasted from about 2300 B.C. to 606 B.C..
BAKHMETEFF: Boris Alexandrovitch Bakhmeteff (1880-1951) was a Russian hydraulician. In 1912, he developed the concept of specific energy and energy diagram for open channel flows.
BARRAGE: French word for dam or weir but commonly used to describe a large dam structure in English.
BAZIN: H. Bazin (1829-1917) was a French hydraulician, engineer and member of the French *Corps des Ponts-et-Chaussées*. He worked as an assistant of Henri P.G. Darcy.
BELANGER: Jean-Baptiste Ch. Bélanger (1789-1874) was a French hydraulic engineer and professor at the *Ecole Nationale Supérieure des Ponts et Chaussées* (Paris). He originally suggested the application of the momentum principle to hydraulic jump flow (Bélanger 1828). In the same book, he presented the first "backwater" calculation for open channel flow.
BÉLANGER EQUATION: Momentum equation applied across a hydraulic jump in a horizontal channel (named after J.B.C. Bélanger).
BELIDOR: Bertrand Forêt de Bélidor (1693-1761) was a teacher at the *Ecole Nationale des Ponts et Chaussées*. His treatise "Architecture Hydraulique" (Bélidor 1737-1753) was a well-known hydraulic textbook in Europe during the 18th and 19th centuries.

BERNOULLI: Daniel Bernoulli (1700-1782) was a Swiss mathematician, physicist and botanist who developed the Bernoulli equation in his *Hydrodynamica, de viribus et motibus fluidorum* textbook (1st draft in 1733, 1st publication in 1738, Strasbourg).
BIDONE: Giorgio Bidone (1781-1839) was an Italian hydraulician. His experimental investigations on the hydraulic jump were published in 1820 and 1826.
BORDA: Jean-Charles de Borda (1733-1799) was a French mathematician and military engineer. He achieved the rank of *Capitaine de Vaisseau* and participated in the US War of Independence with the French Navy. He investigated the flow through orifices and developed the Borda Mouthpiece.
BORDA MOUTHPIECE: A horizontal re-entrant tube in the side of a tank with a length such that the issuing jet is not affected by the presence of the walls.
BOSSUT: Abbé Charles Bossut (1730-1804) was a French ecclesiastic and experimental hydraulician, author of a hydrodynamic treaty (Bossut, 1772).
BOUNDARY LAYER: Thin layer of fluid in the neighbourhood of a solid boundary where friction plays an essential part (i.e. flow region affected by the presence of the boundary). A range of velocities exists across the boundary layer from zero at the boundary to the free-stream velocity at the outer edge of the boundary layer.
BOUSSINESQ: J. Boussinesq (1842-1929) was a French hydrodynamicist. His treatise *Essai sur la théorie des eaux courantes* (1877) remains an outstanding contribution in hydraulics literature.
BOUSSINESQ COEFFICIENT: Momentum correction coefficient named after J. Boussinesq, French mathematician, who first proposed it (Boussinesq, 1877).
BRACCIO: Ancient measure of length (from the Italian "braccia"). One braccio equals 0.6096 m (or 2 ft).
BRESSE: Jacques Antoine Charles Bresse (1822-1883) was a French applied mathematician and hydraulician. He was Professor at the *Ecole Nationale Supérieure des Ponts et Chaussées*, Paris; successor to J.B.C. Bélanger. His contribution to gradually-varied flows in open channel hydraulics is considerable (Bresse, 1860).
BUTTRESS DAM: A special type of dam in which the water face consists of a series of slabs or arches supported on their air faces by a series of buttresses.
BUOYANCY: Tendency of a body to float, to rise or to drop when submerged in a fluid at rest. The physical law of Buoyancy (or Archimedes' principle) was discovered by the Greek mathematician Archimedes. It states that any body submerged in a fluid at rest is subjected to a vertical (or buoyant) force. The magnitude of the buoyant force is equal to the weight of the fluid displaced by the body.
BYEWASH: Ancient name for a spillway.
CASCADE: 1. A steep stream intermediate between a rapid and a water fall. The slope is steep enough to allow a succession of small drops but not sufficient to cause the water to drop vertically (i.e. waterfall). 2. A man-made channel

consisting of a series of steps: e.g., a stepped fountain, a staircase chute, a stepped sewer.

CATARACT: A series of rapids or waterfalls. It is usually termed for large rivers: e.g., the six cataracts of the River Nile between Karthum and Aswan.

CATENA D'ACQUA: Italian term for "chain of water". Variation of the cascade developed during the Italian Renaissance. Water is channelled down the centre of an architectural ramp contained on both sides by stone carved into a scroll pattern to give a chain-like appearance. Waters flow as a supercritical regime with regularly-spaced increase and decrease of channel width, giving a sense of continuous motion highlighted by shock wave patterns at the free-surface. One of the best examples is at Villa Lante, Italy. The stonework was carved into crayfish, the emblem of the owner, Cardinal Gambara.

CAVITATION: Formation of vapour bubbles and vapour pockets within a homogeneous liquid caused by excessive stress (Franc et al., 1995). Cavitation may occur in low-pressure regions where the liquid has been accelerated (e.g. turbines, marine propellers, baffle blocks of dissipation basin). Cavitation modifies the hydraulic characteristics of a system, and it is characterised by damaging erosion, additional noise, vibrations and energy dissipation.

CHADAR: Type of narrow sloping chute peculiar to Islamic gardens and perfected by the Mughal gardens in Northern India (e.g. at Nishat Bagh). These stone channels were used to carry water from one terrace garden to another. A steep slope ($\alpha \sim 20$ to 35°) was selected so as to reflect sunlight to the maximum degree. The chute bottom was very rough to enhance turbulence and free-surface aeration. The discharge per unit width was usually small, resulting in thin sheets of aerated waters. Photographs (e.g. Plumptre, 1993) show a superposition of "white waters" with roll waves, producing a white rough surface in constant move.

CHEZY: Antoine Chézy (1717-1798) (or Antoine de Chézy) was a French engineer and member of the *Corps des Ponts-et-Chaussées*. He designed canals for the water supply of the city of Paris. In 1768 he proposed a resistance formula for open channel flows called the Chézy Equation. In 1798, he became Director of the *Ecole Nationale Supérieure des Ponts et Chaussées* after teaching there for many years.

CHIMU: Indian from the Yuncan tribe, located near Trujillo on the north-west coast of Peru. The Chimu Empire lasted from A.D. 1250 to 1466. It was overrun by the Incas in 1466.

CHOKE: In open channel flow, a channel contraction might obstruct the flow and induce the appearance of critical flow conditions (i.e. control section). Such a constriction is sometimes called a "choke".

CHOKING FLOW: Critical flow in a channel contraction. The term is used for both open channel flow and compressible flow.

CHORD LENGTH: 1. The chord or chord length of an airfoil is the straight line distance joining the leading and trailing edges of the foil. 2. The chord length of a bubble (or bubble chord length) is the length of the straight line

connecting the two intersections of the air-bubble free-surface with the leading tip of the measurement probe (e.g. conductivity probe, conical hot-film probe) as the bubble is transfixed by the probe tip.
COFFERDAM: Temporary structure enclosing all or part of the construction area so that construction can proceed in dry. A diversion cofferdam diverts a stream into a pipe or channel.
COLBERT: Jean-Baptiste Colbert (1619-1683) was a French statesman. Under King Louis XIV, he was the Minister of Finances, the Minister of *Bâtiments et Manufactures* and the Minister of the Marine.
CONTROL: Considering an open channel, sub-critical flows are controlled by the downstream conditions. This is called a downstream flow control. Conversely supercritical flows are controlled only by the upstream flow conditions (i.e. upstream flow control).
CONTROL SECTION: In an open channel, cross-section where critical flow conditions take place. The concept of control and control section are used with the same meaning.
CORIOLIS: Gustave C. Coriolis (1792-1843) was a French engineer and mathematician who first described the Coriolis Force (i.e. effect of motion on a rotating body).
CORIOLIS COEFFICIENT: Kinetic energy correction coefficient named after G.C. Coriolis who introduced first the correction coefficient (Coriolis, 1836).
CRAYA: Antoine Craya was a French hydraulician and professor at the University of Grenoble.
CREAGER PROFILE: Spillway shape developed from a mathematical extension of the original data of Bazin in 1886-88 (Creager, 1917).
CREST OF SPILLWAY: Upper part of a spillway. The term "crest of dam" refers to the upper part of an uncontrolled overflow.
CRIB: 1. Framework of bars or spars for strengthening; 2. frame of logs or beams to be filled with stones, rubble or filling material and sunk as a foundation or retaining wall.
CRIB DAM: Gravity dam built up of boxes, cribs, crossed timbers or gabions, and filled with earth or rock.
CRITICAL FLOW CONDITIONS: In open channel flows, the flow conditions such as the specific energy (of the mean flow) is minimum are called the critical flow conditions. With commonly-used Froude Number definitions, the critical flow conditions occur for $Fr = 1$. If the flow is critical, small changes in specific energy cause large changes in depth. In practice, critical flow over a long reach of channel is unstable.
CULVERT: Covered channel of relatively short length installed to drain water through an embankment (e.g. highway, railroad, dam).
CYCLOPEAN DAM: Gravity masonry dam made of very large stones embedded in concrete.
DAM BREAK WAVE: Flood wave resulting from a sudden dam break. A related form is the wave run-up generated by a tsunami reaching the shoreline.

DANEL: Pierre Danel (1902-1966) was a French hydraulician and engineer. One of the pioneers of modern hydrodynamics, he worked from 1928 to his death for *Neyrpic* known prior to 1948 as *Ateliers Neyret-Beylier-Piccard et Pictet*.
DARCY: Henri Philibert Gaspard Darcy (1805-1858) was a French civil engineer. He performed numerous experiments of flow resistance in pipes (Darcy, 1858) and in open channels (Darcy & Bazin, 1865), and of seepage flow in porous media (Darcy, 1856). He gave his name to the Darcy-Weisbach Friction Factor and to the Darcy Law in porous media.
DEBRIS: Debris comprise mainly large boulders, rock fragments, gravel-sized to clay-sized material, tree and wood material that accumulate in creeks.
DIFFUSION: The process whereby particles of liquids, gases or solids intermingle as the result of their spontaneous movement caused by thermal agitation and in dissolved substances move from a region of higher concentration to one of lower concentration. The term "turbulent diffusion" is used to describe the spreading of particles caused by turbulent agitation.
DIFFUSION COEFFICIENT: Quantity of a substance that in diffusing from one region to another passes through each unit of cross-section per unit of time when the volume concentration is unity. The units of the diffusion coefficient are m^2/s.
DIFFUSIVITY: Another name for the diffusion coefficient.
DIMENSIONAL ANALYSIS: Organisation technique used to reduce the complexity of a study, by expressing the relevant parameters in terms of numerical magnitude and associated units, and grouping them into dimensionless numbers. The use of dimensionless numbers increases the generality of the results.
DIVERSION CHANNEL: Waterway used to divert water from its natural course.
DIVERSION DAM: Dam or weir built across a river to divert water into a canal. It raises the upstream water level of the river but does not provide any storage volume.
DRAINAGE LAYER: Layer of pervious material to relieve pore pressures and/or to facilitate drainage: e.g., drainage layer in an earth fill dam.
DROP: 1. Volume of liquid surrounded by gas in a free-fall motion (i.e. dropping); 2. by extension, small volume of liquid in motion in a gas; 3. a rapid change of bed elevation also called step.
DROPLET: Small drop of liquid.
DROP STRUCTURE: Single step structure characterised by a sudden decrease in bed elevation.
EARTH DAM: Massive earthen embankment with sloping faces and made watertight.
ECOLE NATIONALE SUPÉRIEURE DES PONTS ET CHAUSSÉES, PARIS: French civil engineering school founded in 1747. The name may be directly translated as "National School of Bridge and Road Engineering". Among the directors there were the famous hydraulicians A. Chézy and G. de Prony. Other famous professors included B.F. de Bélidor, J.B.C. Bélanger, J.A.C. Bresse, G.G. Coriolis and L.M.H. Navier.

EMBANKMENT: Fill material (e.g. earth, rock) placed with sloping sides and with a length greater than its height.
ESCALIER D'EAU: See *WATER STAIRCASE*.
FACE: External surface which limits a structure: e.g. air face of a dam (i.e. downstream face), water face (i.e. upstream face) of a dam.
FAWER JUMP: Undular hydraulic jump.
FICK: Adolf Eugen Fick was a 19th century German physiologist who developed the diffusion equation for neutral particle (Fick, 1855).
FILTER: Band of granular material which is graded so as to allow seepage to flow across the filter without causing the migration of the material from adjacent zones.
FLASHBOARD: A board or a series of boards placed on or at the side of a dam to increase the depth of water. Flashboards are usually lengths of timber, concrete or steel placed on the crest of a spillway to raise the upstream water level.
FLASH FLOOD: Flood of short duration with a relatively high peak flow rate.
FLASHY: Term applied to rivers and streams whose discharge can rise and fall suddenly, and are often unpredictable.
FLIP BUCKET: A flip bucket or ski-jump is a concave curve at the downstream end of a spillway, to deflect the flow upwards. Its purpose is to throw the water clear of the hydraulic structure and to induce the disintegration of the jet in the air.
FORCHHEIMER: Philipp Forchheimer (1852-1933) was an Austrian hydraulician who contributed significantly to the study of hydraulic engineering and groundwater hydrology.
FORTIER: André Fortier was a French scientist and engineer. He became later Professor at the Sorbonne, Paris.
FOURIER: Jean Baptiste Joseph Fourier (1768-1830) was a French mathematician and physicist known for his development of the Fourier series. In 1794 he was offered a professorship of mathematics at the *Ecole Normale* in Paris and was later appointed at the *Ecole Polytechnique*. In 1798 he joined the expedition to Egypt lead by the (then) General Napoléon BONAPARTE. His research in mathematical physics culminated with the classic study *Théorie Analytique de la Chaleur* (1822) in which he enunciated his theory of heat conduction.
FREE-SURFACE: Interface between a liquid and a gas. More generally a free-surface is the interface between the fluid (at rest or in motion) and the atmosphere. In two-phase gas-liquid flow, the term "free-surface" includes also the air-water interface of gas bubbles and liquid drops.
FREE-SURFACE AERATION: Natural aeration occurring at the free surface of high velocity flows is referred to as free surface aeration or self-aeration.
FRENCH REVOLUTION (RÉVOLUTION FRANÇAISE): revolutionary period that shook France between 1787 and 1799. It reached a turning point in 1789 and led to the destitution of the monarchy in 1791. The constitution of the First Republic was drafted in 1790 and adopted in 1791.

FRONTINUS: Sextus Julius Frontinus (A.D. 35-103 or 104) was a Roman engineer and soldier. After A.D. 97, he was *curator aquarumi,* in charge of the water supply system of Rome. He dealt with discharge measurements in pipes and canals. In his analysis he correctly related the proportionality between discharge and cross-section area. His book *De Aquaeductu Urbis Romae* ("Concerning the Aqueducts of the City of Rome") described the operation and maintenance of Rome's water supply system.
FROUDE: William Froude (1810-1879) was a English naval architect and hydrodynamicist who invented the dynamometer and used it for the testing of model ships in towing tanks. He was assisted by his son Robert Edmund Froude who, after the death of his father, continued some of his work. In 1868, he utilised Reech's Law of Similarity to study the resistance of model ships.
FROUDE NUMBER: The Froude Number is proportional to the square root of the ratio of the inertial forces over the weight of fluid. It is used generally for scaling free surface flows, open channels and hydraulic structures. Although the dimensionless number was named after William Froude, several French researchers used it before. Dupuit (1848) and Bresse (1860) highlighted the significance of the number to differentiate the open channel flow regimes. Bazin (1865) confirmed experimentally the findings. Ferdinand Reech introduced the dimensionless number for testing ships and propellers in 1852. The number is called the Reech-Froude Number in France.
GABION: A gabion consists of rockfill material enlaced by a basket or a mesh. The word originates from the Italian *gabbia*, meaning large cage.
GABION DAM: Crib dam built up of gabions.
GAS TRANSFER: Process by which gas is transferred into or out of solution (i.e. dissolution or de-sorption respectively).
GAUCKLER: Philippe Gaspard Gauckler (1826-1905) was a French engineer and member of the *Corps des Ponts-et-Chaussées*. He re-analysed the experimental data of Darcy and Bazin (1865), and he presented a flow resistance formula for open channel flows, the Gauckler-Manning Formula, sometimes called improperly the Manning Equation (Gauckler, 1867). His son was *Directeur des Antiquités et des Beaux-Arts* (Director of Antiquities and Fine Arts) for the French Republic in Tunisia where he directed an extensive survey of Roman hydraulic works in Tunisia.
GHAZNAVID (or *GHAZNEVID*): One of the Moslem dynasties (10th to 12th centuries) ruling south-western Asia. Its capital city was at Ghazni (Afghanistan).
GOUJON: Jean Goujon (1510-1568) was a French architectural sculptor during the Renaissance period. One of his masterpieces is the sculptures of the *Nymphes* (Nymphs) on the Fountaine des Innocents in Paris.
GRAVITY DAM: Dam which relies on its weight for stability. The term normally refers to a masonry or concrete dam.
HASMONEAN: Describing the dynasty of the Maccabees, in Israel. The

Hasmonean Kingdom was created following the uprising of the Jews in 166 B.C..

HENNIN: Georg Wilhelm Hennin (1680-1750) was a young Dutchman hired by Tsar Peter the Great to design and build several dams in Russia (Danilveskii, 1940; Schnitter, 1994). He went to Russia in 1698 and stayed until his death in April 1750.

HERO: Greek mathematician (1st century A.D.) working in Alexandria, Egypt. He wrote at least 13 books on mathematics, mechanics and physics. He designed and experimented the first steam engine. His treatise *Pneumatica* described Hero's fountain, siphons, steam-powered engines, a water organ, and hydraulic and mechanical water devices. It influenced directly the waterworks design during the Italian Renaissance. In his book *Dioptra*, Hero stated rightly the concept of continuity for incompressible flow: the discharge being equal to the area of the cross-section of the flow times the speed of the flow.

HIMYARITE: Important Arab tribe of Antiquity, dwelling in southern Arabia (700 B.C. to A.D. 550).

HOHOKAMS: Native Americans in the south-west of the USA (Arizona), they built several canal systems in the Salt River Valley during the period 350 B.C. to A.D. 650. They migrated to northern Mexico around A.D. 900, where they built other irrigation systems.

HYDRAULIC DIAMETER: Is defined as the equivalent pipe diameter: that is, four times the flow cross-section area divided by the wetted perimeter.

HYDRAULIC FILL DAM: Embankment dam constructed of material which are conveyed and placed by suspension in flowing water.

HYDRAULIC JUMP: Transition from a rapid (supercritical flow) to a slow flow motion. Described by Leonardo da Vinci, the first experimental investigation was published by Giorgio Bidone in 1820. The basic theory was developed by Bélanger (1828). It has been verified experimentally by Bakhmeteff and Matzke (1936).

HYPER-CONCENTRATED FLOW: Sediment-laden flow with large suspended sediment concentrations (i.e. typically more than 1% in volume). Spectacular hyper-concentrated flows are observed in the Yellow River Basin (China) with volumetric concentrations greater than 8%.

IDEAL FLUID: Is frictionless and incompressible. A frictionless fluid is non-viscous.

IDLE DISCHARGE: Old expression for spill or waste water flow.

INCA: South-American Indian of the Quechuan tribes of the highlands of Peru. The Inca civilisation dominated Peru between A.D. 1200 and 1532, and was terminated by the Spanish Conquest.

INCEPTION OF AIR ENTRAINMENT: Characteristic flow conditions for which free-surface air entrainment starts.

INFLOW: 1. Upstream flow; 2. incoming flow.

INLET: 1. Upstream opening of a culvert, pipe or channel; 2. a tidal inlet is a narrow water passage between peninsulas or islands.

INTAKE: Any structure in a reservoir through which water can be drawn into a waterway or pipe: by extension, upstream end of a channel.
INTERFACE: Surface forming a common boundary of two phases (e.g. gas-liquid interface) or two fluids.
INTERNATIONAL SYSTEM OF UNITS: See *SYSTÈME INTERNATIONAL D'UNITÉS*.
INVERT: 1. Lowest portion of the internal cross-section of a conduit; 2. channel bed of a spillway; 3. bottom of a culvert barrel.
INVISCID FLOW: Is a non-viscous flow.
IRROTATIONAL FLOW: Is defined as a zero vorticity flow. Fluid particles within a region have no rotation. If a fluid is at rest and is frictionless, any later motion of this fluid will be irrotational. In irrotational flow each element of the moving fluid undergoes no net rotation, with respect to chosen coordinates axes, from one instant to another.
KÁRMÁN: Theodore Von Kármán (1881-1963) was a Hungarian fluid dynamicist and aerodynamicist who worked in Germany (1906-1929), and later in the USA. He was a student of Ludwig Prandtl in Germany. He gave his name to the vortex shedding behind a cylinder – Karman Vortex Street.
KÁRMÁN CONSTANT: "Universal" constant of proportionality between the Prandtl Mixing Length and the distance from the boundary. The constant is named after Theodore Von Kármán. Experimental results indicate that $\kappa = 0.40$.
KEULEGAN: Garbis Hovannes Keulegan (1890-1989) was an Armenian mathematician who worked as hydraulician for the US Bureau of Standards since its creation in 1932.
LAMINAR FLOW: is characterised by fluid particles moving along smooth paths in laminas or layers, with one layer gliding smoothly over an adjacent layer. Laminar flows are governed by Newton's Law of Viscosity which relates the shear stress to the rate of angular deformation: $\tau = \mu * \partial v/\partial y$.
LEFT ABUTMENT: Abutment on the left-hand side of an observer when looking downstream.
LEFT BANK (LEFT WALL): Looking downstream, the left bank or the left channel wall is on the left.
LE NÔTRE: André Le Nôtre (1613-1700) was a French landscape architect. He set the style of formal gardens. His two masterpieces were the gardens of the Château de Vaux-le-Vicomte and of the Château de Versailles near Paris, France.
LEONARDO DA VINCI: Italian artist (painter and sculptor) who extended his interest to medicine, science, engineering and architecture (1452-1519).
LINING: Coating on a channel bed to provide water tightness, to prevent erosion or to reduce friction.
LUMBER: Timber sawed or split into boards, planks or staves.
McKAY: Professor Gordon M. McKay (1913-1989) was Professor in Civil Engineering at the University of Queensland.
MANNING: Robert Manning (1816-1897) was Chief Engineer of the Office of

Public Works, Ireland. In 1889, he presented two formulae (Manning, 1890). One was to become the so-called "Gauckler-Manning Formula" but Robert Manning did prefer to use the second formula that he gave in his paper. It must be noted that the "Gauckler-Manning Formula" was proposed first by the Frenchman P.G. Gauckler (Gauckler, 1867).
MANSART: Jules Hardouin Mansart (1646-1708) was the First Architect of King Louis XIV. He enlarged the Château de Versailles and the park. He was the grand nephew of another French architect, Nicolas François Mansart (1598-1666).
MASONRY DAM: Dam constructed mainly of stone, brick or concrete blocks jointed with mortar.
METRIC SYSTEM: See *SYSTÈME MÉTRIQUE*.
MINIMUM ENERGY LOSS CULVERT: Culvert designed with very smooth shapes to minimise energy losses. The design of a minimum energy loss culvert is associated with the concept of constant total head. The inlet and outlet must be streamlined in such a way that significant form losses are avoided (Apelt, 1983).
MIXING LENGTH: The mixing length theory is a turbulence theory developed by L. Prandtl, first formulated in 1925 (Prandtl, 1925). Prandtl assumed that the mixing length is the characteristic distance travelled by a particle of fluid before its momentum is changed by the new environment.
MOCHICA: 1. South American civilisation (A.D. 200-1000) living in the Moche River Valley, Peru along the Pacific coastline; 2. Language of the Yuncas.
MOMENTUM EXCHANGE COEFFICIENT: In turbulent flows the apparent kinematic viscosity (or kinematic eddy viscosity) is analogous to the kinematic viscosity in laminar flows. It is also called the Eddy Viscosity or the Eddy Coefficient. The momentum exchange coefficient is proportional to the shear stress divided by the strain rate. It was first introduced by the Frenchman J.V. Boussinesq (1877, 1896).
MOOR: 1. A native of Mauritania, a region corresponding to parts of Morocco and Algeria; 2. A Moslem of native North African races.
MORNING-GLORY SPILLWAY: Vertical discharge shaft, more particularly the circular hole form of a drop inlet spillway. The shape of the intake is similar to a Morning-Glory flower (American native plant *Ipomocea*). It is sometimes called a "Tulip intake".
MUGHAL: Also Mughul, Mogul or Moghul. A name or adjective referring to the Mongol conquerors of India and to their descendants. The Mongols occupied India from A.D. 1526 up to the 18th century. However the authority of the Mughal emperor was purely nominal after 1707. The 4th emperor, Jahangir (1569-1627), married a Persian princess, Mehr-on Nesa, who became known as Nur Jahan. His son Shah Jahan (1592-1666) built the famous Taj Mahal between 1631 and 1654 in memory of his favourite wife Arjumand Banu Begum, better known by her title as Mumtaz Mahal or Taj Mahal.
NABATAEAN: Habitant from an ancient kingdom to the east and south-east of Palestine that included the Neguev Desert. The Nabataean Kingdom lasted from

around 312 B.C. to A.D. 106. The Nabataeans built a large number of soil-and-retention dams. Some are still in use today as shown by Schnitter (1994).
NAPPE FLOW: Flow situation above a stepped chute for which the water flows as a succession of free-fall jets.
NORMAL DEPTH: Uniform equilibrium open channel flow depth.
OBVERT: Roof of the barrel of a culvert. Another name is "soffit".
OUTLET: Opening through which water can be freely discharged from a reservoir to the river (e.g. bottom outlet).
PASCAL: Blaise Pascal (1623-1662) was a French mathematician, physicist and philosopher. He developed the modern theory of probability. Between 1646 and 1648, he formulated the concept of pressure and showed that the pressure in a fluid is transmitted through the fluid in all directions. He measured also the air pressure both in Paris and on the top of a mountain overlooking Clermont-Ferrand (France).
PASCAL: Unit of pressure named after the Frenchman B. Pascal: one Pascal equals a Newton per square-metre.
PERVIOUS ZONE: Part of the cross-section of an embankment comprising material of high permeability.
PITOT: Henri Pitot (1695-1771) was a French mathematician, astronomer and hydraulician. He was a member of the French *Académie des Sciences* from 1724. He invented the Pitot tube to measure flow velocity in the Seine River (first presentation in 1732 at the *Académie des Sciences de Paris*).
PITTING: Formation of small pits and holes on surfaces due to erosive or corrosive action (e.g. cavitation pitting).
PLUNGING JET: A liquid jet discharging into a receiving pool of liquid. A plunging jet flow is associated with gas bubble entrainment when the jet velocity exceeds a critical value, called "onset velocity of gas entrainment".
PRANDTL: Ludwig Prandtl (1875-1953) was a German physicist and aerodynamicist who introduced the concept of boundary layer in 1904 and developed the turbulent "mixing length" theory.
PRONY: Gaspard Clair François Marie Riche de Prony (1755-1839) was a French mathematician and engineer. He succeeded A. Chézy as director general of the *Ecole Nationale Supérieure des Ponts et Chaussées*, Paris, during the French Revolution.
REECH: Ferdinand Reech (1805-1880) was a French naval instructor who proposed first the Reech-Froude Number in 1852 for the testing of model ships and propellers.
REHBOCK: Theodor Rehbock (1864-1950) was a German hydraulician and professor at the Technical University of Karlsruhe. His contribution to the design of hydraulic structures and physical modelling is important.
RENAISSANCE: Period of great revival of art, literature and learning in Europe in the 14th, 15th and 16th centuries.
REYNOLDS: Osborne Reynolds (1842-1912) was a British physicist and math-

ematician who expressed first the Reynolds Number in 1883 and later the Reynolds Stress (i.e. the turbulent shear stress).

REYNOLDS NUMBER: Dimensionless number proportional to the ratio of the inertial force to the viscous force.

RIBLET: Series of longitudinal grooves. Riblets are used to reduce skin drag (e.g. on aircraft, ship hull). The presence of longitudinal grooves along a solid boundary modifies the bottom shear stress and the turbulent bursting process. Optimum groove width and depth are about 20 to 40 times the laminar sublayer thickness (i.e. about 10 to 20 μm in air, 1 to 2 mm in water).

RICHELIEU: Armand Jean du Plessis (1585-1642), Duke of Richelieu and French Cardinal, was the Prime Minister of King Louis XIII of France from 1624 to his death.

RIGHT ABUTMENT: Abutment on the right-hand side of an observer when looking downstream.

RIGHT BANK (RIGHT WALL): Looking downstream, the right bank or the right channel wall is on the right.

RIQUET: Pierre Paul Riquet (1604-1680) was the designer and Chief Engineer of the Canal du Midi built between 1666 and 1681. It provides an inland route between the Atlantic and the Mediterranean across southern France.

ROCKFILL: Material composed of large rocks or stones loosely placed.

ROCKFILL DAM: Embankment dam in which more than 50% of the total volume comprises compacted or dumped pervious natural stone.

ROLLER: Large-scale turbulent eddy.

ROLLER COMPACTED CONCRETE (RCC): Roller compacted concrete is defined as a no-slump consistency concrete that is placed in horizontal lifts and compacted by vibratory rollers.

ROLL WAVE: On steep slopes free-surface flows may become unstable. The phenomenon is clearly visible at low flow rates. The waters flow down the chute in a series of wave fronts (called roll waves).

ROUSE: Hunter Rouse (1906-1996) was an eminent hydraulician who was Professor and Director of the Iowa Institute of Hydraulic Research at the University of Iowa (USA).

S.A.F.: St Anthony's Falls Hydraulic Laboratory at the University of Minnesota (USA).

SABAEN: Ancient name of the people of Yemen in southern Arabia. Renowned for the visit of the Queen of Sabah (or Sheba) to the King of Israel around 950 B.C. and for the construction of the famous Marib Dam (115 B.C. to A.D. 575). The fame of the Marib Dam was such that its final destruction (in A.D. 575) was recorded in the Koran.

SCALE EFFECT: Discrepancy between model and prototype resulting when one or more dimensionless parameters have different values in the model and prototype.

SCOUR: Bed material removal caused by the eroding power of the flow.

SEEPAGE: Interstitial movement of water that may take place through a dam, its foundation or abutments.
SENNACHERIB (or *AKKADIAN SIN-AKHKHEERIBA*): King of Assyria (705-681 B.C.), son of Sargon II (who ruled during 722-705 B.C.). He built a huge water supply for his capital city Nineveh (near the actual Mossul, Iraq) in several stages. The final stage included several dams, some equipped with stepped spillways (i.e. River Khosr dams), and over 75 km of canals and paved channels.
SEPARATION: In a boundary layer, a deceleration of fluid particles leading to a reversed flow within the boundary layer is called a separation. The decelerated fluid particles are forced outwards and the boundary layer is separated from the wall. At the point of separation, the velocity gradient normal to the wall is zero:

$$\left(\frac{\partial V_x}{\partial y}\right)_{y=0} = 0$$

SEPARATION POINT: In a boundary layer, intersection of the solid boundary with the streamline dividing the separation zone and the deflected outer flow. The separation point is a stagnation point.
SEQUENT DEPTH: In open channel flow, the solution of the momentum equation at a transition between supercritical and subcritical flow gives two flow depths (upstream and downstream flow depths). These are called sequent depths.
SEWAGE: Refused liquid or waste matter carried off by sewers. It may be a combination of water-carried wastes from residences and industries together with ground water, surface water and storm water.
SEWER: An artificial subterranean conduit to carry off water and waste matter.
SHOCK WAVES: With supercritical flows, a flow disturbance (e.g. change of direction, contraction) induces the development of shock waves propagating at the free-surface across the channel (e.g. Ippen and Harleman, 1956). Shock waves are called also lateral shock waves, oblique hydraulic jumps, Mach waves, crosswaves and diagonal jumps.
SIDE-CHANNEL SPILLWAY: A side-channel spillway consists of an open spillway (along the side of a channel) discharging into a channel running along the foot of the spillway and carrying the flow away in a direction parallel to the spillway crest (e.g. Arizona-side spillway of the Hoover Dam, USA).
SIMILITUDE: Correspondence between the behaviour of a model and that of its prototype, with or without geometric similarity. The correspondence is usually limited by scale effects.
SKIMMING FLOW: Flow regime above a stepped chute for which the water flows as a coherent stream in a direction parallel to the pseudo-bottom formed by the edges of the steps. The same term is used to characterise the flow regime of large discharges above rockfill and closely-spaced large roughness elements.

SLOPE: 1. Side of a hill; 2. inclined face of a canal; 3. inclination from the horizontal.
SPECIFIC ENERGY: Is proportional to the energy per unit mass, measured with the channel bottom as the elevation datum. It is expressed in metres of water. The concept of "specific energy", first developed by B.A. Bakhmeteff in 1912, is commonly used in open channel flows.
SPILLWAY: Opening built into a dam or the side of a reservoir to release (to spill) excess flood waters. Ancient names for a spillway include waster waterway, wastewater weir, idle-discharge outlet, byewash or bywash.
SPLITTER: Concrete block installed on a chute to split the flow, to increase the energy dissipation and to aerate the flow.
STAGNATION POINT: Is defined where the velocity is zero. When a streamline intersects itself, the intersection is a stagnation point. For irrotational flow a streamline intersects itself at right-angle.
STAIRCASE: Another adjective for "stepped": e.g., a staircase cascade.
STILLING BASIN: A stilling basin is structure for dissipating the energy of the flow below a spillway, outlet work, chute or canal structure. In the majority of cases, a hydraulic jump is used as energy dissipator within the stilling basin.
STOP-LOGS: Form of sluice gate comprising a series of wooden planks, one above the other, and held at each end.
STORM WATERS: Waters drained off a catchment area during or after a heavy rainfall.
STRAUB: L.G. Straub (1901-1963) was Professor and Director of the St Anthony Falls Hydraulics Laboratory at the University of Minnesota (USA).
STREAMLINE: Is the line drawn so that the velocity vector is always tangential to it (i.e. no flow across a streamline). When the streamlines converge the velocity increases. The concept of streamline was first introduced by the Frenchman J.C. de Borda.
SUBCRITICAL FLOW: For mean flow velocities less than the critical flow velocity, the flow is defined as subcritical. The Froude Number is less than unity. In practice, subcritical flows are best controlled by the downstream flow conditions.
SUPERCRITICAL FLOW: When the flow depth is less than the critical flow depth, the flow is supercritical and the Froude Number is larger than one. Supercritical flows can be controlled only from upstream (e.g. gate, bottom outlet).
SURFACE TENSION: Property of a liquid surface displayed by its acting as if it were a stretched elastic membrane. Surface tension depends primarily upon the attraction forces between the particles within the given liquid and also upon the gas, solid or liquid in contact with it. The action of surface tension is to increase the pressure within a water droplet or within an air bubble. For a spherical bubble of diameter d_{ab}, the increase of internal pressure necessary to balance the tensile force caused by surface tension equals: $\Delta P = 4*\sigma/d_{ab}$ where σ is the surface tension.
SYSTÈME INTERNATIONAL D'UNITÉS: International system of units adopted

in 1960 based on the mètre-kilogram-second (MKS) system. It is commonly called SI unit system. The basic seven units are: for length, the metre; for mass, the kilogram; for time, the second; for electric current, the ampere; for luminous intensity, the candela; for amount of substance, the mole; for thermodynamic temperature, the Kelvin. Conversion tables are given in Appendix 2.

SYSTÈME MÉTRIQUE: International decimal system of weights and measures which was adopted in 1795 during the French Revolution. Between 1791 and 1795, the *Académie des Sciences de Paris* prepared a logical system of units based on the metre for length and the kilogram for mass. The standard metre was defined as 1 E-7 times a meridional quadrant of earth. The gram was equal to the mass of 1 cm^3 of pure water at the temperature of its maximum density (i.e. 4 Celsius) and 1 kilogram equalled 1,000 grams. The litre was defined as the volume occupied by a cube of 1E+3 cm^3.

TAILWATER DEPTH: Downstream flow depth.

TAILWATER LEVEL: Downstream free-surface elevation.

TOTAL HEAD: The total head is the total energy per unit mass and per gravity unit. It is expressed in metres of water.

TRASHRACK: Screen comprising metal or reinforced concrete bars located in a waterway at an intake to precent the ingress of floating or submerged debris.

TURBULENCE: Flow motion characterised by its unpredictable behaviour, strong mixing properties and a broad spectrum of length scales (Lesieur, 1994).

TURBULENT FLOW: In turbulent flows the fluid particles move in very irregular paths, causing an exchange of momentum from one portion of the fluid to another.

TURRIANO: Juanelo Turriano (1511-1585) was an Italian clockmaker, mathematician and engineer who worked for the Spanish kings Charles V and Philip II. It is reported that he checked the design of the Alicante Dam for King Philip II.

UNIFORM EQUILIBRIUM FLOW: Occurs when the velocity and depth are identically the same at every point, in magnitude and direction, for a given instant in a prismatic canal:

$$\frac{\partial V}{\partial s} = 0 \qquad \frac{\partial d}{\partial s} = 0$$

in which time is held constant and ∂s is a displacement in the flow direction. That is, steady uniform flow (e.g. liquid flow through a long pipe at a constant rate) and unsteady uniform flow (e.g. liquid flow through a long pipe at a decreasing rate).

UPLIFT: Upward pressure in the pores of a material (interstitial pressure) or on the base of a structure.

VAUBAN: Sébastien Vauban (1633-1707), Maréchal de France, extended the feeder system of the Canal du Midi in 1686-1687.

VILLARÉAL DE BERRIZ: Don Pedro Bernardo Villaréal de Berriz (1670-1740) was a Basque nobleman. He designed several buttress dams, some of these being still in use (Smith, 1971).
VITRUVIUS: Roman architect and engineer (94-?? B.C.). He built several aqueducts to supply the Roman capital with water. (Note: there is some uncertainty about his full name: "Marcus Vitruvius Pollio" or "Lucius Vitruvius Mamurra", Garbrecht, 1987.)
V.N.I.I.G.: Institute of Hydrotechnics Vedeneev in St Petersburg (Russia).
VOC: Volatile Organic Compound (e.g. chlorine).
VON KÁRMÁN CONSTANT: see *KÁRMÁN CONSTANT*.
WADI: Arabic word for a valley which becomes a watercourse in rainy seasons.
WAKE REGION: The separation region downstream of the streamline that separates from a boundary is called a wake or wake region.
WARRIE: Australian Aboriginal name for "rushing water".
WASTE WATERWAY: Old name for a spillway. Particularly used in irrigation with reference to the waste of waters resulting from a spill.
WASTEWEIR: A spillway. The name refers to the waste of hydroelectric power or irrigation waters resulting from the spill. A "staircase" wasteweir is a stepped spillway.
WATER: Common name applied to the liquid state of the hydrogen-oxygen combination H_2O. Although the molecular structure of water is simple, the physical and chemical properties of H_2O are unusually complicated. Water is a colourless, tasteless, and odourless liquid at room temperature. One most important property of water is its ability to dissolve many other substances: H_2O is frequently called the universal solvent. Under standard atmospheric pressure, the freezing point of water is 0 Celsius (273.16 K) and its boiling point is 100 Celsius (373.16 K).
WATER CLOCK: Ancient device for measuring time by the gradual flow of water (through a small orifice, into a floating vessel). *Clepsydra* in Greek.
WATERFALL: Abrupt drop of water over a precipice, characterised by a free-falling nappe of waters. The highest waterfalls are the Angel Fall (979 m) in Venezuela (Churún Merú), Tugel Fall (948 m) in South Africa, Mtarazi (762 m) in Zimbabwe.
WATER-MILL: Mill (or wheel) powered by water.
WATER STAIRCASE: Or *ESCALIER D'EAU* is the common architectural name given to a stepped cascade with flat steps.
WEIR: Low river dam used to raise the upstream water level. Measuring weir are built across a stream for the purpose of measuring the flow.
W.E.S.: Waterways Experiment Station of the US Army Corps of Engineers.
WES STANDARD SPILLWAY SHAPE: Spillway shape developed by the US Army Corps of Engineers at the Waterways Experiment Station.
WHITE WATERS: Non technical term used to design free-surface aerated flows.

The refraction of the light by the entrained air bubbles gives the "whitish" appearance to the free-surface of the flow.
WOOD: I.R. Wood is an Emeritus Professor in civil engineering at the University of Canterbury (New Zealand).
YUNCA: Indian of a group of South American tribes of which the Chimus and the Chinchas are the most important. The Yunca civilisation developed a pre-Inca culture on the coast of Peru.

CHAPTER 1

Introduction: Basic concepts in stepped chute hydraulics

1.1 PRESENTATION

Recent advances in technology have permitted the construction of large dams, reservoirs and channels. These progresses have necessitated the development of new design and construction techniques, particularly with the provision of adequate flood release facilities. Chutes and spillways are designed to spill large water discharges over a hydraulic structure (e.g. dam, weir) without major damage to the structure itself and to its environment.

At overflow spillways the flood waters rush as an open channel flow or free-falling jet, and it is essential to dissipate a major part of the flow kinetic energy to avoid damage to the dam toe, the surroundings and eventually the dam itself. Energy dissipation is usually achieved by 1. a high velocity water jet taking off from a flip bucket (or from the dam crest) and impinging into a downstream plunge pool acting as a water cushion (e.g. Chastang Dam, France), 2. a standard stilling basin downstream of the spillway where a hydraulic jump is created to dissipate a large amount of flow energy (e.g. US Bureau of Reclamation designs), or 3. the construction of steps on the spillway to assist in energy dissipation (e.g. Gold Creek Dam, Australia).

In the two first methods, most energy dissipation takes place at the downstream end in the plunge pool and in the stilling basin respectively. At a stepped spillway, the steps increase significantly the rate of energy dissipation taking place in the chute, and eliminate (or reduce greatly) the need for a large energy dissipator at the downstream end. Water flowing over a rough or stepped bottom is highly turbulent and can dissipate a major proportion of its energy.

Although stepped chutes have been used for more than 3,500 years (Chapter 2), the design technique was abandoned in the late 1920s. A recent renewal in interest for stepped spillways has been associated with the development of new construction techniques (e.g. roller compacted concrete) and a number of research projects have been conducted in the last three decades. Characteristics of some model and prototype studies are summarised in Table 1.1 and some examples are shown on Figure 1.1.

A.

B.

Figure 1.1. Stepped spillway models. A. Gabion spillway model (courtesy of Mr. Royet, Cemagref). Model used by Peyras et al. (1992). $\alpha = 45$ degrees, $q_w = 0.22$ m^2/s, h = 0.2 m, W = 0.8 m. Skimming flow regime. B. 1/10-scale model (courtesy of Mr. Royet, Cemagref). Model used by BaCaRa (1991). $\alpha = 51.3$ degrees, h = 0.06 m. Note the few smaller steps at the upstream end, the smooth chute on the left and the pressure tappings on the vertical faces of steps 5, 6 and 7.

C.

D.

Figure 1.1. Stepped spillway models. C. Nihon University spillway model. Model used by Yasuda & Ohtsu (1999). $\alpha = 30$ degrees, h = 0.05 m, W = 0.4 m - $d_c/h = 1$. Skimming flow regime. D. Nihon University spillway model. Model used by Yasuda & Ohtsu (1999). $\alpha = 11.3$ degrees, h = 0.10 m, W = 0.4 m - $d_c/h = 0.2$ nappe flow regime.

Table 1.1. Details of experimental investigations. A. Model studies.

Ref.	Slope	Scale	Step height	Nb of steps	Discharge	Flow regime	Remarks
	(deg.)		h (m)		q_w (m^2/s)		
(1)	(2)	(3)	(4)	(5)	(6)	(7)	(8)
Model studies							
Moore (1943)	N/a		0.15 to 0.46	1		Nappe flow	Drop structure model (USA). W = 0.28 m.
Rand (1955)	N/a		0.2	1	0.01 to 0.08	Nappe flow	Drop structure model (USA). W = 0.5 m.
Horner (1969)						Nappe and skimming flow	(UK).
	22.8 to 40.1		0.102 to 0.203	10			W = 0.3048 m. l = 0.2413 m. Inclined upward steps (δ = 0, +5, 10, 15, 20°).
	22.8		0.051	20			W = 0.3048 m.
	22.8 to 45		0.029 to 0.069	30			W = 0.3048 m. l = 0.06894 m. Inclined upward steps (δ = 0, +5, 10, 15, 20°).
	15.9 to 45		0.305 to 1.067	3			W = 0.6096 m. l = 1.0668 m. Inclined upward steps (δ = 0, +5, 10, 15, 20°).
	22.8		0.449	8			W = 0.6096 m. l = 1.0668 m.
Essery & Horner (1978)	11 to 40		0.03 to 0.05	4 to 18		Nappe and skimming flow	CIRIA tests (UK). Include inclined steps.
Essery et al. (1978)	11 to 45		0.025 to 0.5	4 to 40	0.01 to 0.145	Nappe and skimming flow	CIRIA tests (UK). W = 0.15 m. H_{dam} = 2 m.
Stephenson (1979a,1988)	18.4 to 45		0.15	1 to 4		Nappe and skimming flow	Gabion stepped chute (South Africa). W = 0.38 m
			0.1				W = 0.1 m.
Noori (1984)	5.7		0.004	100	0.007 to 0.09	Skimming flow	(UK). W = 0.5 m.
	11.3		0.013	70	0.025 to 0.2		
Sorensen (1985)	52.05	1/10	0.061	11	0.006 to 0.28		Monksville Dam spillway model (USA). W = 0.305 m. WES crest profile with smaller first steps.
		1/25	0.024	59	0.006 to 0.28	Nappe and skimming flow	
Houston (1987)	72 & 59	1/5 1/10 1/15	0.08 to 0.12		0.2	Skimming flow	W = 0.76 and 1.22 m. Upper Stillwater spillway model.
Baker (1990)	21.8		0.0096 to 0.058	up to 50	up to 0.5	Skimming flow	W = 0.6 m (UK). Wedge shape blocks inclined downward: δ = - 8.3°.
	33.7						Flat blocks parallel to bed.
Diez-Cascon et al. (1991)	53.1	1/10	0.03 & 0.06	50 to 100	0.022 to 0.28	Skimming flow	H_{dam} = 3.8 m (Spain). W = 0.8 m.
BaCaRa (1991)	53.1	1/10	0.06	18	0.026 to 0.207	Skimming flow	H_{dam} = 1.04 m (France). Creager crest profile with 5 smaller first steps. W=1.5 m.
	63.4 59.0 53.1	1/25	0.024	52	0.007 to 0.115	Skimming flow	H_{dam} = 1.2 m (France). Creager crest profile with 3 smaller first steps.
Bayat (1991)	51.3	1/25	0.024		0.006 to 0.07		Godar-e-landar spillway model (Iran). W = 0.3 m.
			0.03				
			0.02				

Table 1.1. Continued.

Frizell & Mefford (1991)	26.6		0.051	47	0.077	Skimming flow	USBR Research laboratory (USA). W = 0.457 m.
Stephenson (1991)	54.5						Kennedy's vale model (South Africa).
Peyras et al. (1991,1992)	18.4, 26.6, 45	1/5	0.20	3, 4, 5	0.04 to 0.27	Nappe and skimming flow	Gabion stepped chute (France). W = 0.8 m.
Beitz & Lawless (1992)	51.3 & 48.0	1/60	0.02	10	6E-4 to 0.093	Nappe and skimming flow	Burton Gorge Dam spillway model (Australia). Smooth crest with smaller first steps.
Frizell (1992)	26.6		0.051	47	0.58	Skimming flow	USBR Research laboratory (USA). W = 0.457 m.
Tozzi (1992)	53.13	1/15	0.0083 0.0166 0.0333 0.05 0.10		0.086 to 0.201	Skimming flow	Model 1 (W = 0.7 m). WES crest profile with smaller first steps (Brazil). H_{dam} = 2.1 m.
Bindo et al. (1993)	51.34	1/21.3	0.038	31 - 43	0.01 to 0.142	Skimming flow	M'Bali spillway model (France). Creager crest profile with smaller first steps (W = 0.9 m).
		1/42.7	0.019	43	0.007 to 0.04	Skimming flow	
Christodoulou (1993)	55		0.025	15	0.02 to 0.09	Skimming flow	W = 0.5 m (Greece). WES crest profile.
Montes (1994)	36.8 & 45		0.03			Nappe and skimming flow	Smooth crest followed by smaller first steps.
Ru et al. (1994)	53.13	1/10			0.003 to 0.32	Nappe and skimming flow	W=0.45m. WES crest followed by small first steps.
			0.02	48			Model B.
			0.04	24			Model C.
			0.06	16			Model D.
			0.08	12			Model E.
Ruff & Frizell (1994)	26.6		0.154		2.6	Skimming flow	H_{dam} = 15.2 m. W = 1.52 m. Inclined downward steps.
Elviro & Mateos (1995)	53.13	1/6 to 1/16					H_{dam} = 7 m.
Kells (1995)	45	1/5		3		Nappe and skimming flow	Gabion steps with impervious horizontal capping.
Zhou (1996)	53.1		0.04	45	0.012 to 0.189	Skimming flow	W=0.2 m. WES crest followed by small first steps.
Tozzi et al. (1998)	52.2	1/15	0.053	50	0.230	Skimming flow	WES ogee crest followed by smaller first steps.
Chamani & Rajaratnam (1999)	59		0.313, 0.625, 0.125		0.073 to 0.205	Skimming flow	H_{dam} = 2.5 m. W = 0.30 m. Ogee crest.
	51.3		0.313, 0.125		0.073 to 0.205	Skimming flow	
Yasuda & Ohtsu (1999)	5.7		0.006 to 0.010	14 to 112	0.008 to 0.080	Skimming flow	W = 0.4 m. (Japan)
	11.3		0.006 to 0.10	14 to 112	0.008 to 0.080	Skimming flow	W = 0.4 m.
	19		0.002, 0.003, 0.006, 0.012,	7 to 352	0.008 to 0.089	Skimming flow	W = 0.4 m.

Table 1.1. Continued.

		0.025, 0.04, 0.05, 0.08				
	30	0.004 to 0.07	8 to 128	0.008 to 0.089	Skimming flow	W = 0.4 m.
	55	0.003 to 0.064	5 to 152	0.008 to 0.089	Skimming flow	W = 0.4 m.
Shvainshtein (1999)	38.7	0.05	8	0.08 to 0.2	Skimming flow	Ogee crest followed by smaller first steps.
	51.3	0.0625	20	0.08 to 0.2	Skimming flow	H_{dam} = 1.14 m. W=0.48 m.
Boes (2000)	30	0.023, 0.046, 0.092	60 to 240		Skimming flow	W = 0.5 m. L = 5.7 m. Nozzle intake.
	50	0.031 0.093	60 to 180		Skimming flow	
Matos (2000)	53.1	0.08	40	0.08 to 0.2	Skimming flow	W = 1 m.
University of Queensland (Present study)	3.2	0.143	1	0.04 to 0.15	Nappe flow NA3	W = 0.25 m. L = 3.2 m. Supercritical intake flow: d_o = 0.02 to 0.04 m.
	3.4	0.143	10	0.04 to 0.17	Nappe flow NA3	W = 0.5 m. L = 24 m. Supercritical intake flow: d_o = 0.03 m.
	3.4	0.071	10	0.04 to 0.17	Transition flow, Skimming flow	W = 0.5 m. L = 24 m. Supercritical intake flow: d_o = 0.03 m.
	21.8	0.10	9	0.005 to 0.5	Nappe flow, Transition flow, Skimming flow	W = 1 m. Broad-crest design. Inflow: smooth sidewall convergent (4.8:1 contraction ratio). Low upstream turbulence level (Tu < 5%).

Table 1.1. Details of experimental investigations. B. Prototype studies.

Ref. (1)	Slope (deg.) (2)	Scale (3)	Step height h (m) (4)	Nb of steps (5)	Discharge q_w (m^2/s) (6)	Flow regime (7)	Remarks (8)
Prototype Studies							
Grinchuk et al. (1977)	8.7	1/1	0.41	12	1.8 to 60	Skimming flow	Dneiper Hydro Plant, Ukraine (former USSR). Full-scale tests. W = 14.2 m. Concrete wedge blocks.
Baker (1994)	18.4	1/1	0.19	--	Q_w = 0.7 to 2 m^3/s	Skimming flow	Brushes Clough spillway. Concrete wedge blocks.
Chanson & Whitmore (1996)	20.6	1/1	1.5	12	Q_w = 27 m^3/s	Nappe flow	Gold Creek Dam spillway (W = 55 m). Flood event: 2 May 1996.
Sanchez-Bribiesca & Gonzalez-Villareal (1996)	51.34	1/1	0.3	--	Q_w = 1,017 m^3/s	Skimming flow	Trigomil Dam during construction (W = 75 m). Flood event: January 1992. Unprotected RCC steps.

Notes: δ: angle between the step and horizontal (δ > 0 inclined upward, δ < 0 inclined downward); H_{dam}: dam height (i.e. crest elevation above downstream toe); W: channel width.

A.

B.

Figure 1.2. Examples of stepped chutes. A. Clywedog Dam, UK (1968) (Concrete, 1993). View from downstream with waters discharging over the stepped spillways in between the dam buttresses. B. Dartmouth Dam, Australia (1977) (courtesy of the Victoria Rural Water Corporation). View from downstream: the dam is on the left. On the right, note the rock-lined stepped cascade and the concrete intake at the top.

C.

D.

Figure 1.2. Examples of stepped chutes. C. RCC Dam: Riou Dam (France, 1990). View from downstream in 1998. Note the short stilling basin. D. Gabion Stepped Weir near Bologna, Italy (1966) (courtesy of Officine Maccaferri). Note the concrete slab protection on the step faces. $\alpha = 45°$, h = 1 m, W = 11 m, H_{dam} = 5 m.

F.

E.

Figure 1.2. Examples of stepped chutes. E. River training near Macerata, Italy (courtesy of Officine Maccaferri). Flow from the background to the foreground. Gabion lining with steps regularly spaced. F. Training of a torrent near La Paz, Bolivia (1986) (Courtesy of Officine Maccaferri). Gabion steps with lumber protection at the edges of the steps.

G.

Figure 1.2. G. Stepped storm water way, Hong Kong, in 1993. Intake of the stepped storm waterway. $\alpha = 50°$, $h = 0.4$ m, $W = 4$ m.

H.

I.

Figure 1.2. Examples of stepped chutes. H. Stepped storm water way, Hong Kong, in 1993. $\alpha = 10^o$, W = 4 m. Note the baffle blocks at the end of the channel. I. Stepped fountain, Peak Tram Cascade, Hong Kong, in 1994.

J.

K.

Figure 1.2. Examples of stepped chutes. J. Stepped fountain, Gap in the Alps (France) (photograph taken in June 1998). K. Series of six check dams to control the solid transport in the Shaar Descent, Saudi Arabia (1982) (courtesy of Officine Maccaferri). Note the concrete capping to protect the weir crest.

L.

M.

Figure 1.2. Examples of stepped chutes. L. Joe Sippel Diversion Weir, Australia (1984) (photograph taken in 1997). M. Murgon Weir, Australia (1935) (photograph taken in 1997). Timber crib structure.

1.2 APPLICATIONS

In the last decades, stepped spillways have become a popular method for handling flood releases (e.g. Fig. 1.2A & B). Examples of application of stepped channels are presented in Figure 1.2 and Table 1.2 (located at the end of Chapter 1). The development of new construction materials (e.g. roller compacted concrete (RCC), reinforced gabions) has increased the interest in stepped spillways. Construction of stepped spillways is compatible with slipforming and RCC placing methods (Fig. 1.2C). Gabion stepped spillways are the most common type of spillways used for gabion dams (Fig. 1.2D). The steps increase signi-

N. 

O. 

Figure 1.2. Examples of stepped chutes. N. Pulteney Weir, Bath, UK (1971). In May 1997 (courtesy of Mr. and Mrs. Burridge). O. Greenup Weir, Australia (1958) (photograph taken in 1998). Timber crib structure.

ficantly the rate of energy dissipation taking place along the spillway face, and reduce the size and the cost of the downstream stilling basin. Steps are used also to dissipate energy in stormwater channels (e.g. Canberra, Hong Kong) and for river training (Fig. 1.2E & F). In Hong Kong Island, the author observed storm-waterways on steep hill slopes designed with stepped channels to assist

P.

Q.

Figure 1.2. Examples of stepped chutes. P. Stepped fountain: Le Bosquet des Rocailles, Château de Versailles (France, 1683). Photograph taken in June 1998. Q. River training used as a cascade in a park, Niigata Prefecture (Japan). Also spelled Nikita prefecture. Ref.: Soil and Water (1993). *Soil and Water Conservation in Taiwan Newsletter*, Soil and Water Conservation Bureau, Taiwan Provincial Government, Taiwan, No. 6, Dec.

R.

S.

Figure 1.2. Examples of stepped chutes. R. Double stepped road gutter, Sumner Road, Brisbane (Australia). Left: looking d/s toward the freeway. Right: looking u/s from the freeway. Double road gutter with stepped profile. Centenary freeway, next to Sumner Rd. freeway entrance, between Darra and Mt. Ommaney, Brisbane. Flat profile. Left: looking d/s toward the freeway; Right: looking u/s from the freeway. 12 Nov. 1996. This road gutter operates at each large storm (visual observations from drivers). S. Bonneville Lock Dam Tailwater Channel, Columbia river project (USA) (courtesy of the US Army Corps of Engineers, Portland District). Looking upstream.

Figure 1.2. Examples of stepped chutes. T. Bolshevik Dam (Russia, 1980) (courtesy of Prof. Y. Pravdivets). H_{dam} = 11.5 m, design flow: 3.3 m^2/s, S_o = 0.12-0.2, W = 12 m. Overlapping precast concrete slabs (1.5 m × 3 m × 0.16 m).

in energy dissipation in addition to usual baffles (Fig. 1.2G & H). Energy dissipation takes place down a tailwater channel in Figure 1.2S. Figure 1.2N shows a stepped weir built downstream of a bridge to protect the pier foundations. Figure 1.2R presents a stepped road gutter, a technique commonly used in Asia, Europe and Australia.

Stepped channels may be used to increase the discharge capacity. In Russia, Soviet engineers developed the concept of overflow earth dam (Grinchuk et al., 1977; Pravdivets & Bramley, 1989). The spillway consists of a revetment of precast concrete blocks laid on a filter and erosion protection layer (Fig. 1.2T). The channel bed is very flexible and allows differential settlements; individual blocks do not need to be connected to adjacent blocks. High discharge capacity can be achieved (Table 1.2). The high degree of safety allows the use of such channels as primary spillways (Pravdivets & Bramley, 1989). Overflow embankment channels were also built in the USA and Australia with RCC and conventional concrete (Chapter 7). In England, a stepped intake of a Morning-Glory type spillway was selected following model tests which showed a larger discharge capacity than with a smooth intake (i.e. Ladybower Reservoir).

In north-eastern Australia, a number of stepped diversion weirs have been built. Most weirs comprise a downstream stepped face which is formed by rows of timber sheets (for the oldest weirs), steel-sheet piles or concrete hollow boxes (filled with rocks and earthfill) with concrete capping between the rows for more recent structures (e.g. Fig. 1.2L, M & O). The stepped profile contributes to the energy dissipation until the tailwater level rises and drowns out the weir.

Stepped cascades are utilised in water treatment plants. Artificial stepped cascades and drop structures can be introduced along or beside rivers and streams to re-oxygenate waters with low dissolved oxygen contents. Near Chicago (USA), five artificial cascades were designed along a waterway system to help the re-oxygenation of the polluted canal (Gasparotto, 1992; Robison, 1994). The waterfalls were landscaped as leisure parks and combined flow aeration and aesthetics. At Petit-Saut Dam, French Guyana, a two steps re-aeration cascade was added to re-oxygenate the turbined waters, despite the associated hydropower loss. In the Netherlands, the Montferland Water Treatment Plant was designed to remove nitrate from ground water by sulphur/limestone denitrification. It includes an aeration cascade to re-oxygenate depleted waters at the end of the process (Hoek et al., 1992). Aesthetical applications of stepped cascades include stepped cascades and fountains in parks and in cities (e.g. Fig. 1.2I, J & P).

Another application is the debris torrent check dam. In mountain areas, debris[1] torrents may have catastrophic impacts. Check dams or debris dams can be used to reduce the impact of debris torrents. Their construction near the headwaters of the creeks may prevent the "initiation" of the debris flow. Checks dams are usually constructed as a succession of drop structures (h = 0.5 to 5 m) to reduce the steep gradient of the creek, to enhance the energy dissipation and to prevent the onset of debris flow (Fig. 1.2K & Q). Some debris dams include culverts and stepped overflow spillway (e.g. Harvey Creek). They are common in Europe (Austria, France, Italy, Poland, Switzerland), North America and the Far East Asia (Japan, Taiwan) (Chapter 7).

1.3 FLOW REGIMES

A stepped chute is an open channel with a series of drops in the invert. For a given geometry, the flow pattern may be either *nappe flow* at low flow rates, *transition flow* for intermediate discharges or *skimming flow* at larger flow rates (Fig. 1.3).

A low flows (i.e. nappe flow regime), the total fall is divided into a number of smaller free-falls (Fig. 1.3A). The water proceeds in a series of plunges from one step to another (Fig. 1.4). The flow from each step hits the step below as a falling jet, sometimes followed by a hydraulic jump. The energy dissipation occurs by jet breakup in air, by jet impact the step, and with the formation of a fully developed or partial hydraulic jump on the step. Over

[1]Debris comprise mainly large boulders, rock fragments, gravel-sized to clay-sized material, tree and wood material that accumulate in mountain creeks (Vandine, 1985).

small dams, a large rate of energy dissipation can take place (Ellis, 1989; Peyras et al., 1992; Chanson, 1995a). The step height must be relatively large for nappe flow. This situation is not often practical but may apply to relatively flat stepped channels or at low flow rates.

For a given step geometry, an increase in flow rate may bring an intermediate flow pattern between nappe and skimming flow: the transition flow regime (Figs 1.3B & 1.5). The transition flow is characterised by a pool of recirculating waters (and sometimes a small air cavity), and significant spray and water deflection immediately downstream of the stagnation point (Fig. 1.3B). On flat slopes, the supercritical flow downstream of the spray region is decelerated up to the downstream step edge (Chapter 4). The transition flow pattern exhibits significant longitudinal variations of the flow properties on each step and as well as from one step to the next one. The flow seems very chaotic and does not present the coherent appearance of skimming flows.

In a skimming flow regime, the water flows down the stepped face as a coherent stream, skimming over the steps and cushioned by the recirculating fluid trapped between them (Fig. 1.3C). The external edges of the steps form a pseudo bottom over which the flow passes. Beneath this, recirculating vortices develop and are maintained through the transmission of shear stress from the water flowing past the edge of the steps. At the upstream end, the flow is smooth and no air entrainment occurs. After a few steps the flow is characterised by a strong air entrainment and by vortices at the step toes (Fig. 1.6). The energy dissipation in the flow appears to be enhanced by the momentum transfer to the recirculating fluid (Rajaratnam, 1990). On large dams, skimming flow situations allow larger flow rates than nappe flows (Chapter 5).

1.3.1 *Prediction of the flow regime*

The type of stepped flow regime is a function of the discharge and step geometry. The author re-analysed a large number of experimental observations of change in flow regimes. The data are summarised in Table 1.3 and all but one were obtained with flat horizontal steps. For each set of data, Column 6 indicates the type of transition: from nappe to transition flow (NA-TRA), from transition to skimming flow (TRA-SK). The indication (NA-SK) suggests that the researchers ignored the transition flow regime.

Overall the results suggest that the upper limit of nappe flow may be approximated as

$$\frac{d_c}{h} = 0.89 - 0.4 * \frac{h}{l} \qquad \text{Transition NA-TRA} \quad (1.1)$$

where d_c is the critical depth, h is the step height and l is the step length, while the lower limits of skimming flow may be estimated as:

$$\frac{d_c}{h} = 1.2 - 0.325 * \frac{h}{l} \qquad \text{Transition TRA-SK} \quad (1.2)$$

Both equations are compared with the experimental data in Figure 1.7. Two issues must be clearly understood:

1. Equations 1.1 and 1.2 were deduced for flat horizontal steps with h/l ranging from 0.05 to 1.7 (i.e. $3.4 < \alpha < 60°$); there is no information on their validity outside of that range and their accuracy is no better than +/- 10%;
2. Equations 1.1 and 1.2 characterise a change in flow regime for uniform or quasi-uniform flows ONLY. For rapidly varied flows, the results are inaccurate. For example, at the upstream end of a stepped chute, the accelerating waters could flow as thick free-falling nappes before becoming a skimming flow regime further downstream (see Chapter 5).

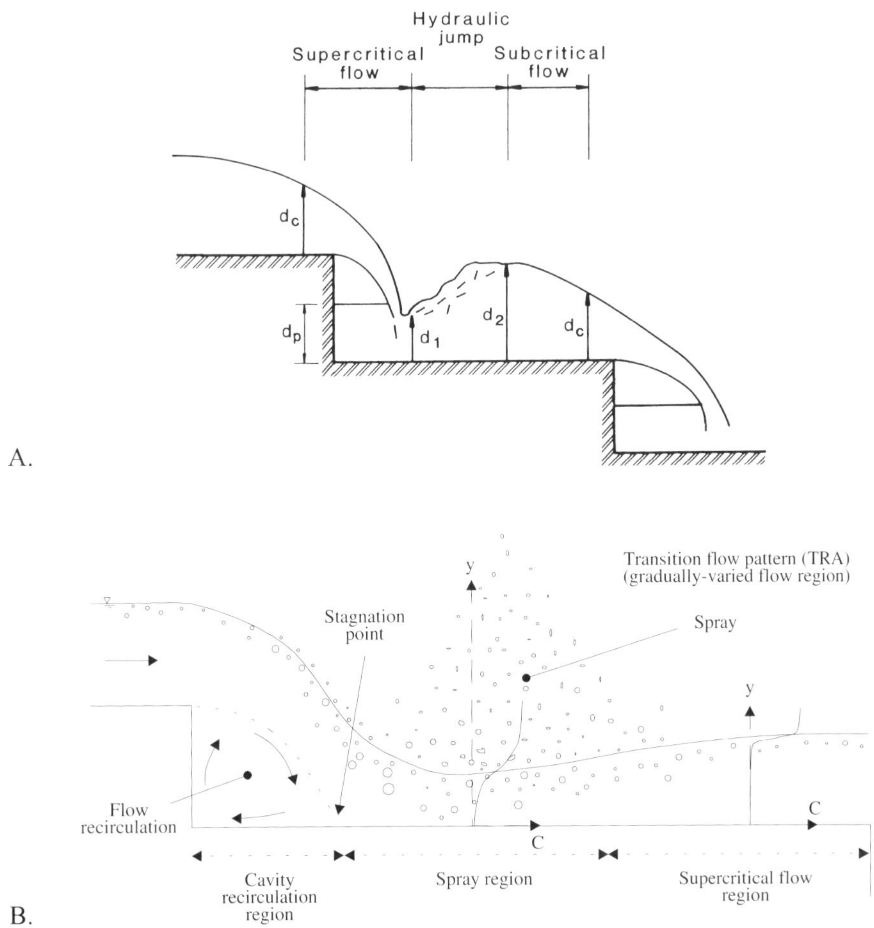

Figure 1.3. Flow regimes above a stepped chute. A. Nappe flow regime with fully-developed hydraulic jump. B. Transition flow regime.

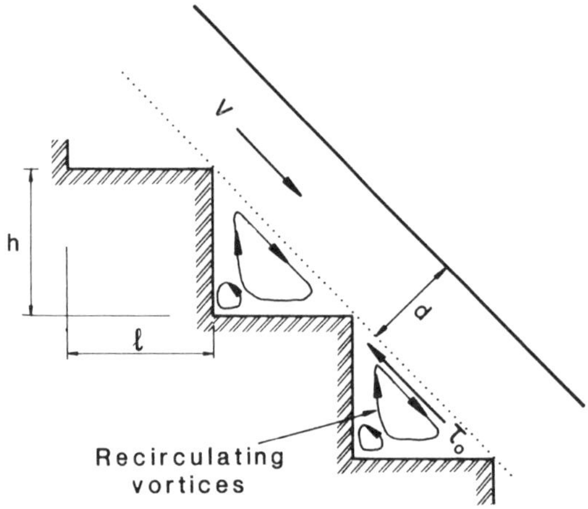

Figure 1.3. Flow regimes above a stepped chute. C. Skimming flow regime with stable cavity recirculation.

Table 1.3. Experimental data - transition flow conditions.

Ref. (1)	Slope α (deg.) (2)	$\frac{h}{l}$ (3)	Step height h (m) (4)	$\frac{d_c}{h}$ Onset (5)	Type of transition (6)	Remarks (7)
Horner (1969)	22.8	0.42	0.029	0.81	NA-SK	30-step model.
	27.8	0.53	0.036	0.80	NA-SK	Horizontal steps.
	36.4	0.74	0.051	0.82	NA-SK	Figure 4.16.
	40.4	0.85	0.069	0.79	NA-SK	
Essery & Horner (1978)	11.3	0.20	0.050	1.15	NA-SK	Flat horizontal steps.
	22.8	0.42	0.029	0.81	NA-SK	
	27.7	0.53	0.036	0.82	NA-SK	
	36.4	0.74	0.050	0.82	NA-SK	
	40.1	0.84	0.057	0.80	NA-SK	
Peyras et al. (1991)	18.4	0.33	0.200	0.74	NA-SK	Gabion. Flat horizontal steps.
	26.6	0.50	0.200	0.67	NA-SK	
	45.0	1.00	0.200	0.61	NA-SK	
Beitz & Lawless (1992)	51.3	1.25	0.020	0.40	NA-SK	Flat horizontal steps.
DNR (1989)	18.4	0.33	0.067	0.56	NA-SK	Pooled steps: d_t/h=0.1.
Montes (1994)	45.0	1.00		0.62	NA-SK	Flat horizontal steps.
	36.9	0.75		0.90	NA-SK	
Kells (1995)	45.0	1.00		0.50	NA-SK	Gabions: horizontal capped (impervious) steps.
Stephenson (1988)	45.0	1.00	7.543	0.74	NA-SK	Flat horizontal steps
	26.6	0.50	5.996	0.77	NA-SK	
	18.4	0.33	4.871	0.79	NA-SK	
	11.3	0.20	3.609	0.83	NA-SK	
Elviro & Mateos (1995)	53.1	1.33		0.51	NA-TRA	
	53.1	1.33		0.72	TRA-SK	
Ru et al. (1994)	53.1	1.33	0.080	0.57	NA-SK	Model E (or B4). Flat horizontal steps.
Ohtsu & Yasuda (1997), Yasuda and Ohtsu (1999)						Flat horizontal steps.
		0.10		1.25	TRA-SK	Lower limit of skimming flow.
		0.34		1.05	TRA-SK	
		0.58		0.93	TRA-SK	
		1.00		0.84	TRA-SK	
		1.43		0.82	TRA-SK	
		0.10		0.76	NA-TRA	Upper limit of nappe flow.
		0.34		0.78	NA-TRA	
		0.58		0.72	NA-TRA	
		1.00		0.60	NA-TRA	

Table 1.3. Continued.

Haddad (1998)	25.6	0.48	0.024	0.98	TRA	Flat horizontal steps. Transition
	13.5	0.24	0.024	1.37	NA-TRA	Upper limit of nappe flow.
	13.5	0.24	0.024	1.46	TRA-SK	Lower limit of skimming flow.
	6.8	0.12	0.024	0.88	NA-TRA	W = 0.75 m.
	6.8	0.12	0.024	1.15	TRA-SK	
	13.5	0.24	0.024	0.88	NA-TRA	
	13.5	0.24	0.024	1.11	TRA-SK	
	25.6	0.48	0.024	0.72	NA-TRA	
	25.6	0.48	0.024	0.98	TRA-SK	
Chamani & Rajaratnam (1999b)	59.1	1.67		0.26	NA-TRA	Flat horizontal steps. h = 0.125, 0.0625, 0.0312 m.
	59.1	1.67		0.22	NA-TRA	
	59.1	1.67		0.20	NA-TRA	
	51.3	1.25		0.36	NA-TRA	h = 0.125 & 0.03125 m
	51.3	1.25		0.32	NA-TRA	
Pinheiro & Fael (2000)	18.4	0.33	0.050	0.95	NA-SK	Flat horizontal steps. W = 0.7 m.
	14.0	0.25	0.050	1.16	NA-SK	
Shvainshtein (1999)	38.7	0.8	0.05	0.72	NA-SK	W = 0.48 m.
	51.3	1.25	0.0625	0.72	NA-SK	Flat horizontal steps.
Boes (2000)	30.0	0.58		0.83	NA-SK	Supercritical inflow.
	50.0	1.19		0.74	NA-SK	Flat horizontal steps.
University of Queensland	3.4	0.06		1.07	NA-TRA	Supercritical inflow. Flat horizontal steps. h = 0.143 & 0.07 m.
	21.8	0.4	0.10	0.53	NA-TRA	Broad-crest. Flat horizontal steps. h = 0.10 m.
	21.8	0.4	0.10	0.97	TRA-SK	

Notes: Type of transition: NA-SK = from nappe to skimming flow; NA-TRA = from nappe to transition flow; TRA-SK = from transition from skimming flow; TRA = transition flow conditions. Data re-analysis by the author.

Figure 1.4. Example of nappe flow regime. Gabion stepped model used by Peyras et al. (1992). (courtesy of Mr. Royet, Cemagref). Note the steps inclined upwards. α = 18.4°, q_w = 0.054 m^2/s, h = 0.2 m, W = 0.8 m, H_{dam} = 4 m.

A.

B.

Figure 1.5. Example of transition flow regime. $\alpha = 22°$, $h = 0.10$ m, $W = 1$ m, $q_W = 0.06$ m^2/s, $d_c/h = 0.71$. A. Sideview. Note the sudden flow bulking downstream of the inception point of air entrainment. B. Detail of the inception point region (flow from top left to bottom right).

A.

B.

Figure 1.6. Examples of skimming flow regime. A. Nihon University model. $\alpha = 19^o$, $W = 0.4$ m, $h = 0.04$ m, $d_c/h = 1.25$. B. Skimming flow over three 1/25 scale models (courtesy of Mr. Royet, Cemagref). From the left to the right: $\alpha = 63.4$, 59 and 53°. $h = 0.024$ m.

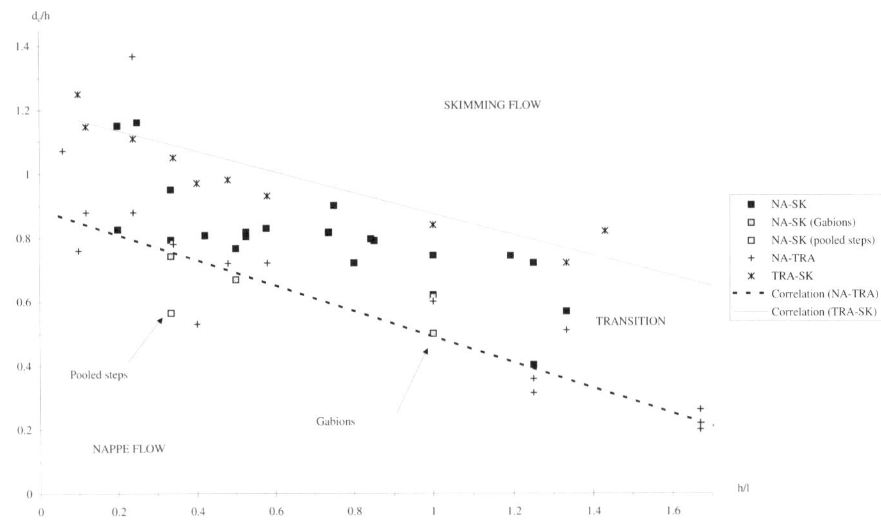

Figure 1.7. Flow conditions for the transition from nappe to skimming flow. Summary of experimental data (Table 1.3).

1.4 AIR ENTRAINMENT ON STEPPED CHUTES

1.4.1 *Introduction*

Stepped chute flows are characterised by a high level of turbulence, and large amounts of air are entrained. Air bubble entrainment is caused by turbulence fluctuations acting next to the air-water free surface. Through this interface, air is continuously trapped and released. Air entrainment occurs when the turbulent kinetic energy is large enough to overcome both surface tension and gravity effects. The turbulent velocity normal to the free surface v' must overcome the surface tension pressure, and be greater than the bubble rise velocity component for the bubbles to be carried away. These conditions are:

$$v' > \sqrt{\frac{8 * \sigma}{\rho_w * d_{ab}}} \tag{1.3}$$

$$v' > u_r * \cos\alpha \tag{1.4}$$

where σ is the surface tension, ρ_w is the water density, d_{ab} is the air bubble diameter, u_r is the bubble rise velocity and α is the channel slope (Ervine & Falvey, 1987; Chanson, 1993a).

Air entrainment occurs when the turbulent velocity v′ satisfies both Equations 1.3 and 1.4. That is, air entrainment at the free surface occurs for turbulent ve-

Figure 1.8. Air entrainment in a stepped chute. $\alpha = 50°$.

locities v′ greater than 0.1 to 0.3 m/s and air bubbles in the range 8-40 mm are the most likely to be entrained for $0 < \alpha < 75°$ (Chanson, 1995a; Fig. 1.6). For steep slopes, the action of the buoyancy force is reduced and larger bubbles are expected to be carried away. On stepped spillways the level of turbulence is high and both conditions (Eqs 1.3 & 1.4) are satisfied (e.g. Fig. 1.8). Hydraulic calculations of stepped chutes must take into account the effects of free-surface aeration (Chanson, 1993c, 1995a).

1.4.2 *Effects of air entrainment*

In stepped channels, the amount of entrained air is an important design parameter. Self-aeration[2] was initially studied because of the effects of entrained air

[2]Natural aeration occurring at the free surface of high velocity flows is referred as free surface aeration or self-aeration.

on the volume of the flow. Air entrainment increases the bulk of the flow which is a design parameter that determines the height of spillway sidewalls.

The presence of air within high-velocity flows may prevent or reduce the damage caused by cavitation (e.g. May, 1987; Chanson, 1997a). Peterka (1953) and Russell & Sheehan (1974) showed that 4 to 8% of air concentration next to the spillway bottom may prevent cavitation damage on concrete surfaces. On stepped spillways, the high rate of energy dissipation reduces the flow momentum in comparison with a smooth chute. The reduction of flow velocity and the resulting increase of flow depth contribute both to lower risks of cavitation as the cavitation index increases. The presence of air within the boundary layer reduces the shear stress between the flow layers and hence the shear force (Chanson, 1994a). The resulting increase of flow momentum must be taken into account when designing a ski jump and a stilling basin downstream of a spillway, a stepped spillway or a rockfill dam with possible overtopping. Air entrainment on spillways and chutes is also recognised for its contribution to the mass transfer of atmospheric gases such as oxygen and nitrogen (Wilhelms & Gulliver, 1989; Toombes & Chanson, 2000). This process contributes to the re-oxygenation of polluted streams and rivers, but it may also cause high fish mortality downstream of large hydraulic structures.

For the designer, the flow bulking is estimated from the total quantity of air entrained while the prevention of cavitation damage requires the knowledge of the air concentration in the fluid layers close to the spillway bottom. The reduction of the drag observed with air entrainment will reduce the energy dissipation above the spillway and hence its efficiency. It must be emphasised that the drag reduction process, observed on smooth-invert chutes, is not completely understood on stepped chutes, although the presence of bubbles across the flow and the bubble size distribution affect the turbulence and the turbulent mixing mechanisms (Chapter 5). The effects of air entrainment on stepped spillway flows are currently under active study in a number of centres around the world (e.g. Table 1.1).

1.4.3 *Air entrainment mechanisms on stepped chutes*

Air entrainment is defined as the entrainment/entrapment of undissolved air that is carried away within the flowing fluid. The resulting air-water mixture consists of both air packets surrounded by fluid and water droplets surrounded by air. Air entrainment may be localised or continuous along an air-water interface. Local aeration includes air entrainment at plunging jet. Interfacial aeration is defined as the aeration process along an air-water interface, usually parallel to the flow direction. Recent reviews of these processes include Wood (1991) and Chanson (1997a).

In a nappe flow, air is entrained at the jet interfaces and by a plunging jet mechanism at the intersection of the lower nappe with the recirculating pool, while de-aeration is often observed downstream (Fig. 1.9A). In the free-falling nappe, interfacial aeration takes place at both the upper and lower nappes. At the lower nappe, the developing shear layer is characterised by a high level of turbulence and significant interfacial air entrainment is observed. The latter may be modelled by a air diffusion model:

$$C = \frac{1}{2} * \left(1 - \operatorname{erf}\left(\frac{y - Y_{50}}{2 * \sqrt{\frac{D_t}{V} * x}}\right)\right) \tag{1.5}$$

where C is the void fraction, y is the vertical co-ordinate, Y_{50} is the location where C = 50%, D_t is an average air bubble diffusivity, V is the jet velocity, x is the horizontal distance measured from the upstream step edge and the function erf is defined as: $\operatorname{erf}(u) = 2/\sqrt{\pi} * \int_0^u \exp(-t^2/2)*dt$ (e.g. Chanson, 1997a; Brattberg et al., 1998). At the upper nappe, the air entrainment process is more complex. The air-water interface is affected by turbulent eddies acting next to the free-surface and by a rapid pressure redistribution downstream of the step brink. A modified solution of the diffusion equation was proposed for the upper nappe of spillway aeration devices:

$$C = \frac{1}{2} * \left(1 - \operatorname{erf}\left(\frac{Y_{50} - y}{2 * \sqrt{\frac{D_t}{V} * x * \left(1 + k * \frac{Y_{50} - y}{x}\right)}}\right)\right) \tag{1.6}$$

where the quantity D_t and Y_{50} are those of the upper nappe and k is a dimensionless parameter taking into account the pressure redistribution (Chanson, 1989a). Equation 1.6 however, neglects the initial free-surface aeration, and it is not valid when the jet core becomes aerated. At the nappe impact additional air is entrained by a plunging jet mechanism and it may become a dominant process at deep-pooled steps. When a hydraulic jump occurs downstream of the nappe impact, additional air is entrained in the jump roller (e.g. Resch et al., 1974; Chanson & Brattberg, 2000).

In the transition flow regime, air bubble entrainment takes place along the jet upper nappe and in the spray region downstream of the stagnation point. The flow is highly turbulent, and air and water are continuously mixed. The air entrainment process in transition flow is not yet fully understood (Chapter 4).

In skimming flows, the aerated flow region follows a region where the free-surface is smooth and glassy. Next to the boundary however, turbulence is generated and the boundary layer grows until the outer edge of the boundary layer reaches the surface (Figs 1.8 & 1.9B). When the outer edge of the boundary layer reaches the free surface, the turbulence can initiate natural free surface aeration. The location of the start of air entrainment is called the point of inception. Downstream of the inception point of free-surface aeration, the flow becomes rapidly aerated and the free-surface appears white as shown by photographs (Krest'yaninov & Pravdivets, 1986; Fig. 1.8). Air and water are fully mixed forming a homogeneous two-phase flow that is similar to that observed in smooth-invert spillway flow (e.g. Chanson, 1997a). The distribution of air concentration may be described by a diffusion model:

$$C = 1 - \tanh^2\left(K' - \frac{y}{2 * D' * Y_{90}}\right) \tag{1.7}$$

where tanh is the hyperbolic tangent function, y is the distance normal to the invert, Y_{90} is the distance where C = 90%, D′ is a dimensionless turbulent diffusivity and K′ is an integration constant (Appendix 6; Chanson, 1997a). D′ and K′ are functions of the mean air content C_{mean} only and they may be estimated as:

$$D' = \frac{0.848 * C_{mean} - 0.00302}{1 + 1.1375 * C_{mean} - 2.2925 * C_{mean}^2} \qquad C_{mean} < 0.7 \tag{1.8}$$

$$K' = 0.32745015 + \frac{0.5}{D'} \tag{1.9}$$

1.4.4 *Basic definitions*

For air-water flow, the characteristic clear-water flow depth d is defined as:

$$d = \int_0^{Y_{90}} (1 - C) * dy \tag{1.10a}$$

where y is the distance measured normal to the bottom, C is the void fraction and Y_{90} is the distance where C = 90% (Fig. 1.10). The depth Y_{90} characterises the air-water flow thickness and it is sometimes used by designers to size the sidewall heights. The depth flow mean air concentration C_{mean} is defined as:

$$(1 - C_{mean}) * Y_{90} = d \tag{1.11}$$

A.

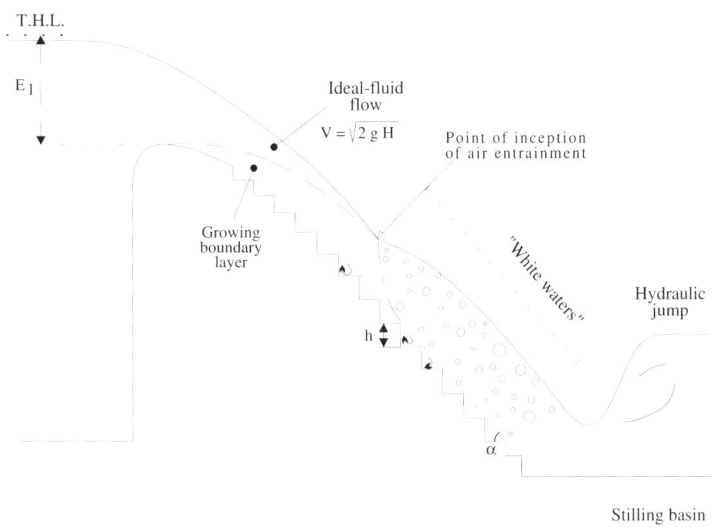

B.

Figure 1.9. Air entrainment on stepped spillways. A. Flow aeration in nappe flow regime with fully-developed hydraulic jump. B. Flow aeration in skimming flow regime.

The average water velocity U_w equals:

$$U_w = \frac{q_w}{d} \tag{1.12}$$

where q_w is the water discharge per unit width.

In a free-falling jet, the clear-water jet thickness is calculated between the lower and upper nappes:

$$d = \int_{Y_{90}^{lower}}^{Y_{90}^{upper}} (1-C) * dy \tag{1.10b}$$

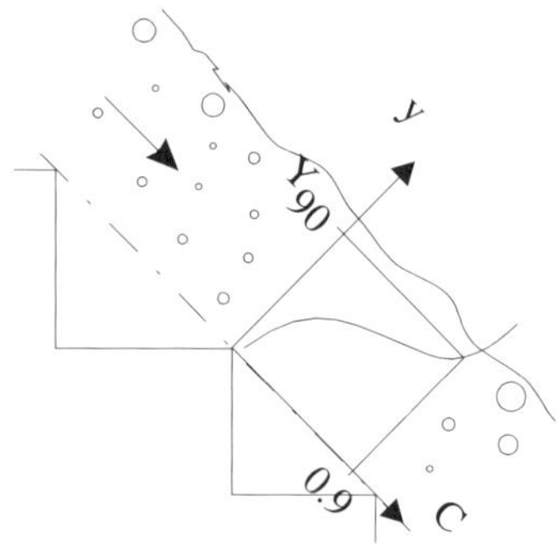

Figure 1.10. Air entrainment on stepped spillways. Definition sketch.

where the subscript (lower and upper) refer to the lower and upper jet interfaces. Basic hydraulic calculations must be performed based upon the clear-water depth (Eq. 1.10) and the flow velocity U_w (Eq. 1.12). For example, flow resistance calculations, residual specific energy H_{res}.

1.5 STRUCTURE OF THE BOOK

The first chapter of this book describes the basic hydraulic characteristics of stepped channels and their applications. The following chapter reviews the history of stepped spillways and irrigation channels. Afterwards, each type of flow regime is analysed separately as their flow features are very different; thus, nappe flow (Chapter 3), transition flow (Chapter 4) and skimming flow (Chapter 5). The flow properties are detailed and design calculations are compared with model and prototype data. The mechanisms of air-water gas transfer are discussed in Chapter 6. Practical examples of stepped chute design are developed in Chapter 7. In Chapter 8, failures and accidents are reviewed and the safety of stepped channels is discussed. Wave phenomena and instabilities are reviewed and discussed in a separate chapter (Chapter 9). In the summary and conclusion (Chapter 10), key issues are summarised and basic questions are answered.

In addition, a detailed list of references are presented. This is followed by a list of physical and chemical properties of fluids (App. 1). The book presents results expressed in SI Units and a table of unit conversions is given in Appendix 2. Several appendices detail particular calculations: nappe trajectory at a drop structure (App. 3), bubble rise velocity calculations (App. 4), modelling form drag and flow resistance in skimming flows (App. 5), void fraction distributions in chute flows (App. 6), chute calculations in skimming flow (App. 7), modelling air-water gas transfer in skimming flows (App. 8). Appendix 9 presents a correction form. Readers who find an error or mistake are

Table 1.2. Typical examples of stepped cascades and chutes.

Name	Ref.	Slope ? (deg.)	Dam height H_{dam} (m)	Max. disch. q_w (m^2/s)	Step height h (m)	Nb of steps	Type of steps	Remarks
(1)	(2)	(3)	(4)	(5)	(6)	(7)	(8)	(9)
Masonry dam spillway								
Gilboa Dam, USA 1926	[GM]		49	7.8	6.1	8	Steps inclined downwards: δ = -2.9 to -5.7°	W = 403.6 m. 9 steps.
Concrete dam spillway								
Kemer Dam, Turkey 1958	[JA]	51.3	114	17.2		~ 31	Precast concrete blocks.	Concrete gravity dam. W ~290 m.
Clywedog Dam, UK 1968	[E2]	60	72	2.8	0.76		Precast concrete beams.	Buttress Dam. Precast beam spillway. W = 182.9 m.
De Mist Kraal Weir, South Africa 1986	[HD]	59	30	29	1	19	Horizontal steps.	RCC Dam. W = 195 m.
Zaaihoek Dam, South Africa 1986	[HD]	58.2	45	15.6	1	40	Horizontal steps.	RCC Dam. W = 160 m.
Monksville, USA 1987	[SO]	52	36.6	9.3	0.61		Horizontal steps.	RCC Dam. W = 61 m.
Olivettes Dam, France 1987	[BO, GO]	53.1	36	6.6	0.6	47	Horizontal steps.	RCC Dam. W = 40 m.
Upper Stillwater, USA 1987	[HO]	72 & 59	61	11.6	0.61		Horizontal steps.	Overflow RCC dam. Flat crest (9.1m long) followed by a 75° slope. W = 183 m.
Wolwedans, South Africa 1989	[GE]	63.4	70	12.4	1			Overflow RCC dam. W = 77.25 m.
M'Bali Dam, RCA 1990	[BI]	51.3	24.5	16	0.8	36	Horizontal steps.	RCC dam. W = 60 m.
Riou Dam, France 1990		59	22	1.15	0.43	42	Horizontal steps. Conventional concrete.	Overflow RCC dam. Broad-crest design. W = 96 m.
New Victoria Dam, Australia 1993	[Wa1]	72 & 51.3	52	5.4	0.6	85	Horizontal steps.	Overflow RCC dam. W = 130 m.
Petit-Saut Dam, Guyana 1994	[DU, GO]	51.3	37	4	0.6		Horizontal steps.	RCC dam.
Nakasujigawa Dam, Japan 1998	[HS]	54.6	71.6	6.6	0.75		Horizontal steps with chamfer.	Concrete gravity dam. W = 175 m.
Earth overflow dam								
Brushes Clough Dam, UK 1859/1991	[BG]	18.43	26	Q_W = 3.66 m^3/s	0.19		Wedge concrete blocks inclined downward: δ = - 5.6°	Earth dam (1859) with new overflow (1991), trapezoidal section (2.2m wide, 1V:2H sideslope).
Dneiper Hydro Plant, Ukraine (former USSR) 1976	[PR2]	8.75		60	0.405	12	Concrete block system. Horizontal steps.	Full scale tests. V = 23 m/s. W = 14.2 m.
Volymia Dam, Russia (former USSR) 1978	[PR1]	26.7	20			~ 16	Concrete stepped block system.	Experimental earth dam in Siberia. Uncontrolled spillway.
Sosnovski Dam, Russia (former USSR) 1978	[PR2]		11	3.3			Concrete stepped block system.	Earth dam. W = 12 m.

Table 1.2. Continued.

Lukhovitsky dams, Russia (former USSR) 1978, 1980 and 1981	[KR]		11 11.5 7.5	3 3.3 2.9			Concrete stepped block system.	3 earthen dams. W = 12, 12 and 7.5 m respectively.
Transbaikal Region Dam, Russia (former USSR) 1986	[MIL]	14	9.4	20			Concrete stepped block system.	Earth dam. Reservoir capacity: 1.5 Mm^3. W = 110 m.
<u>Gabion dam</u>								
Rietspruit outfall, South Africa	[ST]			13		2 to 4	Gabion steps.	3 gabion stepped weirs. W = 50 m.
Cubatao weirs Brazil 1986	[AG]	63	9, 11 and 8		1	9, 11 and 8	Horizontal gabion steps.	Solid transport and flood flow controls. 3 gabion weirs.
<u>Tunnel spillway</u>								
Baltimore Sewer System, USA 1905?	[BA]	43	4.5		0.305	15	Flat steps protected by 0.15m thick granite blocks	'Flight' sewer to dissipate flow energy.
Stoyord Tunnel System, Norway 1993	[WP]	11.3		Q_W = 120 m^3/s	4	22	Pooled steps with 1m high wall.	Flow de-aeration U/S of a vertical shaft. Length: 420 m.
<u>Unlined rock spillway</u>								
Bellfield Dam, Australia 1966	[MIC, T]		54.9	20.9	12.2	3	Unlined cascade. Flat steps.	Earth-cored rockfill-flanked dam. W = 23 m.
Dartmouth Dam, Australia 1977	[T]		180	Q_W = 2755 m^3/s	15	10	Unlined cascade in granite. Flat steps.	Earth and rockfill dam. Channel width varies from 91.4 m at crest up to 350 m.
La Grande 2 Dam, Canada 1982	[PO]	30	134	Q_W = 16140 m^3/s	17.8 9.1 to 12.2	1 11	Smooth profiled step. Pool depth = 8.5 m. Unlined horizontal steps.	W = 122 m. 1st step: concrete lined. 1.8-km long cascade. Unlined excavations. 2nd to 12th steps.
North Dandalup Dam, Australia 1994	[Wa2]	14	62	25	0.6 (RCC) 14 (unlined)		RCC crest and stepped chute, followed by unlined rock cascade	W = 52 m.
Tehri Dam spillway, India	[VI]	15	211.9	Q_W = 11000 m^3/s	50 to 58	4	Smooth pooled steps. Pool depths from 14 to 18 m.	Smooth profiled cascade system. 4 smooth drops. W = 80 to 95 m.
Paunglaung Dam spillway, Myanmar	[LY]	12	120	Q_W = 10000 m^3/s	25 to 35	4	Pooled steps with 10m high walls.	Unlined rock.
<u>Debris dam</u>								
Harvey Creek, Canada	[HU]					16	Flat steps.	Stepped debris spillway and double culvert.
Yuba River Debris Barrier, USA 1904	[E4]	15.4	6.8	9.8	1.8 to 2.6	3	Smooth profiled steps. Concrete facing.	Hydraulic fill dam anchored by rockfill, piling and concrete facing. W = 380 m.
<u>Timber and crib dam</u>								
Arizona Canal Dam, USA 1887	[E3]		10	33		3	Horizontal steps.	Timber cribs filled with loose rock and gravel.
Canyon Ferry Dam, USA 1898	[WE]		8.8		2.3 to 3.05	3	Horizontal steps.	Timber cribs filled with stones. W = 148 m.

Table 1.2. Continued.

Feather River diversion weir, USA 1912	[ET]	26.6	7.9	73	2.4	4	Horizontal steps.	Crib weir. W = 85.3 m.
Murgon Weir, Australia 1935						3	Horizontal steps.	Timber cribs filled with rocks. Still in use today.
Silverleaf Weir, Australia 1953			5.1			4	Horizontal steps	Timber cribs filled with rocks with upstream clay embankment. Refurbished in 1995.
Greenup Weir, Australia		25.4	5.0		1.3	4	Horizontal steps.	Timber cribs filled with rocks. Still in use today.
<u>Irrigation and diversion weirs</u>								
Val Bird Weir, Australia, 1983			4.0	12.4	1.6 to 2	3	Steps inclined downwards: d = 0.95 deg.. 1 flat step and 2 pooled steps (d_t = 0.15 m).	Steps made of sheet-piles connected by concrete slabs.
Joe Sippel Weir, Australia 1984			6.5			3	Pooled steps.	Rockfill structure with concrete slabs steps.
Maranoa River Weir, Australia 1984	[MIK]	14.5	4.7	~ 16	1.5	4	Horizontal steps with nib wall dissipators.	Interlinked concrete boxes filled with rockfill covered by concrete slabs. W = 90 m.
Bucca Weir, Australia 1987	[QW]	63.4	11.8	55.4	1.6	1	Flat crest (4.5m long) and first step.	Overflow diversion dam (W = 130.8 m). RCC structure.
					0.6	17	Inclined upwards steps: ? = +2.9°	
Jordan II Weir, Australia 1992		17.7	5.3		1.4	3	Flat steps.	Overflow reinforced earth embankment.
<u>Flood control and river training weir</u>								
Pulteney Weir, UK 1971	[GR]	18.4	4.05	Q_w = 368 m^3/s	0.457	3	Concrete weir. Horseshoe shape in plan.	Part of Bath flood control scheme.
<u>Fountains</u>								
Peak Tramway Fountain, Hong Kong		20	1.8		0.15	12	Horizontal steps.	Nappe flow regime.
Taipei World Trade Centre Fountain, Taiwan							Horizontal steps.	Nappe flow regime
<u>Re-aeration cascade</u>								
Calumet Waterway System, USA 1991	[GA, RO]						Waterfalls.	5 artificial cascades.
			3.6	11.57 m^3/s	0.91	4	Pooled steps.	Station No. 1.
			3.6	1.59 m^3/s	0.91	4	Pooled steps.	Station No. 2.
			4.56	13.7 m^3/s	1.52	3	Pooled steps.	Station No. 3.
			4.56	13.7 m^3/s	1.52	3	Pooled steps.	Station No. 4.
			3.6	16.4 m^3/s	0.91	4	Pooled steps.	Station No. 5.

Table 1.2. Continued.

Petit-Saut Dam cascade, French Guyana, 1995			5	1s	2	2	Pooled steps.	Labyrinth-type crest with splitters. Methane removal cascade.
Tulip intake of Morning Glory type spillway								
Ladybower Reservoir, UK 1939	[E1, MA]			240 m^3/s	0.46	16	Masonry steps.	Two 24.4m∅ shafts reducing to 4.5m ∅. Improve discharge efficiency of the overflow bellmouth.
Storm waterway								
Hatton Rd, Hong Kong		50	4		0.4	15	Steps and baffle blocks.	W = 4 m.
		10	10		0.2	35		

Notes: [AG] Agostini et al. (1987); [BA] Babbitt (1922); [BI] Bindo et al. (1993); [BG] Baker and Gardiner (1994); [BO] Bouyge et al. (1988); [DU] Dussart et al. (1993); [E1] *The Engineer* (1939); [E2] *Engineering* (1966); [E3] *Engineering News* (1905a); [E4] *Engineering News* (1905b); [ET] Etcheverry (1916); [GA] Gasparotto (1991); [GM] Gausmann and Madden (1923); [GE] Geringer (1995); [GR] Greenhalgh (1974); [HD] Hollingworth and Druyts (1986); [HO] Houston (1987); [HS] Hakoishi and Sumi (2000); [HU] Hungr et al. (1987); [JA]: Japan Dam Association (1963); [KR] Krest'yaninov and Pravdivets (1986); [LY] Lysne (1992); [MA] Mackey (1988); [MIC] Michels (1966); [MIK] Mickelson and Moorwood (1984); [MIL] Miller et al. (1987); [PO] Post et al. (1987); [PR1]: Pravdivets (1992); [PR2] Pravdivets and Bramley (1989); [QW] Queensland Water Resources (1988); [RO]: Robison (1994); [SO] Sorensen (1985); [ST] Stephenson (1979a); [T] Thomas (1976); [VI] Vittal and Porey (1987); [Wa1] Wark et al. (1991); [Wa2] Wark et al. (1998); [WP] Water Power (1992); [WE] Wegmann (1911).

welcome to record the error on the page and to send a copy to the author. Each symbol is explained in the list of symbols at the beginning of the monograph. Immediately after, a glossary clarifies the main technical (and non-technical) terms used in the monograph.

It must be noted that the present study describes primarily results for channels with flat horizontal steps (unless stated). Some researchers discussed specifically experimental results obtained with inclined and pooled steps: e.g. Essery & Horner (1978), Peyras et al. (1992) and Frizell (1992).

CHAPTER 2

Historical development of stepped spillways, channels and cascades: From Antiquity to the 20th century

2.1 INTRODUCTION

Some studies (e.g. Sorensen, 1985) suggested that the design of stepped channels as energy dissipators was a new technique, developed with the recent introduction of new construction materials (e.g. RCC, strengthened gabions). This statement is simply untrue. The design of stepped channels has been known since Antiquity. The oldest stepped spillway was built about 3,300 years ago. Stepped chutes were designed to contribute to the stability of a structure (e.g. overflow weir) and to dissipate flow energy (Table 2.1; Figs 2.1 to 2.3). The world's oldest stepped channels were probably a series of stepped culverts built in Crete around 1500 B.C. (Fig. 2.4). The Roman aqueducts included also a number of stepped cascade designs (Section 2.4). Large stepped fountains and cascades were built for aesthetical purposes. During the 16th to 18th centuries, some *Grandes Cascades* were larger than contemporary waste waterways and even modern spillways.

This chapter describes the history of stepped spillways and waterways, the use of stepped cascades in the aqueducts, and the historical development of the stepped fountains. It explains how the technique of stepped channels was developed independently by several ancient civilisations. At the end the author reviews the hydraulic characteristics of stepped spillways throughout history. Major differences between spillways and waterways are noted. The former is characterised by a steep slope and high-velocity flows, and the kinetic energy of the flow must be dissipated to prevent damage or possible failure of the dam. The latter (waterway) is usually a flatter channel, with lower flow velocities in which energy dissipation is required to prevent scour and erosion of the invert and of the banks. Irrigation channels, storm waterways, sediment control structures are encompassed in the definition.

2.2 HISTORY OF STEPPED SPILLWAYS

The world's oldest stepped spillway is presumably the overflow stepped weir in Akarnania, Greece, built around 1300 B.C. (Fig. 2.3A & B). The weir is an earthfill embankment, 10.5 m high with a 25 m long crest (Knauss, 1995). The downstream slope is stepped (14 steps) with masonry rubbles set in mortar (Fig. 2.3B). The mean slope is about 45°, varying from 39° down to 73° and the step height ranges from 0.6 to 0.9 m (Knauss, 1995). The dam is still standing (Fig. 2.3A).

Other ancient stepped spillways are those of the two Khosr River dams (or Ajilah dams) in Iraq (Table 2.1). The dams were built around 694 B.C. by the Assyrian King Sennacherib. They were designed to supply water to the Assyrian capital city Nineveh, near the actual Mosul. Both dams featured a stepped downstream face and were intended to discharge the river over their crests. Remains of these dams are still in existence (Smith, 1971; Schnitter, 1994). Later, the Romans built stepped overflow dams in their Empire; remains are found still in Syria, Libya and Tunisia (Table 2.1; Figs 2.1A & 2.2A). An impressive structure was the Kasserine Dam (Fig. 2.2). The overflow spillway extended on the 150 m long crest. Smaller structures were Tareglat (Libya), Qasr Khubbaz (Syria) and Oued Gergour (Tunisia). After the fall of the Roman Empire, Moslem civil engineers gained experience from the Nabataeans, the Romans and the Sabaens. They built dams with stepped overflow in Iraq (e.g. Adheim), Iran (e.g. Khajou; Fig. 2.3), Saudi Arabia (e.g. Darwaish) and Spain (e.g. Mestella Weir). The Adheim Dam was a particularly modern design with a vertical upstream face and steep downstream slope similar to modern gravity dams. Although three sluices were included, flood overflow over the downstream stepped face occurred.

Spanish engineers continued to use the Roman and Moslem structures. They designed also new weirs and dams with stepped spillways: e.g. Almansa Dam, Alicante Dam, Barrarueco de Abajo Dam. Between 1400 and 1850 the dam expertise of Spanish engineers was most exceptional by world standards. It was exported to the "New Indies" after the conquest of America. In central Mexico several stepped overflow dams were built during the 18th and 19th centuries (Fig. 2.2C & D). Some were still used at the beginning of the 20th century (Ministerio de Obras Públicas, Transportes y Medio Ambiante, 1993:256-257). One reservoir, the Pabellon Dam, is also called San Blas Dam (Mexico)[1]. Built in stone masonry, its principal function was irrigation water supply and water power[2] (Fig. 2.3D & E). The reservoir is still in use today.

[1]The dam is located in the Hacienda of San Blas, near the town of Pabellon de Hidalgo, about 35 km from Aguascalientes City.

[2]A water mill was built downstream of and on the left of the dam wall. It is no longer in use.

By the middle of the 17th century, French engineers had learned from the Spanish experience. In the south-west of France, the feeder system of the Canal du Midi[3], designed by Riquet and extended by Vauban, included several stepped cascades (Rolt, 1973). The stepped chutes were designed to prevent scouring.

In England several dams were built near furnaces and water mills. Some included stepped weirs and spillways. It is believed that English engineers gained some experience from the Romans who built aqueducts and dams during their occupation.

It is worth mentioning some relatively ancient timber and crib dams with stepped overflows. Several timber crib structures were reported in central Europe prior to A.D. 1600. Two structures, the Kobila Dam and a Russian dam, lasted a few centuries (Table 2.1; Fig. 2.2B). The north-eastern part of America benefited from the experience of northern European settlers, and timber dams were reported as early as A.D. 1600. During the period 1800-1920, timber dams were popular in America, Australia and New Zealand. Most timber dams were 3 to 10 m high but some much bigger ones were built successfully to a height of 30 m. The stepped overflows contributed to the safe passage of large floods without major damage: e.g. the diversion dam on the Feather River (Table 2.6).

2.3 ANCIENT WATERWAY SYSTEMS

Drop structures and stepped profiles were used in early irrigation and storm waterway systems (Table 2.2). The oldest structure was probably a series of

Symbol	Dams	Symbol	Irrigation systems
x	Akarnania Weir	*	Knossos stepped culverts
O	Khosr River dams	ϕ	Wadi Beihan Valley
@	Roman dams	#	Na'aran Channel
M	Moslem dams	π	Peruvian irrigation systems
S	Spanish dams	+	Grandes cascades (Great stepped cascades)
D	Roman dropshaft cascades		

Figure 2.1. Historical development of stepped channels in the world. Legend.

[3]The Canal du Midi was built between 1666 and 1681 to provide an inland route between the Atlantic and the Mediterranean across southern France. The feeder system was later extended in 1686-1687.

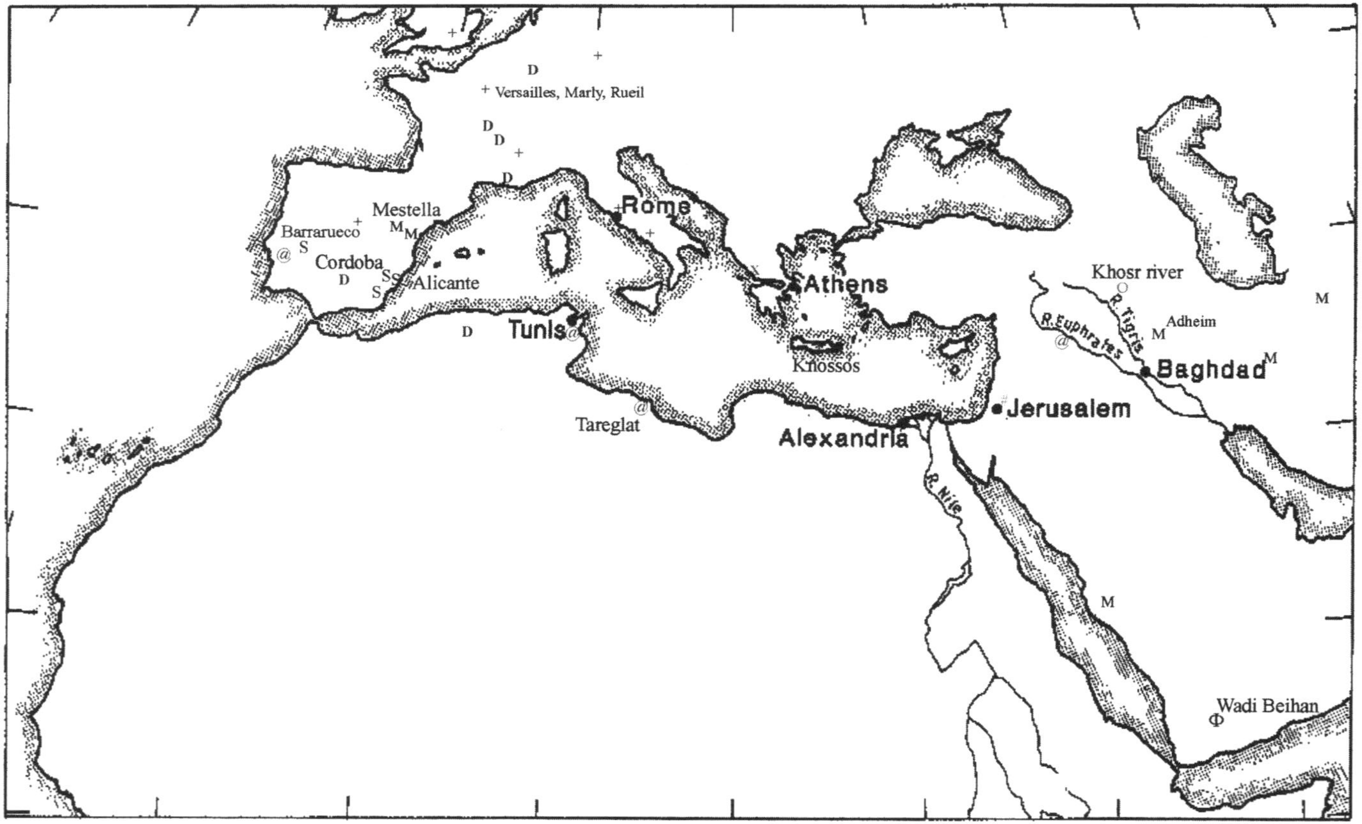

Figure 2.1. Historical development of stepped channels in the World. A. Mediterranean area.

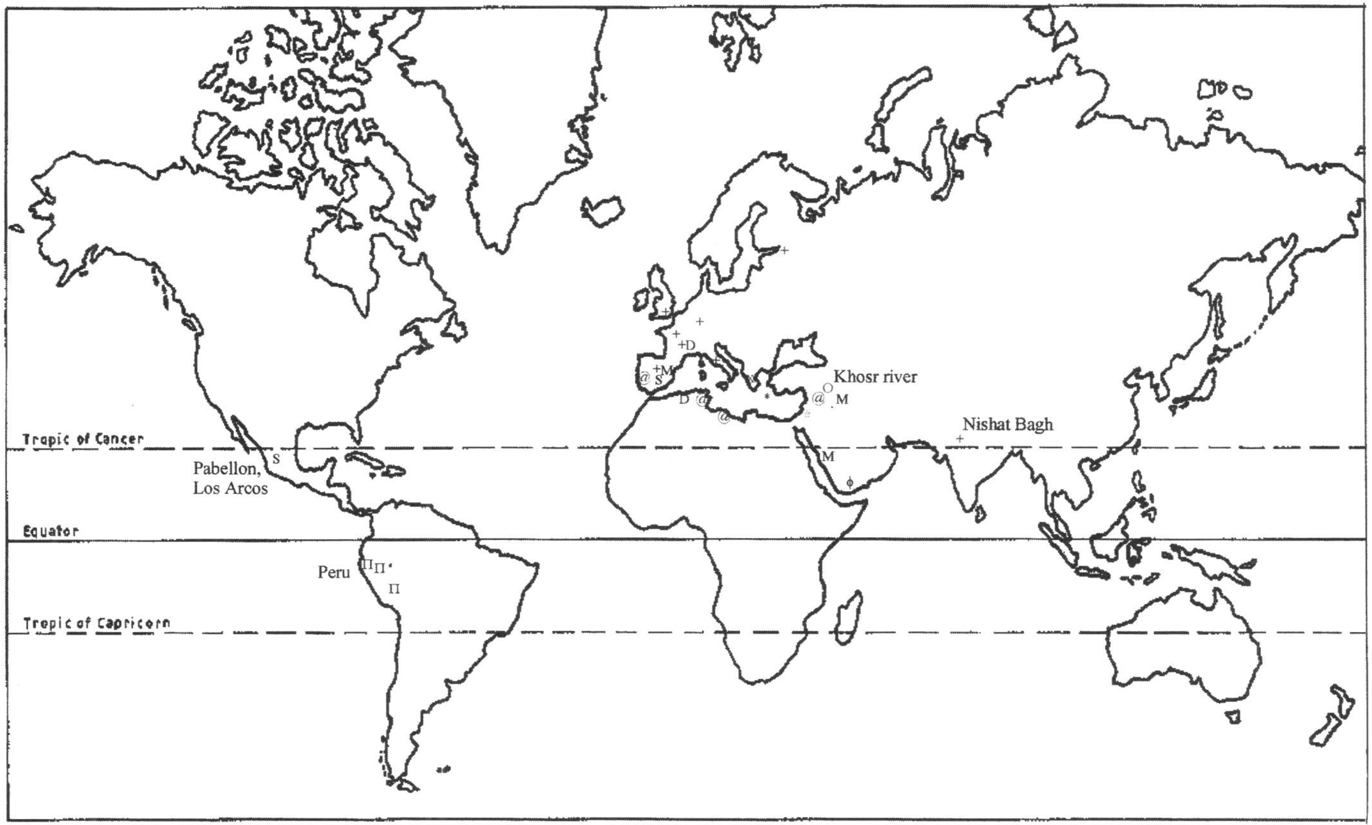

Figure 2.1. Historical development of stepped channels in the World. B. World map.

stepped culverts under a viaduct near Knossos (Crete) (Fig. 2.4). The stepped culverts were designed to carry safely the floodwater down the right slope of the Vlychia Valley beneath a road viaduct leading to the Knossos Palace; the bridge was at least 10 m high and 5.5 m wide (Evans, 1928). Three stepped waterways were excavated although more possibly existed. The steps were made of "squared ashlar masonry" with hard clay mortar. It is believed that the viaduct and stepped culverts were built during the Minoan period around 1500 B.C. They were contemporaneous with the Akarnania Stepped Weir in Greece (Section 2.2).

In Baluchistan, check dams for sediment retention (cultivation of flood plains) were built around 2500 to 1800 B.C.: i.e. the Garbabands (Raikes, 1964-65). One type of construction consisted of upstream and downstream stepped faces with an intermediate earth and rubble filling. This stepped design might have influenced later dam construction in Iran.

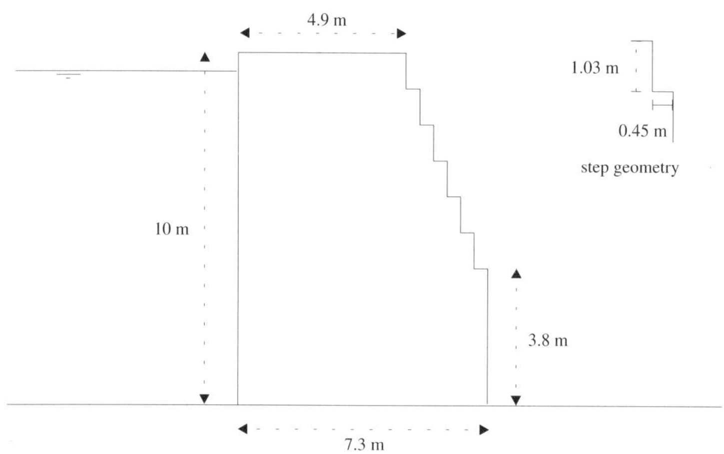

A.

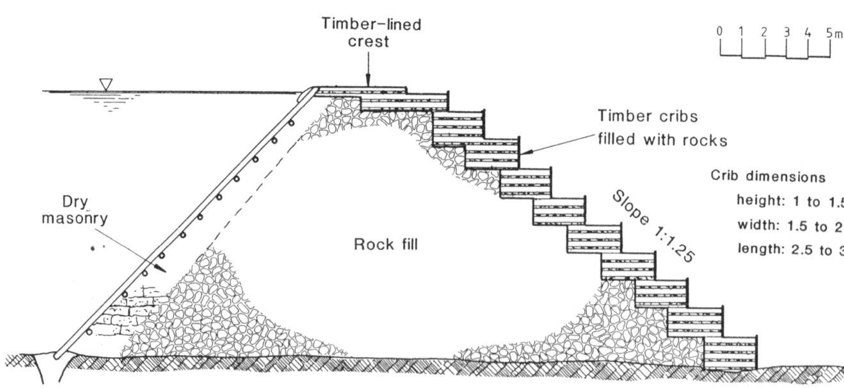

B.

Figure 2.2. Ancient dams with stepped overflow spillway. A. Sketch of the Kasserine Dam, Tunisia (A.D. 100). Details of the spillway. B. Russian dam at the time of Peter the Great, around 1700 (after Gordienko, 1944).

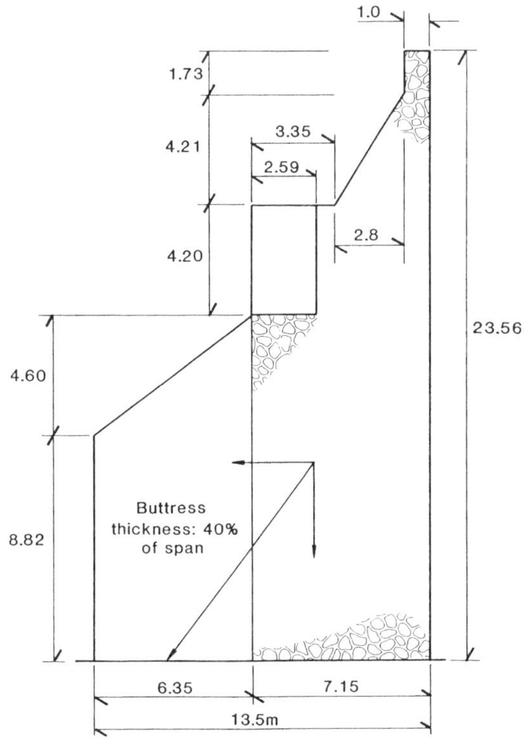

C. Pabellon

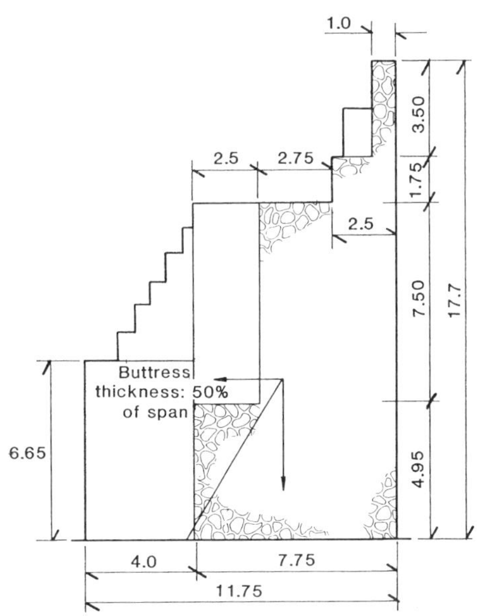

D. Los Arcos

Figure 2.2. Ancient dams with stepped overflow spillway. Spanish dams in Mexico. C. Pabellon Dam 1730? D. Presa de Los Arcos 1780? (after Hinds, 1932).

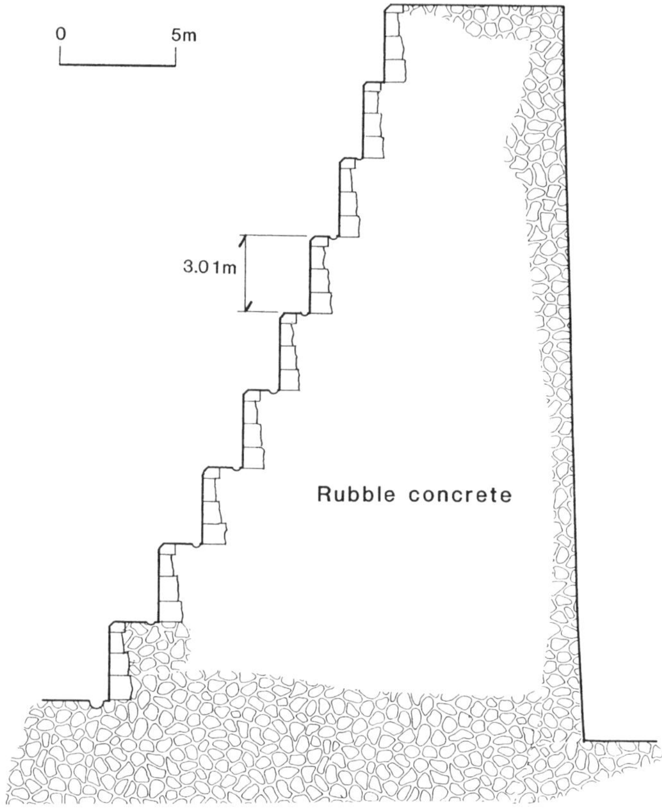

E.

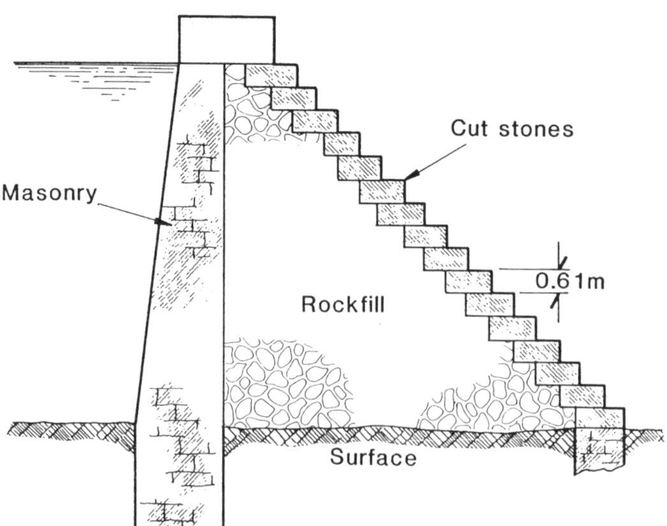

F.

Figure 2.2. Ancient dams with stepped overflow spillway. E. Tytam Dam (or Tai Tam Dam) (Hong Kong, 1887). F. Castlewood Dam, Colorado (USA, 1890). Spillway cross-section (after Schuyler, 1909).

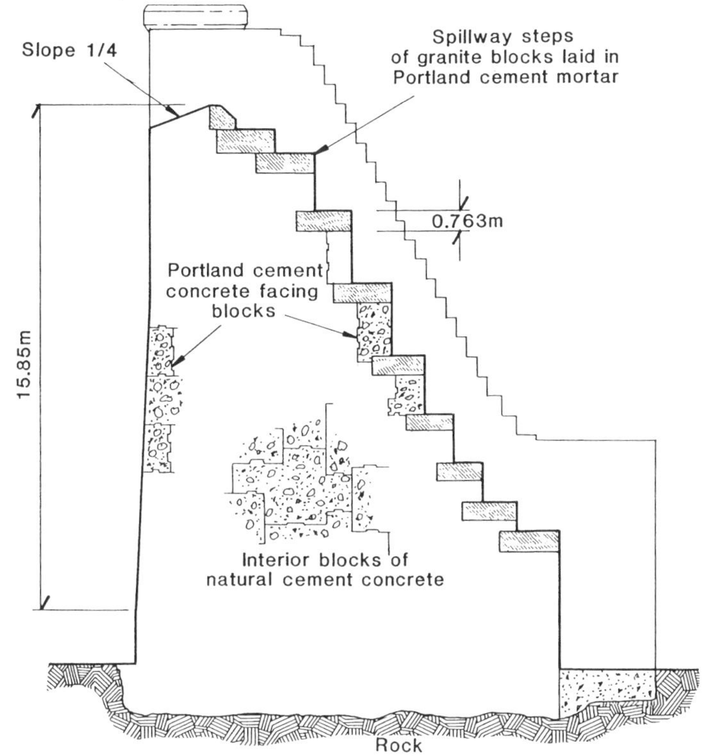

G.

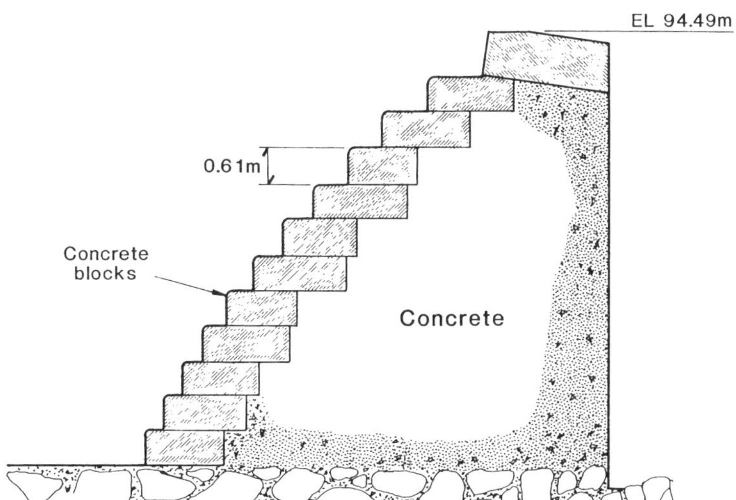

H.

Figure 2.2. Ancient dams with stepped overflow spillway. G. Pedlar River Dam, Virginia (USA), 1905. Spillway cross-section with details of the granite step facing (after Schuyler, 1909). H. Croton Falls Dam (USA, 1911). Spillway cross-section (after Wegmann, 1911).

Figure 2.3. Photographs of ancient stepped spillways. A. Old stepped weir (dam) in Akarnania, Greece 1300 B.C. (courtesy of Professor J. Knauss). Three-quarter view with watermill in foreground and new road ramp over the weir. B. Old stepped weir (dam) in Akarnania, Greece 1300 B.C. (courtesy of Professor J. Knauss). Details of the steps.

C.

D.

Figure 2.3. Photographs of ancient stepped spillways. C. Khajou Bridge-weir[4], Isfahan (Iran, 1660) (courtesy of Mr. A.R. Yazdany). View of the downstream stepped chutes. Note low-flow channels, the stepped chutes being used only for large flood flows. D. Pabellon Dam (Mexico, 1730?) (Hinds, 1953, with permission of the American Society of Civil Engineers). View from the left bank. Note the old water mill in the foreground and the overflow spillway on the right half of the dam.

[4]Other names include Pul i-Khadjn, Khadjoo, Khwaju, Hasanabad or Khvaju.

E.

F.

Figure 2.3. Photographs of ancient stepped spillways. E. Pabellon Dam (Mexico, 1730?) (Gomez-Perez, 1942, with permission of the American Society of Civil Engineers). Waters discharging over the spillway with the old water mill on the right of the picture. F. Ascutney Mill Dam, Windsor (USA, 1834-35) (courtesy of the American Society of Civil Engineers). View of a spill from the right bank. Note the smooth crest followed by the granite steps.

G.

H.

Figure 2.3. Photographs of ancient stepped spillways. G. Tillot Dam (France, 1834). Built as a feeder for the Canal de Bourgogne. Design flow rate: 19 m^3/s, h = 0.4 to 1.2 m, 17.7° mean slope, W = 4.4 down to 2.2 m. View from upstream in January 1997. H. Ternay Dam (France, 1861). Unlined rock cascade at low flow ($Q_w \sim 2$ m^3/s) in June 1998.

I.

J.

Figure 2.3. Photographs of ancient stepped spillways. I. Malmsbury (Australia, 1870) (courtesy of Mr. Rod. Smyth). Australia's first large stepped spillway. Eastern Spillway, Stilling basin in foreground, stepped cascade and upstream drop structure in background (photograph taken on 5 December 1997). J. Pas-du-Riot (France, 1873). View of the masonry spillway (photograph by the author in November 1994).

K.

L.

Figure 2.3. Photographs of ancient stepped spillways. K. Gold Creek Dam (Australia, 1890). View from downstream of the stepped spillway (photograph taken in 1996). L. Gold Creek Dam (Australia, 1890). Stepped spillway in operation in 1996; view from the left bank.

M. N.

Figure 2.3. Photographs of ancient stepped spillways. M. La Tâche Dam (France, 1891). View of the unlined rock spillway in December 1994 after the dam refurbishment. N. Cantref Dam (Wales, 1892) (courtesy of Mr. L. Stuart Davies, Welsh Water). h = 0.3048 m, a = 34°, W = 9.14 m. Design flow rate: 157 m^3/s.

Evidence of stepped canals were found in the Middle-East and in the Americas (Fig. 2.1; Table 2.2). Usually drop structures were introduced to dissipate the flow energy (e.g. Na'aran Channel), and to accommodate with a steep topography (e.g. in Peru). A related form of stepped channel is the terrace irrigation system found in high mountain areas (e.g. in the Cusichaca Valley, Peru). In terrace irrigation, the water falls from one terrace onto the next one and the flow energy must be dissipated at the end of each fall. One of the most ingenious systems was the Quishuarpata Canal (Peru). The canal included two 100 m long steep chutes. Each chute was designed with small steps (h = 1 to 3 cm) along the chute course and large steps (i.e. drop structures) near the downstream end. No stilling ponds were used. The techniques used for the canal highlight the hydraulic expertise of ancient Peruvian engineers (pre-Inca and Inca): they designed canals able to sustain supercritical flows with steep slopes, drop structures and hydraulic jumps at the downstream end of the canals.

Table 2.1. Historical stepped spillways (before A.D. 1850).

Name (1)	Year (2)	Ref. (3)	Dam height (m) (4)	Slope α (deg.) (5)	Construction (6)	Comments (7)
Akarnania Weir, Greece	1300 B.C.	[KN, MU]	10.5	39 to 73	Earthfill overflow weir with downstream stepped face: rubble masonry set in mortar.	Water supply. Equipped with a water mill. Used by the Byzantines. h = 0.6 to 0.9 m.
River Khosr dams, Iraq	694 B.C.	[SM, FO, TH]	2.9		Masonry of limestone, sandstone mortared together.	Built by the Assyrian King Sennacherib to supply water to his capital city Nineveh. Discharge over the dam crest. Lower dam. Called Ajilah Dams [SC2].
			> 1.4	30		Upper dam. 5 steps.
Kasserine Dam, Tunisia	A.D. 100?	[SAL, SM]	10	57	Cut and fitted masonry blocks with mortared joints used to face a rubble and earth core.	Roman dam 220 km SW of Tunis, Tunisia. 4.9m broad crest, followed by 6 steps and a final drop. Discharge over the dam crest. W = 150 m.
Oued Guergour Dam, Tunisia	A.D. 100?	[SAL]	3.6		Rubble masonry dam.	Roman dam between Kasserine and Haïdra. 3 steps (h = 0.55 to 2 m).
Qasr Khubbaz, Syria	A.D. 100/200	[SM, ST]	6.1		Masonry dam with limestone slabs.	Roman dam on the Euphrates river. Reservoir capacity: 9,000 m^3 of water.
Tareglat Dam, Libya	A.D. 200/300	[VI]	> 2		Masonry gravity dam reinforced by buttressing.	Roman dam near Al Khums. Soil-and-water retention dam. Step height: h = 0.6 m.
Monte Novo Dam, Portugal	A.D. 300?	[QU, SC1]	5.7		Curved gravity wall reinforced by two downstream buttresses. Masonry construction.	Roman or post-Roman dam 17 km Est of Evora. Water supply to a watermill. 6 steps.
Dama Dam, Saudi Arabia	A.D. 700?	[FA]	7.5 (?) 10.3 after height.	78	Both outer walls made of dry masonry with earth or rubble core in between.	Moslem dam near Taif. 5 steps (h ~ 1.45 m).
Darwaish Dam, Saudi Arabia	A.D. 700?	[FA]	10	71	Both outer walls made of dry masonry with earth or rubble core in between.	Moslem dam near Taif. 5 steps (h ~ 1.6 m).
Tha'laba Dam, Saudi Arabia	A.D. 700?	[FA]	9	80	Both outer walls made of dry masonry with earth or rubble core in between.	Moslem dam near Taif. Horizontal crest (7.3m long) followed by 5 flat steps (h ~ 1.8 m).
Robella Dam, Spain	A.D. 900?	[SC2]	3.5		Rubble masonry. Shaped crest followed by 5 steps.	Moslem stepped weir. 5 steps (h = 0.37 to 0.7 m).
Mestella Dam, Spain	A.D. 960	[SM]	2.1	27	Rubble masonry and mortar core faced with large masonry blocks and mortared joints.	Stepped weir built by the Moslems. Maximum discharge: around 4,000 m^3/s (?). W = 73 m. 5 steps (h = 0.35 and 1 m).
Khan Dam, Uzbekistan	1000?	[SC2]	15.2	81	Gravity dam. Granite ashlar masonry.	Moslem dam (Ghaznavid dynasty) 100 km North of Samarkand. 2.3m wide crest followed by 7 steps (h = 1.5 to 3.3 m).
Adheim Dam, Iraq	1300?	[SM]	15.2	51	Gravity dam. Cut masonry blocks connected with lead dowels poured into grooves.	Built by the Moslems during the Sassanian period. 7.5m wide crest followed by steps.

Table 2.1. Continued.

Name (1)	Year[a] (2)	Ref. (3)	Dam height (m) (4)	Slope α (deg.) (5)	Construction (6)	Comments (7)
Almansa Dam, Spain	1384?	[SM, WE]	15	40	Curved gravity dam. Rubble masonry with a facing of large masonry blocks.	Discharge over the dam crest. Broad crest followed by 14 steps and an overfall.
Golestan Dam, Iran	1450?	[HA]	16		Gravity dam. Masonry.	Mongolian dam near Mashhad. Overflow spillway: flat crest followed by 3 steps.
Torogh Dam, Iran	1450?	[HA]	20		Gravity dam. Masonry.	Mongolian dam near Mashhad. Stepped rocklined spillway.
Ashburnham Furnace Dam, UK	1563?	[BI]	7		Earth dam with gated masonry spillway. Stones with brick and brickwork repairs.	W = 3.7 m. 3 steps (h ~ 1.5 m).
Kobila Dam, Slovenia	1586	[BR]	10		Timber cribs filled with rocks. Destruction in 1948[a].	Overflow spillway. Crest length: 20 m.
Alicante Dam, Spain	1594	[SM]	41	79	Curved gravity dam. Rubble masonry set in mortar, faced with masonry blocks.	Overflow spillway (W = 80 m). 7 steps: h = 2.7 to 5 m, l = 0.6 to 0.9 m.
Fariman Dam, Iran	1600?	[FA]	21		Gravity dam with a central buttress. Masonry.	Safavid Dynasty. 2 steps (h ~ 6.5 & 14.7 m. Dam heightening in 1930s by 1.5 m.
Akhlamad Dam, Iran	1650?	[FA]	15		Gravity dam. Masonry.	Safavid Dynasty. Flat crest (14.4m long) followed by 2 steps (h = 5 & 10 m).
Khajou Bridge-weir, Iran	1660?		3		Masonry bridge-weir equipped with low-flow deep channels and stepped overflows.	Built by Shah Abbas II. 132.m long weir. 10 steps. Still in use today.
St Ferréol Dam, France	1671	[RO]	32		Waterfalls, cascades, cataracts.	Earth dam with masonry spill weir followed by stepped cascades. Water supply for the Canal du Midi.
Russian dam	1700?	[GO]	12	39	Rockfill dam with timber crib spillway.	Flat crest (3.9 m) followed by 2 first small steps (h = 0.3 & 0.76 m) and 9 steps (h ~ 1.3 m). Design discharge: 0.53 m^3/s.
Barrarueco de Abajo, Spain	1700?	[GD]	6.3	~ 25	Buttress dam. Masonry construction.	Overflow spillway. 5 steps inclined downward.
Kamenskii Dam, Russia	1730?	[DA]			Timber crib dam filled with clay.	Designed by G.W. Hennin. Overflow spillway (W = 23.5 m) with 5 steps. Design discharge ~ 5.2 m^2/s.
Laisota Dam, Spain	1730?	[SM]	4.3		Masonry buttress dam (twin-arch).	Designed by Don Pedro Bernardo Villareal de Berriz.
Pabellon Dam, Mexico	1730?	[GP, HI, SM]	24		Buttress dam. Rubble masonry set in mortar.	Spanish construction (also called San Blas Dam). Discharge over the crest. 3 steps. Reservoir still in use today.
Presa de los Arcos, Mexico	1780?	[HI, SM]	18		Buttress dam. Rubble masonry set in mortar.	Spanish dam across the Rio Morcinique. Overflow spillway. 4 steps.

Table 2.1. Continued.

Name	Year[a]	Ref.	Dam height (m)	Slope α (deg.)	Construction	Comments
(1)	(2)	(3)	(4)	(5)	(6)	(7)
Puentes Dam, Spain	1791	[SM, WE]	50	51	Gravity dam. Rubble masonry core set in mortar and faced with large cut stones. Dam failure in 1802[b].	Dam across the Rio Guadalentin. Discharge over the crest. 4 steps followed by an uniform slope. Step height: h = 4.175 m.
Penarth Weir, UK	1818	[BI]	2.4		Masonry crest. Horizontal steps with sloping faces.	Stepped weir on the River Severn below Newtown. 2 steps (h ~ 1.2 m). W = 42.5 m.
Ascutney Mill Dam, USA	1834-35	[BA, SC3]	12.8		Arched gravity dam made of cut granite with rubble filled core. Downstream stepped buttressing wall.	Overflow stepped spillway (W = 30.48 m) across the Connecticut River. Smooth concrete crest followed by stone stepped profile. Water supply to watermills and later hydropower (1898). Still in use today
Tillot Dam, France	1834		9.2	17.7	Rubble masonry gravity dam with 4 downstream buttresses. Flat crest followed by 2 serries of 4 steps each. Converging channel.	W = 5.5 m down to 2.2 m. h = 0.6 to 1.2 m. Design flow rate: 19 m^3/s. No energy dissipator. Feeder of the Canal de Bourgogne.

Notes: [a]: in chronological order of construction (A.D.). [b]: see Chapter 8.
References: [BA] Battison (1975); [BI] Binnie (1987); [BR] Breznik (1984); [DA] Danilevslkii (1940); [FA]: Fahlbusch (1987); [FO] Forbes (1955); [GD] Garcia-Diego (1977); [GO]: Gordienko (1944); [GP] Gomez-Perez (1942); [HA]: Hartung & Kuros (1987); [HI] Hinds (1932,1953); [KN] Knauss (1995); [MU] Murray (1984); [QU] Quintela et al. (1987); [RO] Rolt (1973); [SAL] Saladin (1886); [SC1] Schnitter (1991); [SC2] Schnitter (1994); [SC3] Schodek (1987); [SM] Smith (1971); [ST] Stein (1940); [TH] Thompson & Hutchinson (1929); [VI] Vita-Finzi (1961); [WE] Wegmann (1911).

2.4 STEPPED CASCADES IN ROMAN AQUEDUCTS

The hydraulic expertise of Romans has contributed significantly to the advances of science and engineering in Antiquity and up to the end of the Middle Ages. The aqueducts in Rome, France and North Africa for example, leave an indelible trace of this savoir-faire (e.g. Ashby, 1935; Rakob, 1974). A Roman aqueduct was a long canal supplying a township in water for public heath purposes (i.e. public baths, latrines). It was designed with flat longitudinal slope; i.e. typically 1 to 3 m/km in average (Hodge, 1992). Short sections however had a steep-gradient: i.e. up to 78% (Chanson, 2000). Current knowledge and field observations suggest primarily three types of design: smooth-invert chutes, stepped channel and the dropshaft cascade. The two latter designs are of interest (Figs 2.5 to 2.7).

Stepped cascades

Roman engineers used both single drops and stepped cascades (Table 2.3). The Brévenne Aqueduct included a number of steep chutes (e.g. Chanson, 2000). One chute was definitely a stepped design: Chevinay (Fig. 2.5). The steps were made of rockfill covered by stone slabs. The step dimensions are similar to modern precast concrete block systems developed by the Russians (Section 7.2). Another large cascade was found at Andriake in Turkey.

Different step geometries were used: the flat horizontal step (e.g. Beaulieu), the inclined downward flat step (e.g. Chevinay) and the pooled step (e.g. Andriake) (Fig. 2.5). Such a wide range suggests that the Roman engineers had a strong experience, if not expertise, in stepped chute design. Pooled step and inclined downward step designs are not usual even by modern standards (e.g. Peyras et al., 1991; Chanson, 1995a). What was the main purpose of the stepped cascade design? At Chevinay, Beaulieu and Cherchell, the cascades were designed to dissipate the kinetic energy of the flow. At Andriake, the cascade is located at the downstream end of a series of arcades. Was the Andriake cascade built for energy dissipation purpose, to treat the water (re-aeration), for aquatic life (in the step pools) or a combination of the above ? The answer is as yet unknown.

A related design is of course the watermill cascade at Barbegal in the South of France (Fig. 2.7; Table 2.3). The available hydraulic power was large: i.e. about 25 to 50 kW ! There, a component of the dissipated energy was transferred to the water wheels.

Dropshaft cascades

A dropshaft cascade is basically a subterranean chute: it consists of a series of dropshafts (Fig. 2.6). The design of Roman dropshafts included an unusual feature, namely a deep wide shaft pool. The pool of water acted as a cushion at the point of nappe impact, preventing scour at the shaft bottom. The shaft pool facilitated further the entrainment of air bubbles by the plunging jet, maximising the bubble residence time and associated air-water gas transfer. The design contributed successfully to an enhancement of the dissolved oxygen (DO) content. The best documented dropshaft cascades are those of Brisecou (Montjeu), Recret (Yzeron), Cherchell and Valdepuentes aqueducts. The latter aqueduct had two large cascades (Cerro de los Pinos, upstream of the Valdepuentes Bridge and Madinat-al-Zhara), but dropshafts were also found in other places.

Two dropshaft shapes were used: rectangular at Vaugneray, Recret and Montjeu (France), and circular at Cherchell (Algeria), Rusicade (Algeria) and Valdepuentes (Spain). The former shape was used at the older Yzeron Aqueduct, possibly because of the ease of construction. The circular shape was used in newer aqueducts (e.g. Cuicul, Cherchell) suggesting that it was possibly a design evolution.

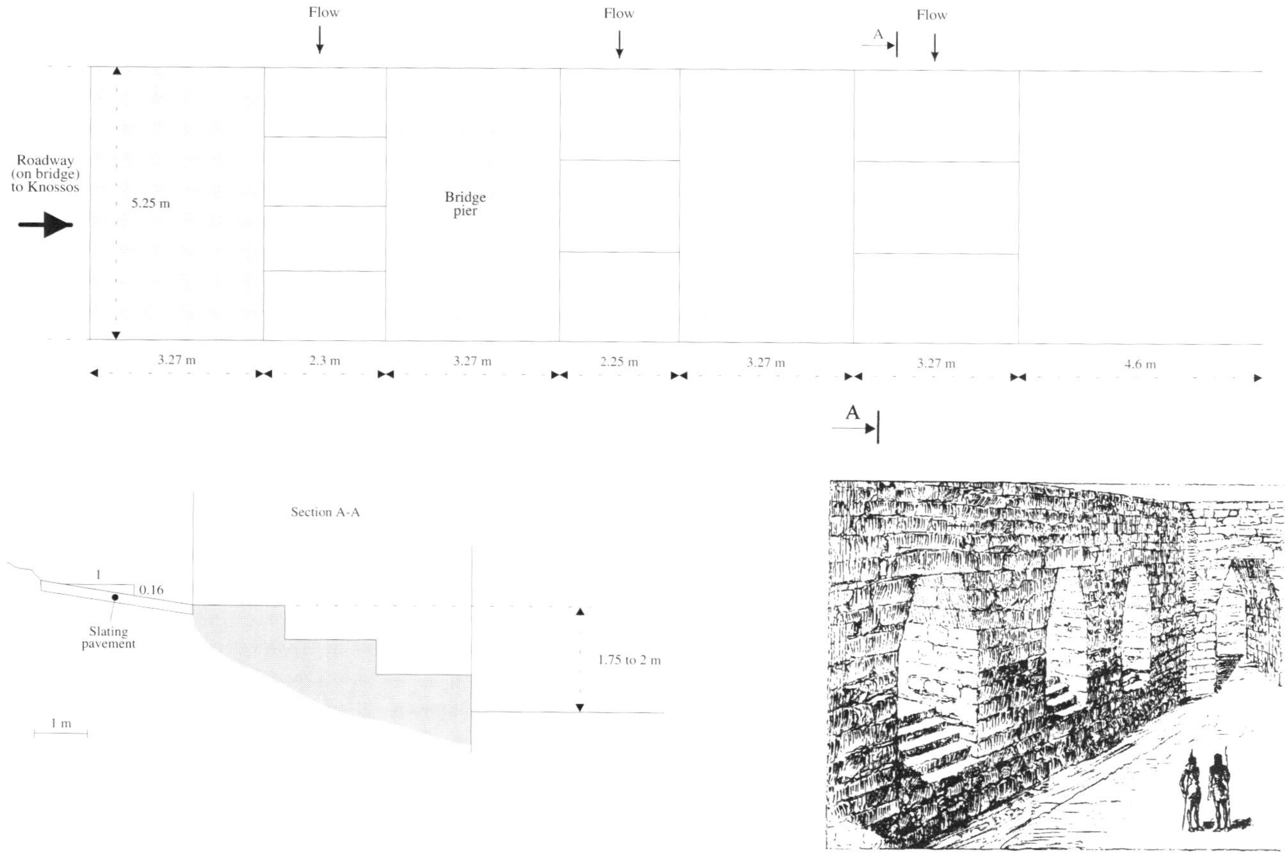

Figure 2.4. Stepped culverts underneath the road to Knossos in the Vlychia Valley (after Evans, 1928).

Table 2.2. Ancient irrigation and storm waterway systems with stepped cascades (before A.D. 1850).

Name	Year	Ref.	Disch q_w (m^2/s)	Slope α (deg.)	Construction and hydraulic design	Comments
(1)	(2)	(3)	(4)	(5)	(6)	(7)
Knossos, Crete	1500 B.C. (?)	[EV]		11	Stepped culverts. Ashlar masonry with clay mortar.	Storm waterway under road viaduct to palace.
					3 steps.	ΔH = 1.75 to 2 m. h = 0.67 m. W = 2.25 to 3.1 m.
					4 steps.	ΔH = 1.75 to 2 m. h = 0.5 m. W = 2.3 m.
Wadi Beihan Valley, Yemen	1000 B.C. to A.D. 200	[LE]				Himyarit irrigation system in Qataban.
Hajar Bin Humeid			2.2 to 4.4	0.03 to 0.11	Paved channel.	Main canal. Maximum water depth: 1.5 m.
			0.4 to 0.7	35 to 60	Stepped intake and drop structures at downstream end.	Secondary canals. W = 0.5 to 5 m. Maximum flow depth < 0.5 m. h = 0.15 to 0.3 m.
Na'aran Channel, Jericho, Israel	103 B.C. to 76	[NE]			Field stones bonded by lime mortar. Energy dissipation by drop structures.	Part of the Wadi Qelt irrigation system built by the Hasmonean King Alexander Janneus. W = 0.6 m.
Salado and Gila valleys, south Arizona, USA	300 B.C. to A.D. 650	[HO, CR]			Earth and stone channels with plastered lining. Stepped profile.	Irrigation systems built by the Hohokams. Re-used by the Mormons of Mesa City in late 19th century.
Moche Valley, Peru	A.D. 200-1500	[FA1, FA2]				Irrigation systems built by the Mochica civilisation, later extended by the Chimus and the Incas. Over 100 km of canals.
Cerro Orejas			0.012		Terrace irrigation system with free-falling nappes.	W = 0.2 to 0.3 m. h < 2 m.
Quishuarpata Canal, Peru	A.D. 1000? to 1532	[FA1]	0.5		Invert made of irregularly shaped granite blocks with granite faced walls.	Artificial irrigation channel (6 km long) in the Hualancay river valley, near Cuzco. W ? 0.8 m.
Chute 1				23.3	Steep narrow steps. Drop structures near the end.	Pre-Inca design. 120 m long. h = 1 to 3 cm.
Chute 2A				32.6	Steep narrow steps.	Chutes 2A and 2B: 150 m long together. h = 1 to 3 cm.
Chute 2B				31	Steep narrow steps. Drop structures near the end.	h = 1 to 3 cm.
Cusichaca Valley, Peru	A.D. 1200 to 1532	[FA1]			Terrace irrigation system with free-falling nappes and granite linings.	Inca irrigation system. h = 4 m.
Inca Irrigation systems, Peru Ollantaytambo	A.D. 1200 to 1532	[CO]		25 to 70[a]	Terrace irrigation systems. Vertical channels and drop structures.	High altitude staircase farms (up to 3,400 m altitude). h = 2.4 to 4.3 m.
Vigevano, near La Sforzesca, Italy	A.D. 1500?	[RI]		26.6	Staircase waterfall (130 steps).	Irrigation channel described by Leonardo da Vinci. h=0.152m.

Notes: h: step height or drop height; [a]: mountain slope.

References: [CO] Cook (1916); [CR] Crouch (1991); [EV] Evans (1928); [FA1] Farrington (1980); [FA2] Farrington & Park (1978); [HO] Hodge (1893); [LE] LeBaron Bowen & Albright (1958); [NE] Netzer (1983); [RI] Richter (1939).

Table 2.3. Stepped cascades and drops in Roman aqueducts.

Steep Section (1)	Δ H (2)	L (3)	S_o = sin α (4)	$(Q_w)_{max}$ (m^3/s) (5)	Remarks (6)
STEPPED CASCADES					
Chabet Illelouine, Cherchell Aqueduct (Alg.)	19	--	--	0.076	Downstream of Oued Bellah.
Beaulieu Aqueduct (Aix-en-Provence, Fra.)	37	--	--	--	Petite cascade. Horizontal and inclined stepped faces: h ~ 0.5 m.
Chevinay, Brévenne Aqueduct (Lyon, Fra.)	87	200	0.40	0.116	Steps inclined downwards: h = 0.92 m, δ = - 11°.
Andriake (Lycia, Turk.)	11	18	0.52	--	Series of 5 pooled steps. W = 1.78 m, h = 2.82 m, d_t = 0.78 m.
DROPS					
Claudia Aqueduct (Rome, Italy)	--	--	--	2.2	Single drop: h = 1.1 m. Near bridge below Vicavaro.
WATER MILL					
Barbegal (Arles, Fra.)	20	60	--	0.26	2 rows of 8 mill wheels each.

Notes: h: step height; $(Q_w)_{max}$: maximum discharge estimate; W: channel width; ΔH: total head loss; δ: angle between the step and horizontal (i.e. δ < 0 inclined downward).
References: Chanson (2000).

The dropshaft cascades were built for large drops in invert elevation: e.g. an overall drop of 200 m at Valdepuentes (Madinat-al-Zhara; Table 2.4). The design had had an excellent reliability record and some cascades were used for centuries[5]. It must be understood that the design was not obvious: a dropshaft cascade was a complex hydraulic structure that included the construction of numerous shafts and interconnection channels in difficult topographic conditions. Two types of dropshaft cascades were built: i.e. a flat invert slope in between shafts and a steep slope. The former design was most common: e.g. at Autun, Recret and Cuicul (Fig. 2.6A). Steep inverts were built at Cherchell and Valdepuentes (Fig. 2.6B) and the connection canals operated with supercritical flow conditions. At Valdepuentes, the invert slope was S_o = 5% between shafts; at Cherchell, a steep chute (S_o = 62%) was located upstream of each shaft. The Valdepuentes Aqueduct was further equipped with three dropshafts with a 90° angle between the inflow and outflow conduits[6]. This type of design was found only at Valdepuentes and Montjeu (France).

The dropshaft cascades might have been used for a rapid vertical drop in invert elevation, kinetic energy dissipation and flow aeration (Chanson, 2000). In the first application, a dropshaft allows the connection between two flat conduits, located at different elevations, along a very short distance; i.e. the

[5]For example, the dropshaft cascades of the Valdepuentes Aqueduct were later re-used by the Moslems.

[6]Upstream of the Valdepuentes stream.

Table 2.4. Dropshaft cascades in Roman aqueducts.

Steep Section	ΔH	L	$S_o = \sin\alpha$	$(Q_w)_{max}$ (m^3/s)	Remarks
(1)	(2)	(3)	(4)	(5)	(6)
Vaugneray, Yzeron Aqueduct (Lyon, Fra.)	22	375	0.058	0.058	Vaugneray branch of Yzeron Aqueduct. About 8 rectangular dropshafts (1.9· 1.1 m^2, h~2.5 m). W=0.4 m.
Recret/Grézieu-la-Varenne, Yzeron Aqueduct (Lyon, Fra.)	38	490	--	0.150	About 15 rectangular dropshafts (1.2· 1.2 m^2, h~2.5 m). W=0.55 m.
Chabet Illelouine, Cherchell Aqueduct (Alg.)	12.3	32	0.36	0.076	4 series of steep chutes (W=0.94 m) followed by circular dropshaft (∅=2.4 m, h=0.8 m).
Moulin Romain, Gunugu Aqueduct (Tun.)	20	--	--	--	4 to 5 circular dropshafts (∅ = 0.8 m, h=3.5 to 4 m). W=0.86 m.
Beaulieu Aqueduct (Aix-en-Pr., Fra.)	37	--	--	--	Combination of steep chutes and dropshafts.
Brisecou Forest, Montjeu Aqueduct (Autun, Fra.)	140	770	0.13	--	Series of rectangular dropshafts (h up to 4.4 m(?)). Possibly up to 24 shafts. At least one 90-degree bend shaft.
Grand thermes, Cuicul Aqueduct (Tun.)	3	85	0.034	--	Series of 4 circular dropshafts ∅ =0.8 m, h=0.4 to 1 m)) on urban distribution line (W = 0.45 m).
Mechernich-Lessenich, Köln Aqueduct (Germ.)	84	--	0.045	0.23	One rectangular dropshaft (1.2 × 0.9 m^2) installed on a steep section.
Cerro de los Pinos, Valdepuentes Aqueduct (Cordoba, Spa.)	120	400	0.29	0.255	Upstream of Valdepuentes bridge. Series of 34 circular dropshafts (∅=0.55 to 0.6 m, h~3 m). Three unusual 90° bend shafts: *spiramina*.
Madinat-al-Zhara, Valdepuentes Aqueduct (Cordoba, Spa.)	200	--	--	0.255	Series of 7 circular dropshafts.

Notes: h: drop in invert elevation; W: channel width; ΔH: total head loss.
References: Chanson (2000), Villanueva (1993).

shaft length. The second application of dropshaft is the dissipation of the kinetic energy of the flow. Such a design is still used today; e.g. storm water systems in Tokyo, the sewer system in Paris. A third application is the flow re-aeration of the flow resulting from the substantial air bubble entrainment taking place in the shaft pool.

2.5 HISTORICAL DEVELOPMENT OF WATER STAIRCASES AND STEPPED FOUNTAINS

In decorative architecture, waterfalls and cascades provide a focal point as well as sounds generated by cascading waters. Although a fountain is originally a spring of water issuing from the earth, the public fountain is a source of public water in cities. The water staircase design contributes to a lot of splashing and white waters, and it is well adapted to a steep topography: e.g. La Grande Cascade de Saint-Cloud on the hill slope of St Cloud overlooking Paris and the River Seine; the water staircase at Château du Touvet on the slopes of the Chartreuse Mountains.

History of stepped fountains and cascades

Over the centuries fountains and cascades have been used as recreational and artistic features (Table 2.5). Greek and Roman architects designed both public and domestic fountains several thousands of years ago. The Moslems developed also a strong artistic sense for water gardens. The Mongols gained expertise from them after the conquest of Persia. Their descendants, the 'Mughals', imported the Moorish garden expertise in India where they built superb water gardens in Kashmir.

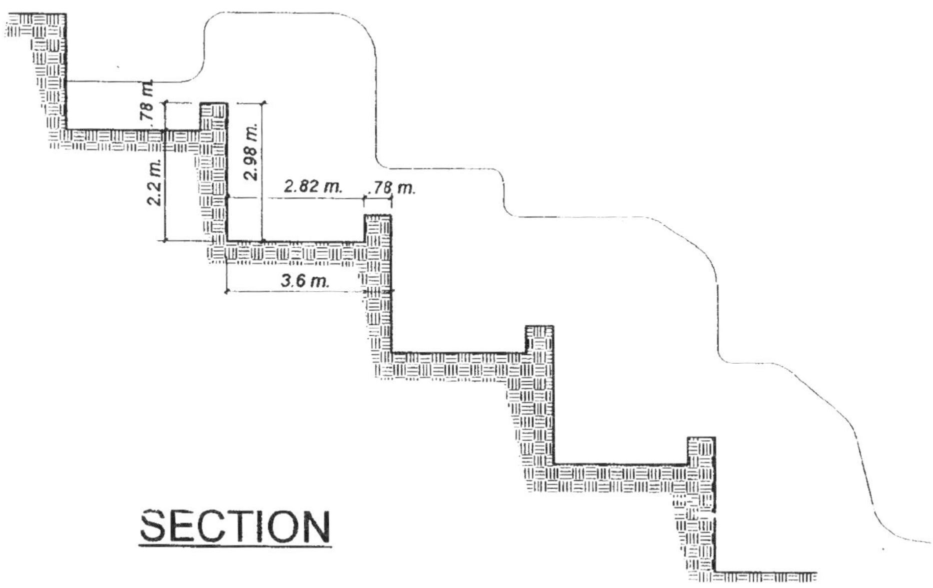

Figure 2.5. Stepped cascades in Roman aqueducts. A. Andriake cascade (courtesy of Dennis Murphy).

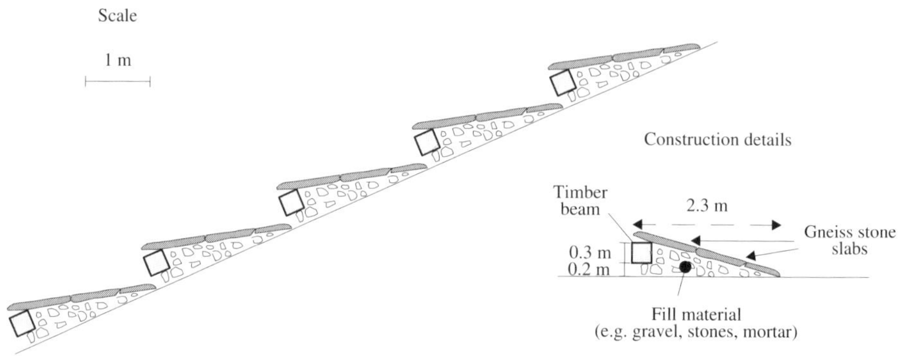

Figure 2.5. Stepped cascades in Roman aqueducts. B. Cascade at Chevinay, Brévenne Aqueduct (Lyon). Inclined downward steps. Author's proposal following the work of P. Leveau et J. Fage.

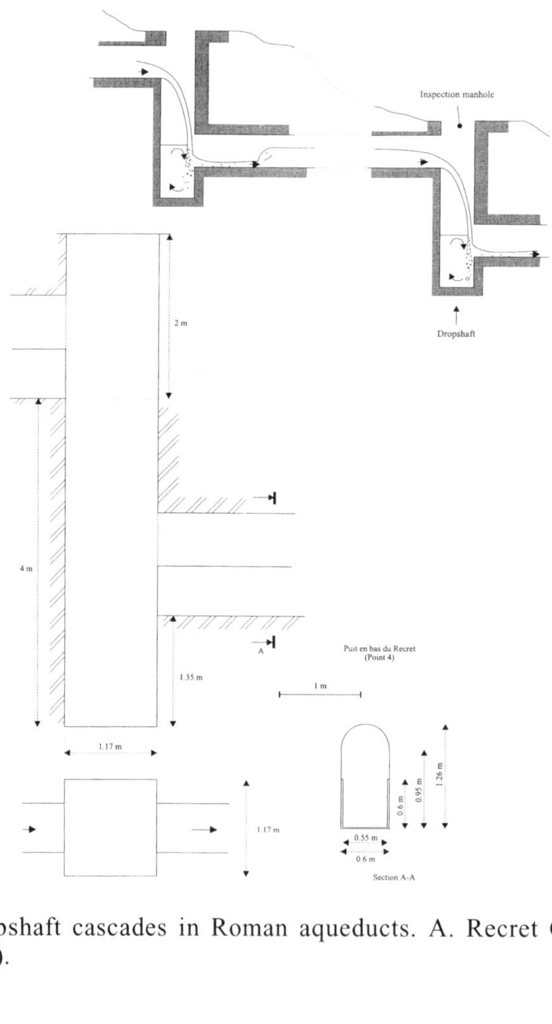

Figure 2.6. Dropshaft cascades in Roman aqueducts. A. Recret Cascade, Yzeron Aqueduct (Lyon).

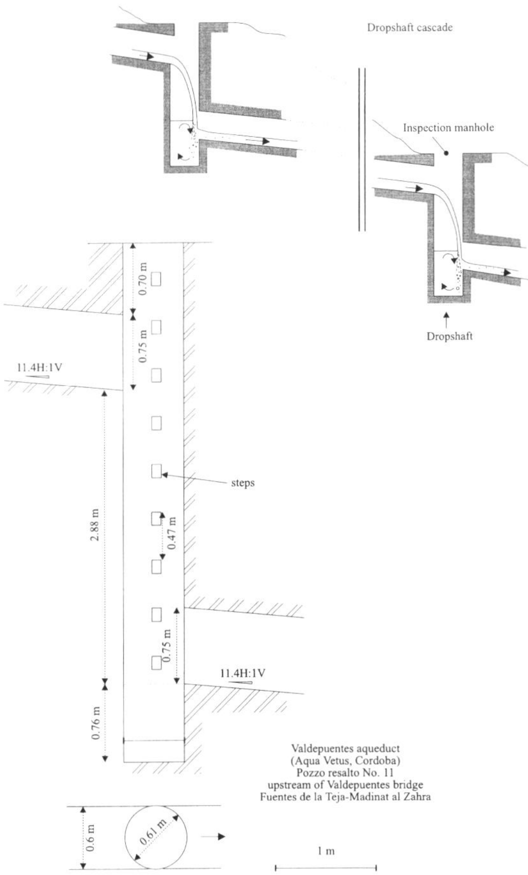

Figure 2.6. Dropshaft cascades in Roman aqueducts. B. Cerro de los Pinos Cascade, Valdepuentes Aqueduct (Cordoba).

In Europe, the tradition of stepped cascades appeared (or re-appeared) in Italy around the Renaissance period; e.g. Villa d'Este, Tivoli. The French gardeners became the masters of stepped cascade design during the 17th century; e.g. Rueil, Marly, Sceaux, St-Cloud, Versailles (Fig. 2.8). Their work influenced all the European countries for the next 200 years. Among the French designs[7], the gardens of Le Château de Marly were most renowned (Fig. 2.8). The park included several magnificent cascades (e.g. La Rivière) which were copied all over Europe including at Chatsworth (UK), Peterhof (Russia), La Granja (Spain) and Wilhemshöhe (Germany) (Chanson, 1998a). At Peterhof one cascade is even called 'Marly'!

[7]The French classical garden is called sometimes the formal water garden in reference to the abundant use of water pieces: e.g. 'bassin', 'buffet d'eau', 'canal', 'cascade', 'douve', 'escalier d'eau', 'étang', 'grille d'eau', 'jet d'eau', 'pièce d'eau'.

Figure 2.7. Barbegal Water Mills (photograph taken in June 1998). $\Delta H = 20$ m, $L = 60$ m. View from downstream. The inflow channel arrived at the top of the hill through a manmade cut into the rock. The mill was equipped with 2 series of 8 water wheels.

It is worth highlighting some design similitude between the Mughal and French gardens. Plumptre (1993) noted a similarity of garden architecture between Taj Mahal completed in 1654, and Vaux-le-Vicomte (completed in 1656). The parallel can be further extended to stepped cascade design: e.g. Nishat Bagh (around 1640) and Rueil (1638), Achabal (around 1620) and Marly (1687). The famous French and Mughal gardens were completed only few years apart albeit the absence of direct contact between the French Kingdom and the Mughal Empire. While the Taj Mahal and Nishat Bagh were the ultimate achievements in Mughal art, Vaux-le Vicomte and Rueil were the precursors of the 17th/18th century European water gardens and cascades (Table 2.5).

After the 18th century, the interest for the great cascades seemed to fade away[8], except in the cities where stepped fountains have been built since the Roman times (Table 2.5; Chapter 7). There are exceptions; e.g. a modern "great cascade" surrounded the olympic cauldron at the Sydney 2000 Olympic Games opening ceremony (Table 7.1). In modern times, the concept of multi-purpose cascades was re-developed. Near Chicago (USA), five re-oxygenation cascades were landscaped as leisure parks combining aesthetics and water quality en-

[8]The "Grande Cascade" of Rueil and "La Rivière" of Marly were replaced by lawns in 1720 and 1728 respectively because of high maintenance costs.

Table 2.5. Characteristics of stepped cascades, water staircases and public fountains.

Name (1)	Slope (deg.) (2)	Cascade height (m) (3)	Step height (m) (4)	Nb of steps (5)	Type of steps (6)	Remarks (7)
Domestic staircase fountains						
Beaurepaire, France (AD 140-200)		0.32	~0.015	7	Flat steps. Rectangular prismatic staircases.	Romain indoor fountain [LA]. Marble construction.
Nissan-lez-Ensérune, France (AD 140-200)		0.30	~0.02	7	Flat steps. Rectangular prismatic staircases.	Romain indoor fountain [LA]. Marble construction.
Garden cascades						
Villa d'Este, Tivoli, Italy 1550-1568					Water staircases, stepped cascades and fountains.	Designed by P. LIGORIO [JE].
Water staircases					Flat steps.	Curved shape.
Ovato fountain					Flat steps followed by drop.	Also called Oval Fountain or Fountain of Tivoli.
Avenue of hundred jets						
Neptune fountain						Built in 1927.
Teatro d'Acqua, Villa Aldobrandini, Frascati, Italy around 1560-1600				9	Water staircase (pooled steps) overhanging a theatre.	Also called the Great Water Theatre. Built for Pope Clement VIII.
Achabal, India around 1600-1620					Large water staircase.	Built by Emperor Jahangir. W ~ 21 m.
Grande Cascade de Rueil, France 1638 (demolition in 1720)					About 30 steps.	Designed by T. Francini for Cardinal de Richelieu. Replaced in 1720 by a lawn. [HE]
Nishat Bagh ('Garden of Gladness'), India around 1620-1640					Long water staircase [PL].	Created by Yamin Al-Aula Asaf Khan, brother of Nur Jahan during the Mughal period.
Villa Garzoni, Collodi, Italy around 1633-1692					Water staircase overhanging a semi-circular arena.	Created by R. Garzoni.
La Grande Cascade de Saint-Cloud, France						
Upper cascade, 1667	~ 12.5		0.2	18	6 series of 3 flat steps (l=0.4 m) and 1 long step (l=1.9 m).	Designed by A. Lepautre. 2 parallel cascades (W = 2 m) separating drops and waterfalls.
Lower cascade, 1697				6	Pooled steps.	Designed by J. Mansart.
Les Grandes Cascades de Sceaux, France 1677	~ 6.3		0.3, 0.85 &0.95	27	Pooled steps with rounded crest.	Designed by A. le Nôtre for the Minister J.B. Colbert.
Bosquet des Rocailles, Versailles, France 1683				4, 7 and 8	5 8-step fountains, 8 7-step fountains, and 4 4-step fountains. Pooled steps.	Also called Bosquet de la Salle de Bal. Designed by J. Mansart. Nappe flow regime.
Les Cascades de Marly, France						Designed by J. Mansart for King Louis XIV.
La Rivière, 1687 (demolition in 1728)	~ 6		~ 0.5	63	Pooled steps made of read and green marble.	About 300 m long. Replaced in 1728 by Le Tapis Vert.
La Cascade de la Rivière, 1687			0.5 to 1.5	4	Pooled steps.	
Les Nappes				6	Pooled steps.	Landscaped with rocks.

Table 2.5. Continued.

La Cascade Champêtre	~ 20 to 25			42	4 series of 9 pooled steps and a large intermediary pool.	Made of white marble. Also called La Cascade Rustique or Les Cascades Champestres.
L'Abreuvoir, 1698				1	Pooled step.	
Chatsworth's Great Cascade, Derbyshire, UK 1696					Horizontal steps with intermediary pools.	Designed by Grillet based upon La Rivière de Marly (see above). Nappe flow regime [JE].
Wilhelmshöhe, Hesse-Kassel, Germany 1701-1713 (originally called Kassel-Weissenstein)						Designed by G. Guerniero [PL]. About 240 m long (W = 10.7 m).
Peterhof's Grand Cascade, St Petersburg, Russia 1704-1715				7	Pooled steps in marble stones. Followed by waterfalls.	Designed by J.B.A. le Blond for Tsar Peter I the Great. Inspired by Marly's cascades. [JE].
La Granja, Segovia, Spain 1720						Built for King Philip V of Spain.
La Cascada				11	Pooled steps. Masonry.	Built by R. Frémin. Inspired by La Rivière de Marly. W ~ 10 m.
Carrera de Caballos				5	Cascading pools.	Include 114 water jets. From the Fountain of Andromeda to the Media Luna and the Ria.
Studley Royal, Fountains Abbey, UK 1725		3		5?		100m long masonry dam [BI]. Nappe flow.
Palazzo Reale, Caseta, Italy 1752, 1770?					Great cascade (12 steps) and water staircases.	Designed by L. Vanvitelli for King Charles III of Naples. Inspired by La Granja. [PL]
L'Escalier d'Eau, Château du Touvet, France 1763					5 successive cascades with rounded step crests.	Designed by P. de Marcieu. Nappe flow regime.
Water staircase No. 1	26	2	0.17	12	Flat steps.	W = 3.6 m. Surrounded by 4 lateral water staircases.
Water staircase No. 2	23	2.1	0.23	9	Flat steps.	W = 3.6 m. Surrounded by 2 lateral water staircases.
Water staircase No. 3	7.25	1.9	0.31	6	Pooled steps.	W = 3.4 m.
Water staircase No. 4	32	2.4	0.27	9	Flat steps.	W = 3.4 m.
Water staircase No. 5	34	1.7	0.28	8	Flat steps.	W = 3.4 m.
<u>Ancient public fountains</u>						
Perge, Turkey AD 200-300					Pooled channel.	Colonnaded Avenue: L = 150 to 00 m, W = 2 m.
					Pooled channel.	Downstream of the Nympheum.
Fontaine des Innocents, Paris, France 1549			0.5 to 1	7	4 semi-circular 7-step cascades. Pooled steps.	Designed and sculpted by J. Goujon. Nappe flow regime.
Cascade du Trocadéro, Paris, France 1878		11.2	1.1 & 3.5	8	7 pooled steps followed by a 3.5m high drop.	Designed by J. Bourdais and G. Davioud for the Universal Exposition in Paris. Demolished in 1935.
Fountain of Porte de St-Cloud, Paris, France 1936					2 circular stepped cascades.	Designed by R. Pommier, J. Billard and P. Landowski. Concrete sculptures and fountains.

References: [BI] Binnie (1987); [HE] Helot-Lécroart (1985); [JE] Jellicoe and Jellicoe (1971); [LA] Lavagne (1998); [PL] Plumptre (1993); Chanson (1998a).

hancement. In Japan and Taiwan, stepped river training is combined with attraction parks. Another use is the fish ladders designed as both aesthetical cascades and fishways in North America and Japan.

Discussion
Most historical stepped cascades were designed with pooled steps. The pooled step design allows a better control of the flow rate and prevent the occurrence of hydraulic jump. Urban fountains are typically designed with a small number of large steps (h = 0.2 to 1 m). Figure 2.9 presents the step height versus the number of steps. The staircase fountains have a larger number of smaller steps, although some grandes cascades are exceptions (e.g. Sceaux; Marly; Fig. 2.9).

The design of some cascades was quite difficult, such as; the unsteady aerated flows on the chadars at Nishat Bagh, the parallel water staircases at Château du Touvet (Fig. 2.8I) the water supply system at Château de Versailles. It is believed that some water garden designers had a solid hydraulic expertise.

2.6 DISCUSSION

2.6.1 *Dissemination of the stepped-chute expertise*

From Antiquity up to the beginning of the 20th century, the Romans, Moslems and Spanish contributed successively to the dissemination of the art of dam-building and stepped channel design. Their contribution in maintaining and developing the expertise in stepped channel design was most important. Dams with stepped overflow spillway were built very early in the Mediterranean area, including the Middle East. The construction technique then spread around the Mediterranean Sea in Roman times (Fig. 2.1). During their expansion period (7th to 9th centuries A.D.), the Moslems learned from the Sabaens, Nabataeans and Romans. The Moslem conquerors of the Hispanic peninsula brought their water traditions with them from the eastern Mediterranean. Seven hundred years of Moorish control of Iberia left a strong hydraulic influence and a solid tradition of irrigation structures. Following the reconquest of Spain, the Catholic Spanish benefited from the Roman and Moslem precedents. Later the Spanish conquered the New World and transferred deliberately their technology in turn.

Spain occupied a exceptional place in the development of large dams and stepped spillways. Most European countries and European settlements in America benefited from their expertise in the fields of dam and stepped spillway design. Clear evidence of the Spanish influence were found certainly in France, Mexico and United States (Table 2.6; Figs 2.1 to 2.3, 2.10)[9].

[9]It is worth noting the existence of some pre-European hydraulic expertise in northern, central and southern America (e.g. Smith, 1971). For example, irrigation stepped channels were used in Mexico and Peru before the Spanish conquest (e.g. Table 2.2).

A.

B.

Figure 2.8. Historical stepped fountains and grandes cascades. A. Pooled channel flanked by colonnaded avenue, Perge, Turkey looking downstream in October 1997 (courtesy of D. Murphy and M. Mengel). B. La Fontaine des Innocents, Paris, designed by J. Goujon (France, 1549).

C.

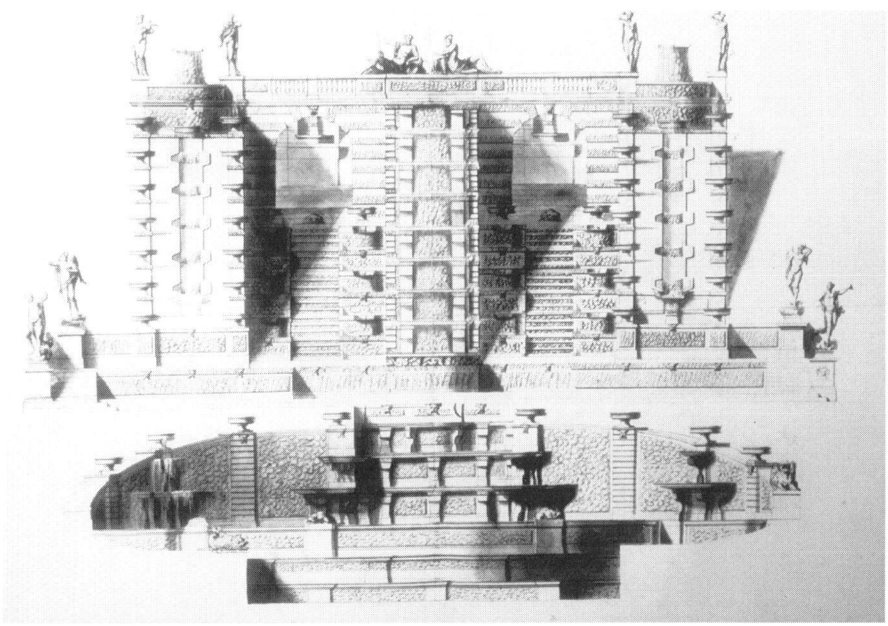

D.

Figure 2.8. Historical stepped fountains and grandes cascades. C. La Grande Cascade, Rueil-Malmaison (France, 1638) (courtesy of the Musée de l'Histoire de Rueilmalmaison). The cascade required regular maintenance work and was replaced by a lawn in 1720. D. La Grande Cascade at Saint-Cloud (France, 1667-1697). Hand drawing by N. Tessin in 1687 of the upper cascade (courtesy of Swedish National Art Museums).

E.

F.

Figure 2.8. Historical stepped fountains and grandes cascades. E. La Grande Cascade at Saint-Cloud (France, 1667-1697) in November 1994. F. Les Grandes Cascades, Sceaux (France, 1677).

Figure 2.8. Historical stepped fountains and *grandes cascades.* G. La Rivière de Marly (France 1687) (courtesy of *Archives Nationales*, *Service Photographique*, Cote 0*/1/1472/p8). Built in 1687, La Rivière was replaced in 1728 by Le Tapis Vert. There are 63 steps made of red and green marble, length about 300 m, mean slope: 6°, h = 0.5 m. (Bottom) Le Tapis Vert, looking downstream towards at Le Petit Parterre, La Pièce des Gerbes, La Grande Pièce d'Eau, Les Nappes, and Le Grand Abreuvoir de Marly.

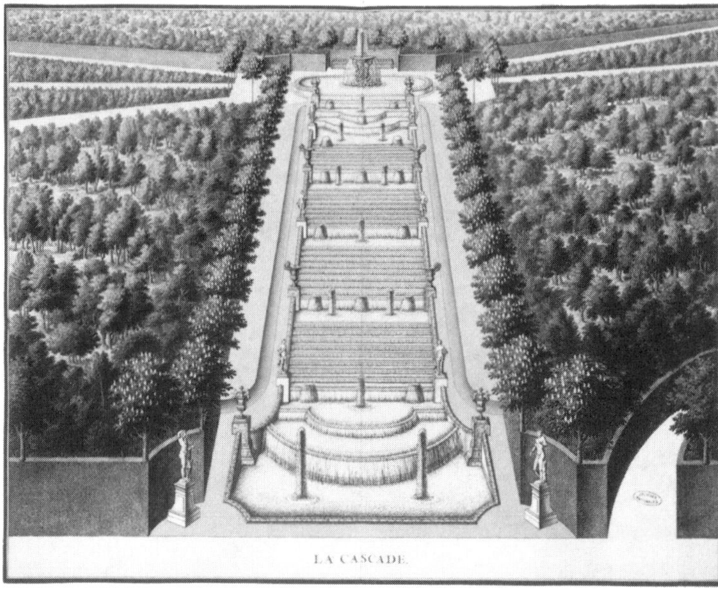

H.

I.

Figure 2.8. Historical stepped fountains and grandes cascades. H. La Cascade Champêtre (or Cascade Rustique), Marly (France) (courtesy of *Archives Nationales*, *Service Photographique*, Cote 0*/1/1471/p27). I. Escalier d'eau, Château du Touvet (France, 1763) (courtesy of Mr. F. Botton).

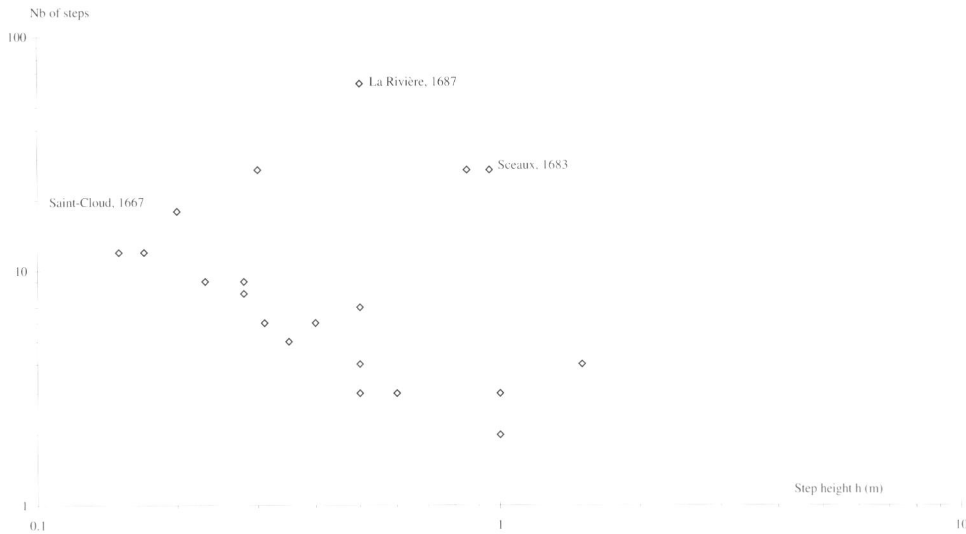

Figure 2.9. Characteristics of stepped cascades: step height versus number of steps.

With most ancient stepped waterways systems (Table 2.2; Fig. 2.1), it is believed that the hydraulic expertise was developed locally through evolution rather than technology transfer. In Yemen and Israel, the inhabitants benefited from some local "savoir-faire": i.e. the Sabaen and Nabataean expertise respectively. In Peru, Indian civilisations used stepped channels and drop structures prior to the Spanish conquest: e.g. Chimus and Incas. Note the different evolution processes: i.e. a dissemination of design expertise for spillways and cascades versus local evolution for irrigation structures.

Parallel with the historical development of stepped fountains

In terms of size and characteristics, the great cascades of the 16th to 18th century period were very similar to the contemporary stepped spillways (Section 2.5; Table 2.5). La Rivière de Marly was over 300 m long with a 6° slope and 0.5 m step height. It was definitively longer than any stepped weirs at the time. The cascades of La Granja, Palazzo Reale and Peterhof were also larger than most contemporary waste waterways. It is worth discussing the parallel development of water staircases and stepped spillways.

Greeks and Romans were renowned for their urban cascades and fountains. Later the Moslems built famous water gardens in their empire. Romans and Moslems contributed to the dissemination of design expertise both in stepped fountains and in stepped spillways since Antiquity. After the Moslem Era, the exchange of ideas between landscape architects and hydraulic engineers (in the field of stepped chutes) vanished. The Moor successors (e.g. the Spaniards and the Mughals) did not continue the dual expertise. The Spanish contribution to stepped spillway design was most important, but the design of stepped cascades was abandoned in Spain. In India, the Mughals kept only the stepped cascade tradition. Famous stepped cascades were built by the Italian[10], French and Mughal princes during the 16th and 17th centuries (Table 2.5), at a time when stepped spillways and weirs were built in Spain (e.g. Alicante Dam). However, no major stepped spillway was built in Italy, France or India during the same period.

The French landscape designers who built the gardens of Versailles and Marly (and their stepped cascades) were superb surveyors and good arithmeticians. They were knowledgeable in water supply engineering (e.g. the Aqueduc de Marly supplying waters to Versailles), but they never applied their expertise to dam engineering or river training engineering.

In the reverse sense, King Philip V of Spain selected a French designer, Frémin, to build the stepped cascade of La Granja despite the Spanish expertise in stepped spillways. In Russia, Tsar Peter the Great selected another Frenchman, J.B.A. Leblond, to design Peterhof's Cascade, while he hired Dutchman G.W. Hennin to design and build dams with stepped overflows. The author has found no evidence suggesting any contact or exchange between Hennin and Leblond.

The lack of exchange between architects and engineers seems to have survived up to the present time. Today most hydraulic engineers ignore the modern expertise in stepped cascades (Section 7.2; Table 7.1). Conversely, most architects are unaware of the recent developments in stepped spillway hydraulics.

2.6.2 *Design concepts and construction methods*

In most early dams (Table 2.1; Figs 2.2 & 2.3), the waters were discharged over the dam crest, and the stepped spillway design was selected to contribute to the stability of the dam and simplicity of shape. Later design engineers realised the advantages of stepped channels for energy dissipation purposes and to prevent scouring. During the 19th century, overflow stepped spillways were selected frequently, with nearly one third of the dams built in the USA being equipped with a stepped cascade (e.g. Schuyler, 1909; Wegmann, 1922) (Tables 2.1 & 2.6; Figs 2.3 & 2.10). Most structures were masonry and concrete dams with a downstream stepped face reinforced by granite blocks (e.g. Goulburn Weir). Some dams were equipped with stepped rocklined spillways: e.g. Ternay and La Tâche Dams. Others had a lateral spillway (e.g. Pas-du-Riot Dam). Earth embankments were usually equipped a lateral spillway (e.g. Val House Dam, Gold Creek Dam).

[10]In Italy, Pirro Ligorio who designed the Villa D'Este garden was strongly influenced by Roman influence: e.g. the water garden at Hadrian's Villa.

Figure 2.10. Stepped spillway designs in the early 20th century. A. Upper Coliban (Australia, 1903). Spillway chute with 6.7 m high drop followed by 5 steps and final 2 step cascade. The dam is on the right handside (photograph taken on 30 January 2000).

The development of the stepped spillway was marked by two milestones: the Gold Creek Dam cascade (1890, Australia) and the New Croton Dam (1906, USA). Completed in 1885, the Gold Creek Dam was initially equipped with an unlined cascade spillway. After heavy scour during flood spills, a concrete[11] stepped cascade (h = 1.5 m, l = 4 m) was built in 1890 (Fig. 2.3). It is believed that the Gold Creek cascade was the world's first concrete stepped spillway and the ancestor of modern RCC stepped spillways (Chanson & Whitmore, 1998). Completed in 1906, the New Croton Dam spillway[12] is probably the first stepped chute designed specifically to maximise energy dissipation (Fig. 2.10). It is still in use despite a major accident in 1955 (Chapter 8).

During the first part of the 20th century, interest in stepped cascades declined: "Cascades were formerly used a great deal (... but they) have seldom lived up to their expectations, and have seldom justified their high cost. (...) Considerably cheaper means of dissipating energy are known today" (Schoklitsch, 1937: 913). New progress in the energy dissipation characteristics of hydraulic jumps (e.g. Bakhmeteff & Matzke, 1936) favoured the design of

[11]Non-reinforced concrete. The strength of the concrete at Gold Creek was probably: $\sigma_c \sim$ 0.5 to 2 MPa (Chanson & Whitmore, 1998).

[12]The stepped cascade of the New Croton Dam appeared in the movie "Daylight" (Universal Studios, 1996), starring Sylvester Stallone. Low overflow is visible at the beginning of the movie, when a convoy of three trucks crosses the bridge over the cascade. The sequence lasts for about 45 seconds.

B.

C.

Figure 2.10. Stepped spillway designs in the early 20th century. B. New Croton Dam (USA, 1906). Spillway design discharge: 1,550 m^3/s, $h = 2.13$ m, stepped slope: 53°. View from the right bank in 1999 (courtesy of Mrs. J. Hacker). Note the bridge connecting the dam crest to the right bank over the spillway weir. C. Llwyn-on Dam spillway (Wales, 1926) (courtesy of Mr. L. Stuart Davies, Welsh Water).

Table 2.6. Stepped spillway construction between 1850 and 1950.

Name	Year	Ref.	Dam height (m)	Slope α (deg.)	Construction	Comments
(1)	(2)	(3)	(4)	(5)	(6)	(7)
Nijar Dam, Spain	1850	[SH, SM]	31		Curved gravity dam in masonry.	Overflow dam. Flat crest followed by 7 steep steps (h = 1.2 to 5.3 m).
Brent Reservoir	1852	[BI]	8.8		Earth dam (built in 1835, heightened in 1852). Spillway: stone above brick masonry.	Lateral overflow stepped spillway. 7 steps: 1st large step (h ~3 m) followed by 6 identical steps (h ~ 0.7 m).
Dale Dyke Dam, UK	1863	[BIB, HU]	29		Earth dam (failure in 1864).	Lateral spillway (W = 7.3 down to 3.5 m). Maximum discharge: 47 m^3/s. 5 steps (h = 0.4 to 0.74 m) followed smooth channel.
Gouffre d'Enfer Dam, France (also called Furens Dam, Furan Dam or Rochetaillée Dam)	1866	[SCH, SM]	60		Curved gravity dam in masonry. Unlined rock spillway channel.	Lateral spillway followed by unlined rock cascade. Refurbishment of spillway intake in 1990s. Cascade still in used today.
Bilberry Dam, UK	1867	[BIB]	16.1		Earth dam with puddle clay corewall (1854-1856?). Lateral overflow spillway with stepped channel.	Maximum discharge: 49 m^3/s. Lateral stepped channel (W = 3.048 m). Reservoir still in use today.
Ternay Dam, France	1867		41		Curved gravity dam in masonry. Unlined rock spillway.	30 steps (h ~ 0.3 to 0.8 m). Refurbishment of spillway intake in 1990s. Still in use today.
Val House Dam, UK	1868	[BIB, HU]	17.4	30	Earth dam. Spillway: concrete floor and ashlar steps.	Lateral overflow spillway with stepped crest (60 m long, 3 steps, h = 1.5 m) followed by flat slope, then stepped channel (W = 12.2 m, 27 steps, h = 1.5 m). Maximum discharge: 160 m^3/s.
Quinson Dam (or Verdon Dam), France	1868	[CO, WE]	20	83 & 51	Gravity dam. Stone blocks with metal clamp links and lead joints. Cut-stone facing.	Overflow dam (W = 40 m). Maximum discharge: 1,200 m^3/s. Flat crest (4.6 m) followed by smooth slope (83°) and 10 steps (51°, h = 0.75 m). In use up to 1971 (new dam completed in 1974).
Ulley Dam, UK	1870?	[HU, SH]	12.6	12.7	Earth embankment. Spillways: ashlar pitching set on concrete.	Two lateral overflow stepped spillways (W = 1.83 m, h = 0.69 m, l = 3.05 m).
Malmsbury Reservoir, Australia	1870		24		Earth embankment with masonry spillways at each end.	Right bank spillway: masonry chute with a series of 8 steps (W ~ 20 m). Still in use today.
Pas-du-Riot Dam, France	1873		36		Curved gravity dam made of Cyclopean masonry. Masonry spillway.	Maximum discharge: 65 m^3/s. Lateral spillway: 7 steps (h ~ 2.5 to 3 m). Trapezoidal section (base width ~ 3 m). Nappe flow regime. Still in use today.
Hijar Dams, Spain	1880	[SH]	43	53?	Curved gravity dams in masonry.	2 nearly-identical dams on the Martin river. Step height: h = 2 m.
Upper Barden Reservoir, UK	1882	[BIA]	42		Earth embankment with masonry spillways at each end.	Maximum discharge: 43 m^3/s. Flat steps at upstream end.

Table 2.6. Continued.

Le Pont Dam, France	1882		27		Curved gravity dam (granite rubble masonry) with 7 buttresses. Pooled masonry steps with rounded crest.	Design by H. Bazin. Maximum discharge capacity: 195 m^3/s (Maximum flood: 139 m^3/s in 1910). Steps: h = 2.8 to 6.97 m, 13° mean slope. W = 29.5 down to 10 m. Pool bed in unlined rock.
Bridgeport Dam, USA	1886	[SH, WE]	13		Gravity dam made of rubble masonry laid in mortar. Downstream stepped face. Upstream earth embankment built afterwards to reduce dam leakage.	Overflow spillway (W = 24.8 m, h = 0.29 to 0.95 m). Maximum discharge capacity: 78 m^3/s.
Arizona Canal Dam, USA	1887	[EN1]	10		Timber cribs filled with loose rock and gravel. Dam failure in 1905[b].	Maximum discharge: 33 m^2/s. 3 horizontal steps.
Tytam Dam (or Tai Tam Dam), Hong Kong	1887	[SH]	29	65	Gravity dam (40% stone, 60% concrete). Downstream stepped face covered by ashlar blocks.	Broad crest (6.4 m) followed by 9 steps (h = 3.05 m).
Castlewood Dam, USA	1890	[SH]	21	45	Rockfill dam with masonry walls. Spillway made of stone blocks laid in cement mortar. Reinforced by an upstream earth embankment in 1900.	Maximum discharge: 115 m^3/s. Broad crest (2.5 m) followed by 16 steps (h = 0.61 m). W = 30.5 m.
Gold Creek Dam, Australia	1890		26	21	Earthfill dam with puddle clay corewall (1885). Concrete stepped spillway (1890) over eroded unlined rock spillway.	Broad-crest (W = 55 m) followed by 12 steps (h = 1.5 m, l = 4 m). Maximum spillway capacity: 200 m^3/s. Crest level lowered by 1.2 m in 1975 and by another 0.4 m in 1998. Still in use today.
Goulburn Weir, Australia	1891	[EV]	15	~ 30	Concrete gravity weir with horizontal steps made of granite blocks.	Design discharge: 1,970 m^3/s. Record discharge 1,982 m^3/s on 7 June 1917. W = 141 m. 12 steps (h = 0.5 m). Gate refurbishment in 1986[b]. Still in use today.
La Tâche Dam (also called Chartrain Dam), France	1891		51	32.6	Curved gravity dam in masonry. Unlined rock spillway cut in the left hill slope.	10 steps (h = 2.4 to 5.4 m) followed by 9 masonry steps (h = 0.2 to 0.3 m). W = 2.2 m. Refurbishment of spillway intake in 1994. Still in use today.
Cantref Dam, Wales UK	1892		23.8	20 to 34	Earth dam with puddle clay corewall. Masonry stepped spillway.	Maximum discharge capacity: 157 m^3/s. h = 0.3048 m. W = 9.14 m. Still in use today. Increase of spillway capacity in the 1980s by addition of two siphon spillways and a reinforced earth spillway.
Winterburn Dam, UK	1893		24		Earth embankment. Masonry stepped spillway.	Maximum discharge capacity: 110 m^3/s. h = 2.26 m. Reservoir still in use today (maximum spillway capacity increased to 270 m^3/s).
Titicus Dam, USA	1895	[WE]	41	~ 63	Earth dam with masonry overflow spillway (masonry of rubble with cut stone laid in regular courses).	Maximum spillway capacity: 472 m^3/s. W = 61 m. 13 steps (h = 0.61 to 3.7 m).

Table 2.6. Continued.

Beacons Dam, UK	1897		16.5		Earth dam with puddle clay corewall. Masonry stepped spillway.	Maximum discharge capacity: 113 m^3/s. h = 0.3048 m. W = 9.14 m. Replaced by a smooth chute in the 1980s.
Cornell University Dam, USA	1897	[SH]	9.1		Curved gravity dam in masonry with downstream stepped face.	Spillway: dam overflow (W = 41 m) and natural spillway ledge (W = 40.5 m). Maximum spillway discharge: 136 m^3/s.
L'Echapre Dam, France	1897		42	24.3	Curved gravity dam in masonry. Unlined rock (gneiss) spillway.	9 steps (h = 3.2 to 4.1 m). W = 3 m. Spillway refurbishment in 1989. Still in use today.
Junction Reefs Dam, Australia	1897		19		Multiple arch dam (concrete and brick). Unlined rock stepped cascade.	h = 2 to 4 m. Still in use today.
Canyon Ferry Dam, USA	1898	[WE]	8.8		Timber cribs filled with stones. Reinforced by upstream earth embankment	W = 148 m. 3 steps (h = 2.3 to 3.05 m).
Leamington Weir, UK	1903	[LA]			Curved gravity weir (masonry).	5 steps. To protect Leamington suspension bridge.
Upper Coliban Reservoir, Australia	1903		30		Earth dam with lateral spillway.	Spillway chute includes a drop (6.7m) and 2 series of steps (h = 1.2 m, W ~ 32 m). Refurbishment of spillway intake in 1993. Still in use.
Yuba River Debris Barrier, USA	1904	[EN2]	6.8	15.4	Debris dam: hydraulic fill anchored by rockfill, piling and concrete facing.	Maximum discharge: 9.8 m^2/s. W = 380 m. Smooth profiled steps with concrete facing. h = 1.8 to 2.6 m.
Pedlar River Dam, USA	1905	[SH]	22.4	62	Concrete dam with spillway steps of granite blocks laid in cement mortar.	Maximum discharge: 7.7 m^2/s. W = 45.7 m. 10 steps (h = 0.76 to 2.3 m). Ventilated steps.
Urft Dam, Germany	1905	[KE, SH]	58		Curved gravity dam (slate masonry) with upstream earth embankment.	Maximum discharge: 200 m^3/s. W = 100 m. Steps cut in natural rock covered by concrete (h = 1.52 m).
New Croton Dam, USA	1906	[W7, GO]	90.5	53	Masonry gravity dam. Spillway built in block stone masonry. Spillway damage in 1955[b].	Maximum discharge: 1,550 m^3/s. W = 305 m. h = 2.13 m. Still in use today.
Wachusett Dam, USA	1906	[WE]	62.5	18.4	Rubble masonry gravity dam. Separate overflow weir with granite facing.	Maximum discharge: 1,135 m^3/s. W = 137.2 m. h = 0.61 m.
Zuni Dam, USA	1908	[SH, WE]	21.3	28.7	Rockfill dam with upstream hydraulic earthfill and downstream stepped face. Partial failure of dam in 1909.	Spillway: unlined rock spillway and dam overflow (h = 0.46 m). Dam failure caused by settlement of left abutment.
Croton Falls Dam, USA	1911	[WE]	52.7		Cyclopean masonry dam with reinforced concrete block facing. Horizontal steps made of granite.	Overflow weir: W = 213 m. 12 steps with rounded step edges: h = 0.61 m, l = 0.305 to 0.91 m.
Feather River Diversion Weir, USA	1912	[ET]	7.9	26.6	Crib weir with horizontal lumber steps.	Maximum discharge: 73 m^2/s. W = 85.3 m.

Table 2.6. Continued.

Klingenberg Dam, Germany	1913?	[EN3]	13.25		Curved gravity dam.	Maximum discharge: 100 m^3/s. W = 37.4 down to 12 m. 12 steps (h = 0.6 to 2.2 m).
Lahontan Dam, USA	1915	[RH, DG]	49	22.9 (right) 26.6 (left)		Maximum discharge: 850 m^3/s. Spillway refurbishment in 1940[b]. Two stepped spillways (h = 3.05 & 3.35 m).
Warren Dam, Australia	1916	[JO]	17.4[a]	35[a]	Concrete gravity dam. Overtopped in 1917 ([b]). Spillway re-designed in 1926.	Maximum spillway discharge: 100 m^3/s. W = 35 m. h = 0.37 m.
Hetch-Hetchy Dam, USA	1923	[KE]	69.4[a]	57[a]	Gravity dam (granite masonry). Dam heightening in 1937-38.	Now called O'Shaughnessy Dam. Uncontrolled siphon-spillway followed by stepped chute[c]. Maximum spillway discharge: 566 m^3/s. W = 55 m. h = 2 m.
Gilboa Dam, USA	1926	[GM]	49		Masonry dam.	Maximum discharge: 3,130 m^3/s. Inclined downwards steps (? = -5.7°): h = 6.1 m. W = 403.6 m.
Llwyn-on Dam, UK	1926		22.1		Earth dam with puddle clay corewall. Masonry stepped spillway.	Maximum discharge capacity: 235 m^3/s. h = 0.3048 m. W = 18.29 m. Still in use today.
St. Francis Dam, USA	1926	[JA]	62.5		Concrete gravity dam. Dam failure in 1928[b].	Step height: h = 0.4 m. W = 67 m.
Eildon Weir, Australia	1927	[KN]	42.7[a]	56 and 22[a]	Rockfill dam with concrete spillway section. Spillway re-designed in 1936. New dam in 1955.	Maximum discharge: 566 m^3/s. Step height: h = 1.83 m. 14 steps (56°) followed by a flat smooth slope and a 22° stepped channel. W = 208 m.
Mühleberg Weir, Austria	1930?	[SC]	20		2 flat steps followed by baffle blocks.	Step heights: h = 8.16 and 8.9 m.
Paderno Power Plant, Austria	1930?	[SC]	18	17.7	Pooled steps.	11 step cascade. Step height: h = 1.6 m. Pool height: d_t = 1 m.
Ryburn Dam, UK	1933	[WO]	26.9	60	Concrete overflow dam with d/s stepped face (concrete steps).	Maximum discharge: ~ 120 m^3/s. Steps: h = 1.86 m, l = 1.07 m. W = 36.6 m.

Notes: [a]: original dam; [b]: see Chapter 8; [c]: stepped geometry selected to improve the new concrete bond during the dam heightening.

References: [BI] Binnie (1987); [BIA] Binnie (1913); [BIB] Binnie (1981); [CO] Colas des Francs (1975); [DG] Douma and Goodpasture (1940); [EN1] *Engineering News* (1905a); [EN2] *Engineering News* (1905b); [EN3] Engels (1914); [ET] Etcheverry (1916); [EV] Evans (1984); [GM] Gausmann & Madden (1923); [GO] Goubet (1992); [HU] Humber (1876); [JA] Jansen (1983); [JO] Johnson & Templar (1976); [KE] Kelen (1933); [KN] Knight (1938); [LA] Labrum (1994); [RH] Rhone (1990); [SCH] Schnitter (1994); [SC] Schoklitsch (1937); [SH] Schuyler (1909); [SM] Smith (1971); [W7] Wegmann (1907); [WE] Wegmann (1911); [WO] Wood (1933).

hydraulic jump stilling basins downstream of the spillway chutes. Stilling basins allowed larger energy dissipation and smaller structures. Altogether, the concept of downstream stilling basins contributed to cheaper construction costs. In the 1970s design engineers regained interest for stepped spillways. The trend was initiated by the introduction of new construction materials (e.g. roller compacted concrete (RCC), gabions with polymer coated wire) and new design techniques (e.g. embankment overtopping protection systems). Over the past two decades, several dams have been built with overflow stepped spillway (Tables 1.2 & 7.3; Chapter 7).

2.6.3 *Design methods*

Since Antiquity, the design of stepped spillways and channels was recognised to reduce flow velocities. During the Renaissance period, Spanish and Italian hydraulicians (e.g. Juanelo Turriano[13], Leonardo da Vinci) developed these principles to stepped weirs. Leonardo da Vinci stated the basic ideas: the flow, "the more rapid it is, the more it wears away its channel"; if a waterfall "is made in deep and wide steps, after the manner of stairs, the waters (...) can no longer descend with a blow of too great a force" (Fig. 2.11). He illustrated his conclusion with a staircase waterfall[14] "down which the water falls so as not to wear away anything" (Richter, 1939). This method was well known during the 19th century. In his textbook, Humber stated: "The byewash[15] will generally have to be made with a very steep mean gradient, and to avoid the excessive scour which could result if an uniform[16] channel were constructed, it is in most cases advisable to carry the byewash down by a series of steps, by which the velocity will be reduced. This is very well illustrated in the case of the Rotherham Works" (Humber 1876: 133). For the Malmsbury Dam Cascade (Fig. 2.3), Sankey (1871) recommended a wide spillway crest followed by a reduction in channel width associated with the introduction of a series of steps. He stated that the protection of the spillway (in terms of erosion and scouring) required that the "breadth" and drop height to be related as:

$$W \sim \frac{1}{\sqrt{h}} \tag{2.1}$$

Downstream of the channel contraction, "the floor dropped by a series of steps for 60 feet and apron beyond (...) so constructed (...) as to ensure the velocity

[13]Interestingly an architectural stepped weir was sketched in a codex attributed to him (Garcia-Diego, 1976), and a similar design was used for the Pulteney Weir (or Horseshoe Weir) completed in 1971 as part of improvement works and flood control scheme for the city of Bath.

[14]The staircase water falls at Vigevano, near La Sforzesca, Italy, included 130 steps with the following step geometry: h = 0.15 m, h/l = 0.5.

[15]channel to carry waste waters: i.e. spillway.

[16]in the meaning of an uniform smooth channel bed (i.e. not stepped).

D.

Figure 2.10. Stepped spillway designs in the early the 20th century. D. Claerwen Dam spillway (Wales, 1952) (courtesy of Mr. L. Stuart Davies, Welsh Water).

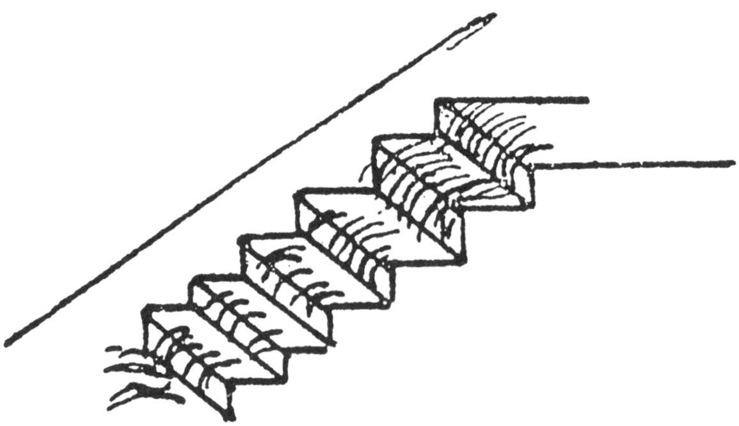

Figure 2.11. Staircase cascade sketched by Leonardo da Vinci (after Richter, 1939). Comparison with Figure 3.1A shows the accuracy of Leonardo da Vinci's drawing.

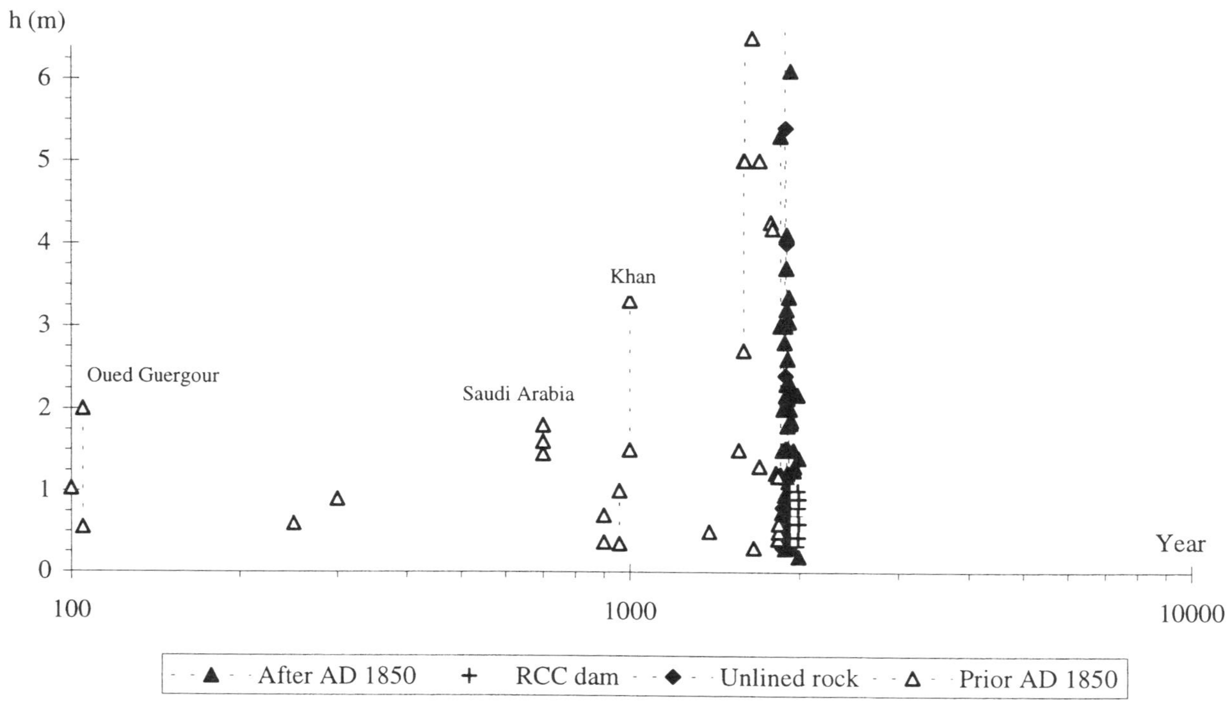

Figure 2.12. Evolution of stepped spillway design over the ages. A. Step height h.

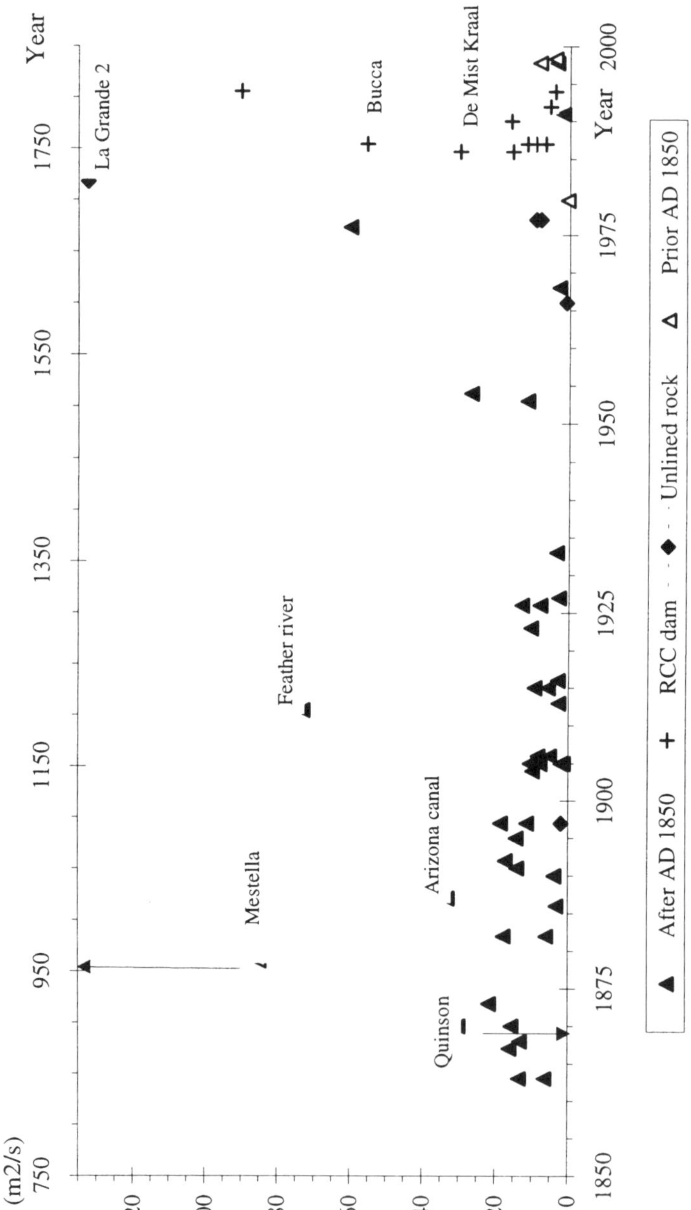

Figure 2.12. Evolution of stepped spillway design over the ages. B. Maximum discharge capacity q_w.

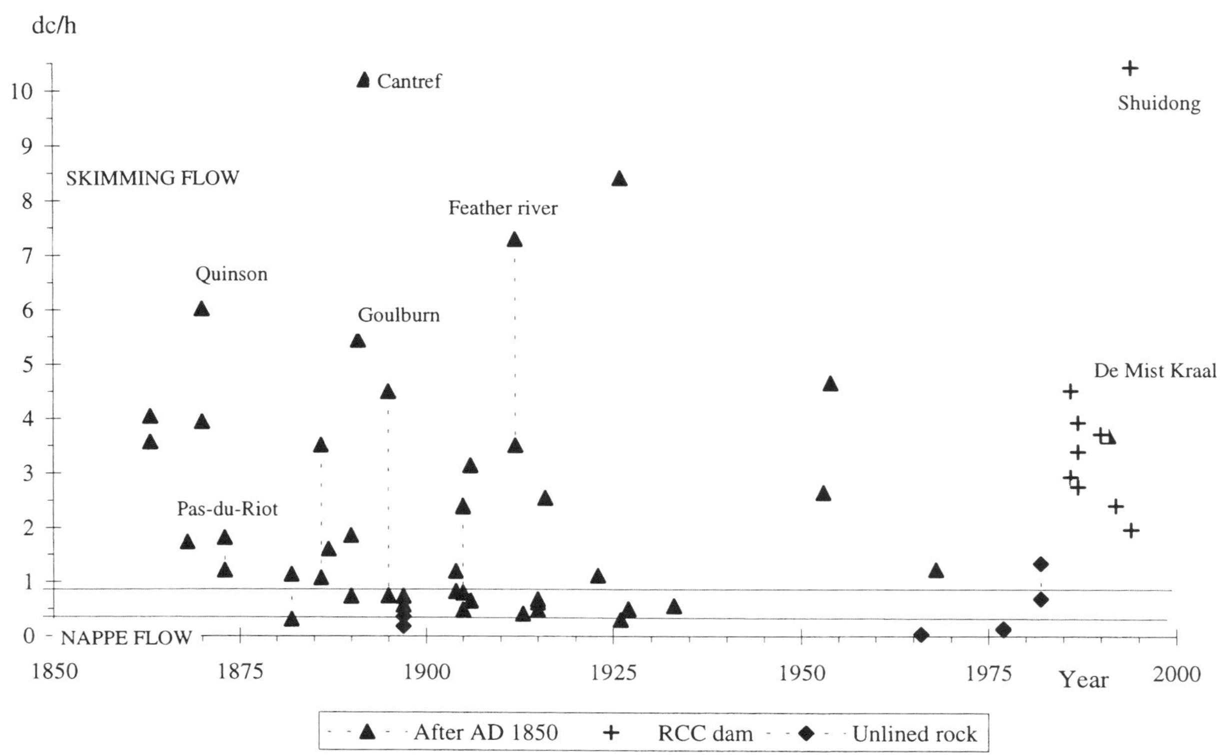

Figure 2.12. Evolution of stepped spillway design over the ages. C. Ratio d_c/h.

of the current being arrested and the quiet discharge of the flood into the Coliban (River)" (Sankey, 1871: 28).

For a given chute geometry (width, step height, step length), small flows behave as a succession of free-falling jets and nappe impact (i.e. nappe flow regime), while larger flows skim over the pseudo bottom formed by the step edges (i.e. skimming flow regime), (Chapter 1). There are fundamental differences between the hydraulics of nappe and skimming flows, and a good observer, like Leonardo da Vinci, could easily distinguish between the two regimes. It is likely that some ancient engineers recognised the basic flow patterns (e.g. nappe or skimming flow). There is, however, evidence that hydraulic engineers did not use hydraulic design guidelines (e.g. flow resistance, head loss), even at the beginning of the 20th century. It is only recently that new progress on the hydraulics of stepped channels has been achieved: e.g. Essery & Horner (1978), Sorensen (1985), Rajaratnam (1990), Peyras et al. (1991), Chanson (1995a).

2.6.4 *Step dimensions and design flow rates*

A continuity in the design of stepped spillways is a point of interest, as the characteristics of ancient stepped spillways are similar and comparable to present designs. Figure 2.12 presents the step height h, the maximum discharge capacity q_w and the ratio d_c/h for a wide range of dams as a function of year of completion. Since Antiquity, the step heights increased slightly up to 1930 (Fig. 2.12A). Recently, unlined rock spillways have been designed with larger step heights, typically 10 m high. The most recent concrete and gabion structures use smaller step heights, ranging from 0.2 to 1 m (Figs 1.2 & 7.12; Section 7.6).

Interestingly, most contemporary stepped spillways are designed for a maximum discharge capacity no greater than some ancient designs (Fig. 2.12B). New stepped spillway designs might be cheaper to build but they do not provide necessarily larger discharge capacities! It is worth mentioning several 19th century dams with stepped spillways which were recently refurbished: e.g. Gold Creek, La Tâche, Pas-du-Riot and Ternay Dams (Table 2.6). In each case, the spillway intake was modified to increase the maximum discharge capacity. But the stepped chute remained unchanged, suggesting a strong confidence in the ancient stepped channel designs (see Section 8.3.2).

In the past, most small weirs and drop structures were designed for a nappe flow regime. However, Figure 2.12C suggests that some ancient stepped spillways were designed for a skimming flow regime: i.e. $d_c/h > 0.4$ to 0.8 (see Section 1.3). One example, the Mestella Weir (A.D. 960), was able to sustain very large skimming flows: i.e. q_w up to 85 m^2/s and d_c/h up to 26.

Stepped weirs and channels have been used for more than 3,500 years, since the early structures in Greece and Crete. Despite recent advances in technology, the design characteristics (geometry, discharge) of stepped spillways show some continuity from Antiquity up to now (Fig. 2.12). In consideration of the recent renewal of interest for stepped spillways, the concept of stepped chute is not a new technique (nor a revolution) but simply an evolution of design.

CHAPTER 3

Hydraulics of nappe flow regime

3.1 INTRODUCTION

For a given flat step geometry, low flows behave as a series of free-falling jets with nappe impact onto the downstream step: i.e. nappe flow regime (Section 1.3). At the upstream end of each step, the flow is characterised by a free-falling nappe, an air cavity and a pool of recirculating fluid (Figs 3.1 & 3.2). Air bubble entrainment occurs at both the upper and lower jet interfaces, with additional air being entrained by a plunging jet mechanism at the intersection of the lower nappe with the recirculating pool.

In this book, the nappe flow regime is defined as a *succession of free-falling nappes and jet impacts* from one step onto the next one when the nappe cavity is *fully aerated*. Nappe ventilation is essential for a proper operation of nappe flows. Examples of problems and accidents associated with unventilated nappes are discussed in Section 3.2.3.

Although a lot of research has been conducted on skimming flow down steep chutes, the hydraulics of nappe flow has received less attention. There are detailed studies on basic flow properties (Horner, 1969; Peyras et al., 1991, 1992; Chanson, 1994b); energy dissipation (Chamani & Rajaratnan, 1994; Chanson & Toombes, 1997a); and the air entrainment processes (Chanson & Toombes, 1997b; Toombes & Chanson, 2000).

3.1.1 *Nappe trajectory calculations*

For a horizontal step, a simple expression of the nappe trajectory derives from the motion equation. Considering an aerated nappe, the flow direction at the brink of the step is nearly horizontal and the trajectory equations of the nappe centreline are:

$$x = V_b * t \tag{3.1a}$$

$$y = h + \frac{d_b}{2} - \frac{1}{2} * g * t^2 \tag{3.1b}$$

A. 

B. 

Figure 3.1. Photographs of nappe flow regime. A. Peak Tramway Fountain, Hong Kong in, 1993. $\alpha = 20°$, h = 0.15 m, horizontal steps. Flow from the right to the left. B. Brushes Clough spillway, UK in, 1994 (courtesy of Mr. Gardiner, NWW). $\alpha = 18.4°$, h = 0.19 m, inclined steps ($\delta = -5.6°$). Flow from the right to the left.

where V_b is the flow velocity at the brink of the step, t is the time, h is the step height and d_b is the flow depth at the brink (Figs 3.2 & 3.7) (App. 3). The length of the drop L_d and the nappe impact conditions derive from the trajectory equations:

$$\frac{L_d}{h} = \left(\frac{d_c}{h}\right)^{3/2} * \sqrt{\frac{h}{d_b}} * \sqrt{1+2*\frac{h}{d_b}} \tag{3.2}$$

$$\frac{d_i}{d_c} = \left(\left(\frac{d_c}{d_b}\right)^2 + 2*\frac{h+\frac{d_b}{2}-d_p}{d_c}\right)^{-1/2} \tag{3.3}$$

$$\frac{V_i}{V_c} = \sqrt{\left(\frac{d_c}{d_b}\right)^2 + 2*\frac{h+\frac{d_b}{2}-d_p}{d_c}} \tag{3.4}$$

$$\tan\theta = \sqrt{2} * \sqrt{\frac{d_b}{d_c}} * \sqrt{\frac{h+\frac{d_b}{2}-d_p}{d_c}} \tag{3.5}$$

where d_i and V_i are the nappe thickness and flow velocity at the intersection of the falling nappe with the receiving pool, d_p is the pool height and θ is the angle of the falling nappe with the horizontal (Fig. 3.7). The pool of water behind the overfalling jet (Figs 3.2 & 3.7) is important as its weight provides the pressure force parallel to the step surface which is required to change the jet direction from an angle θ to the step bottom to parallel to the horizontal face of the step. The Momentum Equation resolved along the step surface is:

$$\frac{1}{2}*\varrho_w*g*d_p^2 - \frac{1}{2}*\varrho_w*g*d_1^2 = \varrho_w*q_w*(V_1 - V_i*\cos\theta) \tag{3.6}$$

assuming that the edges of the jet fluid have not disintegrated into spray and neglecting the shear forces on the surfaces. Assuming that the velocity entering the control volume is approximately the same as leaving it, it yields:

$$\frac{d_p}{d_i} = \sqrt{1+2*\frac{V_i^2}{g*d_i}*(1-\cos\theta)} \tag{3.7}$$

Using the same method as White (1943) the application of the momentum equation at the base of the overfall leads:

$$\frac{d_1}{d_c} = \frac{2 * Fr_b^{-2/3}}{1 + \frac{2}{Fr_b^2} + \sqrt{1 + \frac{2}{Fr_b^2} * \left(1 + \frac{h}{d_c} * Fr_b^{2/3}\right)}} \tag{3.8}$$

where Fr_b is the Froude Number immediately upstream at the overfall brink[1]. Although Rajaratnam & Chamani (1995) criticised the assumptions underlying Equation 3.8, good agreement is observed with laboratory data for subcritical inflows (i.e. $Fr_b = 1$) and $d_c/h < 0.45$ (App. 3). Equations 3.1 to 3.8 are valid for any type of nappe flows: that is, both sub- and super-critical inflow conditions.

3.2 BASIC FLOW CHARACTERISTICS

3.2.1 *Nappe flow sub-regimes*

Three types of nappe flow can be distinguished: 1. nappe flow with fully-developed hydraulic jump at low flow rates (sub-regime NA1), 2. nappe flow with partially-developed hydraulic jump (sub-regime NA2), and 3. nappe flow without hydraulic jump (sub-regime NA3) (Fig. 3.2).

C.

Figure 3.1. Photographs of nappe flow regime. C. Gold Creek Dam (Australia, 1890). h = 1.5 m, l = 4 m, W = 55 m. View from right bank in, 1996 (Q_w = 27 m^3/s).

[1]That is, $Fr_b = 1$ for subcritical inflow conditions.

D.

D.

Figure 3.1. Photographs of nappe flow regime. D. Cantref dam (Wales, 1892) (courtesy of Mr. L. Stuart Davies, Welsh Water). h = 0.3048m, α = 34°, W = 9.14m, design flow rate: 157 m^3/s. Low overflow in October 1996.

E.

Figure 3.1. Photographs of nappe flow regime. E. Nappe flow (NA1) on Lahontan Dam left spillway during a small overflow (courtesy of the US Bureau of Reclamation).

Nappe flows on horizontal steps and inclined upward steps are characterised often by the presence of hydraulic jumps (sub-regimes NA1 and NA2). The dominant hydraulic feature is the occurrence of critical flow conditions at each and every step edge. The flow energy is dissipated by jet breakup in air, by jet impact and mixing, on the step, and by the formation of a hydraulic jump on the step (Fig. 3.2A & B). Such nappe flows operate as a succession of drop structures. In the sub-regime NA1, the flow at each step is supercritical in the free-falling nappe up to the hydraulic jump, subcritical downstream of the jump and critical at the step edge (Fig. 3.2A). With larger flow rates, the hydraulic jump starts interfering with the downstream step edge (sub-regime NA2, Fig. 3.2B).

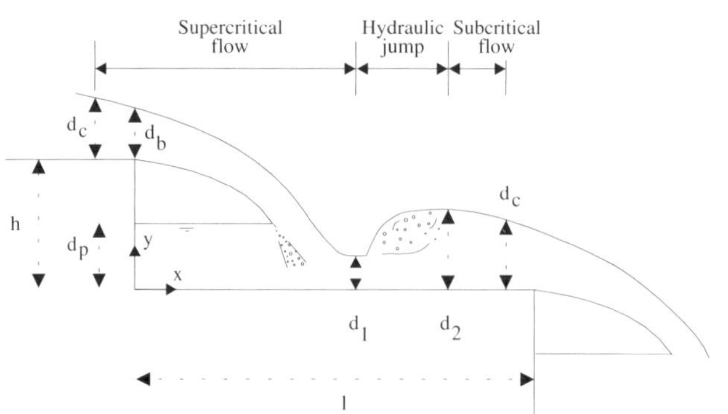

Figure 3.2. Nappe flow sub-regimes. A. Nappe flow with fully developed hydraulic jump (sub-regime NA1).

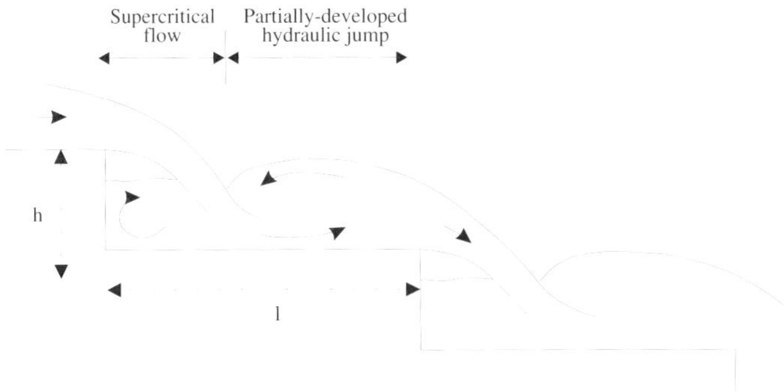

B.

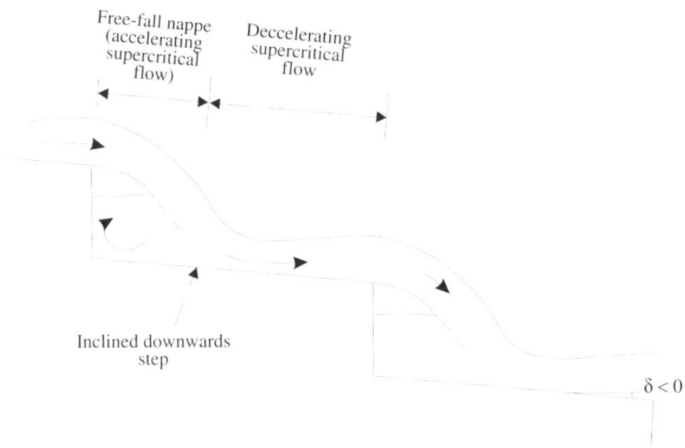

C.

Figure 3.2. Nappe flow sub-regimes. B. Nappe flow with partially-developed hydraulic jump (sub-regime NA2). C. Nappe flow without hydraulic jump (sub-regime NA3).

A nappe flow without hydraulic jump (sub-regime NA3) occurs for larger discharges, before the apparition of transition flow. The sub-regime NA3 is observed on flat horizontal steps and often on channels with inclined-downward steps (Fig. 3.2C). No hydraulic control (i.e. critical flow) takes place and the flow is supercritical at any position. Three dimensional flow patterns may be observed: e.g. shock waves, sidewall waves (Chanson & Toombes, 1997a, b).

The hydraulic analysis of nappe flows differs depending upon the occurrence or not of a hydraulic jump. In the former case, critical flow occurs at each step brink and the flow may be analysed as a series of drop structures (Section 3.3). For nappe flow without hydraulic jump, the analysis of the supercritical flow is more complex (Section 3.4).

3.2.2 *Transition between nappe flow sub-regimes*

For flat horizontal steps, Chanson (1994b) developed a condition for nappe flow regime with fully developed hydraulic jump (sub-regime NA1) based upon the drop length and hydraulic jump length:

$$\left(\frac{d_c}{h}\right) < 0.0916 * \left(\frac{h}{l}\right)^{-1.276} \qquad \text{Nappe flow sub-regime NA1} \quad (3.9)$$

where h and l are the step height and length respectively. Equation (3.9) was obtained for: $0.2 \le h/l \le 6$. Experimental observations are summarised in Table 3.1. The data are compared with Equations 1.1 and 3.9 in Figure 3.3. The results indicate that the sub-regime NA1 occurs only for very low discharges on steep slopes (i.e. $h/l > 0.5$) (Fig. 3.3). The data of Horner show that the disappearance of hydraulic jump occur at larger flow rates on inclined upward steps. The data exhibit however some scatter for the smaller slopes (i.e. $h/l <$ 0.4) (Fig. 3.3).

3.2.3 *Nappe ventilation and oscillations*

When the falling nappe intersects the receiving pool of water, air is entrained at the intersection between the pool and the underside of the jet. The entrained air is drawn from the air cavity beneath the nappe. If the cavity between the nappe and the vertical step is not ventilated, the pressure in the cavity falls below atmospheric, with subpressure and nappe oscillations occurring.

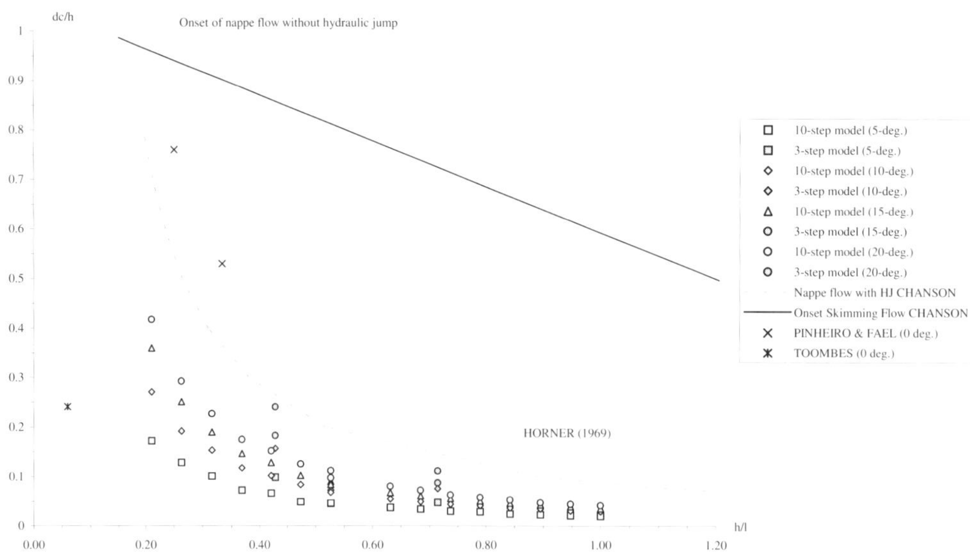

Figure 3.3. Transition between nappe flow sub-regimes NA2 and NA3. Comparison between Equation 3.9 and experimental data (Table 3.1).

Table 3.1. Flow conditions for the transitions between nappe flow sub-regimes.

Ref. (1)	W m (2)	h m (3)	α deg. (4)	δ deg. (5)	h/l (6)	$\frac{dc}{h}$ onset (7)	Type of transition (8)	Remarks (9)
Horner (1969)	0.61	0.051	11.9	5	0.21	0.17	NA2-NA3	10-Steps model.
	0.61	0.064	14.7	5	0.26	0.13	NA2-NA3	Flat steps.
	0.61	0.076	17.5	5	0.32	0.10	NA2-NA3	Un-controlled inflow.
	0.61	0.089	20.2	5	0.37	0.07	NA2-NA3	
	0.61	0.102	22.8	5	0.42	0.07	NA2-NA3	
	0.61	0.114	25.3	5	0.47	0.05	NA2-NA3	
	0.61	0.127	27.8	5	0.53	0.04	NA2-NA3	
	0.61	0.127	27.8	5	0.53	0.05	NA2-NA3	
	0.61	0.152	32.3	5	0.63	0.04	NA2-NA3	
	0.61	0.165	34.4	5	0.68	0.03	NA2-NA3	
	0.61	0.178	36.4	5	0.74	0.03	NA2-NA3	
	0.61	0.191	38.3	5	0.79	0.03	NA2-NA3	
	0.61	0.203	40.1	5	0.84	0.02	NA2-NA3	
	0.61	0.216	41.8	5	0.89	0.02	NA2-NA3	
	0.61	0.229	43.5	5	0.95	0.02	NA2-NA3	
	0.61	0.241	45.0	5	1.00	0.02	NA2-NA3	
	0.61	0.051	11.9	10	0.21	0.27	NA2-NA3	
	0.61	0.064	14.7	10	0.26	0.19	NA2-NA3	
	0.61	0.076	17.5	10	0.32	0.15	NA2-NA3	
	0.61	0.089	20.2	10	0.37	0.12	NA2-NA3	
	0.61	0.102	22.8	10	0.42	0.10	NA2-NA3	
	0.61	0.114	25.3	10	0.47	0.08	NA2-NA3	
	0.61	0.127	27.8	10	0.53	0.07	NA2-NA3	
	0.61	0.127	27.8	10	0.53	0.07	NA2-NA3	
	0.61	0.152	32.3	10	0.63	0.05	NA2-NA3	
	0.61	0.165	34.4	10	0.68	0.05	NA2-NA3	
	0.61	0.178	36.4	10	0.74	0.04	NA2-NA3	
	0.61	0.191	38.3	10	0.79	0.04	NA2-NA3	
	0.61	0.203	40.1	10	0.84	0.04	NA2-NA3	
	0.61	0.216	41.8	10	0.89	0.03	NA2-NA3	
	0.61	0.229	43.5	10	0.95	0.03	NA2-NA3	
	0.61	0.241	45.0	10	1.00	0.03	NA2-NA3	
	0.61	0.051	11.9	15	0.21	0.36	NA2-NA3	
	0.61	0.064	14.7	15	0.26	0.25	NA2-NA3	
	0.61	0.076	17.5	15	0.32	0.19	NA2-NA3	
	0.61	0.089	20.2	15	0.37	0.15	NA2-NA3	
	0.61	0.102	22.8	15	0.42	0.13	NA2-NA3	
	0.61	0.114	25.3	15	0.47	0.10	NA2-NA3	
	0.61	0.127	27.8	15	0.53	0.09	NA2-NA3	
	0.61	0.127	27.8	15	0.53	0.08	NA2-NA3	
	0.61	0.152	32.3	15	0.63	0.07	NA2-NA3	
	0.61	0.165	34.4	15	0.68	0.06	NA2-NA3	
	0.61	0.178	36.4	15	0.74	0.06	NA2-NA3	
	0.61	0.191	38.3	15	0.79	0.05	NA2-NA3	
	0.61	0.203	40.1	15	0.84	0.05	NA2-NA3	
	0.61	0.216	41.8	15	0.89	0.04	NA2-NA3	
	0.61	0.229	43.5	15	0.95	0.04	NA2-NA3	
	0.61	0.241	45.0	15	1.00	0.04	NA2-NA3	
	0.61	0.051	11.9	20	0.21	0.42	NA2-NA3	
	0.61	0.064	14.7	20	0.26	0.29	NA2-NA3	
	0.61	0.076	17.5	20	0.32	0.23	NA2-NA3	
	0.61	0.089	20.2	20	0.37	0.17	NA2-NA3	
	0.61	0.102	22.8	20	0.42	0.15	NA2-NA3	
	0.61	0.114	25.3	20	0.47	0.12	NA2-NA3	
	0.61	0.127	27.8	20	0.53	0.11	NA2-NA3	
	0.61	0.127	27.8	20	0.53	0.10	NA2-NA3	
	0.61	0.152	32.3	20	0.63	0.08	NA2-NA3	
	0.61	0.165	34.4	20	0.68	0.07	NA2-NA3	
	0.61	0.178	36.4	20	0.74	0.06	NA2-NA3	
	0.61	0.191	38.3	20	0.79	0.06	NA2-NA3	

Table 3.1. Continued.

	0.61	0.203	40.1	20	0.84	0.05	NA2-NA3	
	0.61	0.216	41.8	20	0.89	0.05	NA2-NA3	
	0.61	0.229	43.5	20	0.95	0.04	NA2-NA3	
	0.61	0.241	45.0	20	1.00	0.04	NA2-NA3	
	0.61	0.457	23.2	5	0.43	0.10	NA2-NA3	3.Steps model.
	0.61	0.762	35.5	5	0.71	0.05	NA2-NA3	Flat steps.
	0.61	0.457	23.2	10	0.43	0.16	NA2-NA3	Un-controlled inflow.
	0.61	0.762	35.5	10	0.71	0.08	NA2-NA3	
	0.61	0.457	23.2	15	0.43	0.18	NA2-NA3	
	0.61	0.762	35.5	15	0.71	0.09	NA2-NA3	
	0.61	0.457	23.2	20	0.43	0.24	NA2-NA3	
	0.61	0.762	35.5	20	0.71	0.11	NA2-NA3	
Pinheiro & Fael (2000)	0.70	0.050	18.4	0	0.33	0.53	NA2-NA3	10 steps. Flat steps. Un-controlled inflow.
	0.70	0.050	14.0	0	0.25	0.76	NA2-NA3	
University of Queensland	0.50	0.143	3.4	0	0.06	0.24	NA2-NA3	10 steps. Flat steps. Supercritical inflow.

Notes: α: pseudo-bottom slope; δ: step slope (positive upwards).
Type of transition: NA1-NA2 = from nappe flow subregime NA1 to NA2; NA2-NA3 = from nappe flow subregime NA2 to NA3.
Data re-analysis by the author.

Nappe oscillations

Thin nappes of liquid discharging over long free-flow spillways may be subjected to nappe oscillations (Fig. 3.4). These fluttering instabilities, also called "Kelvin-Helmholtz" instabilities, induce oscillations of the water sheet and air movement below the nappe strong enough to be heard miles away (Pariset, 1955; Thomas, 1976). The danger to the structure is usually small unless the fluttering frequency is close to the natural frequency of the structure.

Nappe oscillations are controlled by the motion of the air trapped in the space beneath the nappe. The natural frequency of the air-water system depends on the mass of water and on the volume of trapped air. To a first approximation, the possible frequencies of oscillation F of a nappe leaving an horizontal step can be estimated as:

$$\frac{F}{\sqrt{\frac{g}{d_c}}} = 0.715 * \frac{I + \frac{1}{4}}{\left(-1 + \sqrt{1 + 1.022 * \frac{h}{d_c}}\right)} \tag{3.10}$$

where h is the step height, and I is an integer: I = 1, 2, 3, 4, 5 or more (App. 3). For an oscillating nappe, the number of wavelengths in the nappe approximates the integer I. On Figure 3.5, Equation 3.10 is compared with model and prototype observations (Table 3.2).

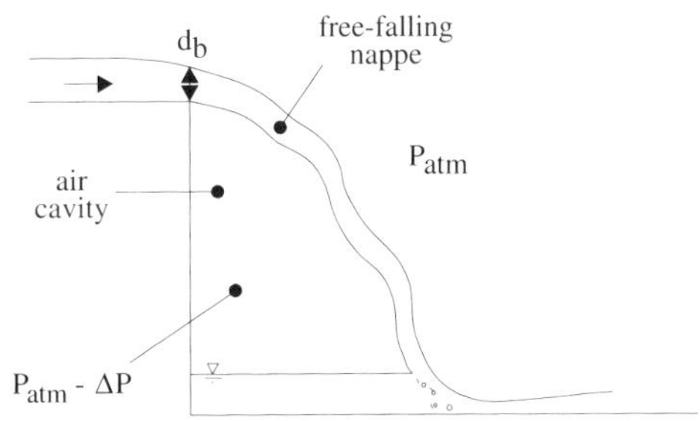

Figure 3.4. Sketch of oscillation of thin free-falling nappe.

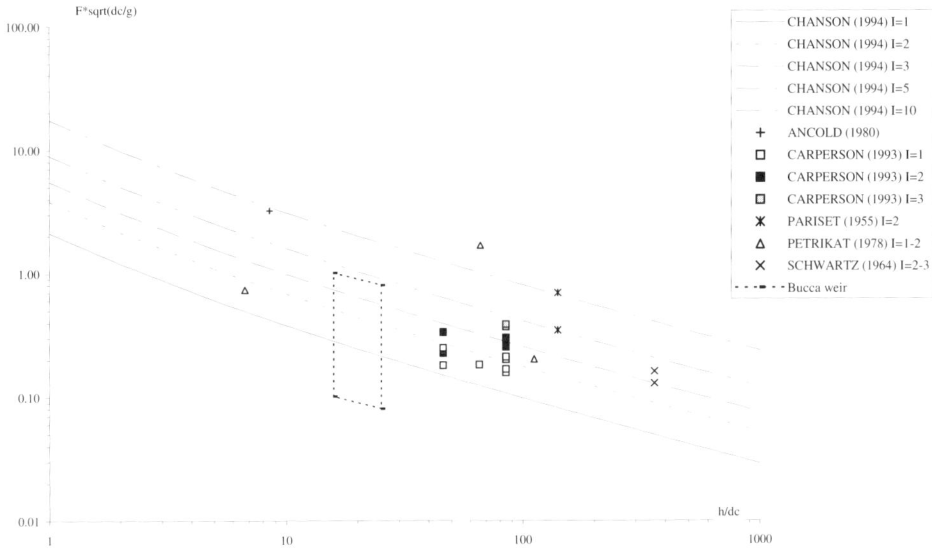

Figure 3.5. Free-falling nappe oscillations. Comparison between Equation 3.10 and experimental data (Table 3.2).

Based on model studies, Schwartz (1964a) indicated that the fluttering instabilities may occur for discharges smaller than 2.1 m^2/s. Nappe splitters may be used to control the oscillation of the nappe (Petrikat, 1958, 1978; Schwartz, 1964a). Splitters are placed on the crest at regular intervals. Each splitter creates a gap in the nappe, allowing the passage for air. Splitters need to be about 0.2 to 0.75 m high. Some were successfully installed at several overfall dams in Australia and New Zealand; e.g. Repulse, Scrivener, Devils Gate, Waitaki Dams (Thomas, 1976). One stepped weir, Bucca Weir, was designed to operate with skimming flow. At low overflows ($q_w \sim 0.05$ to 0.1 m^2/s), deflecting nappes were observed immediately downstream of the crest, associated with loud noises and audio vibrations. The situation was remedied by the installa-

Table 3.2. Observations of free-falling nappe oscillations.

Reference	q_W	h	F	I	Nb of Exp.	Remarks
	m^2/s	m	Hz			
(1)	(2)	(3)	(4)	(5)	(6)	(7)
Pariset (1955)	0.0066	1.65	20	2	--	Model data. W = 2.75 m. $0.03 < d_t < 0.11$ m.
			10	--	--	$0.22 < d_t < 0.5$ m.
Schwartz (1964a)	0.0023	0.9	10	3	1	Model data. Figure 8.
			8	4	1	Figure 9.
Petrikat (1978)	0.0041	0.797	48	1	--	Model data. W = 0.6 m. $0 < d_t < 0.41$ m.
	0.0205	2.8	4	2	1	Prototype data. W = 30 m.
	0.6028	2.25	3 to 5	2	--	Upper weir at Hameln (Ger).
Metropolitan Board (1980)	0.025 to 0.034	0.38	48	--	--	Model data. W = 0.61 m.
Bucca Weir, Australia	0.05 to 0.1	0.6	1 to 10	--	--	Prototype observations. Deflecting nappes downstream of weir crest.
Casperson (1993)	0.0075	1.175	4.3	1	1	Dunedin fountain No. 1 (NZ). W = 6 m.
	0.003 to 0.0075	0.825	4.2 to 6.7	1	12	Dunedin fountain No. 2 (NZ). W = 8 m.
			7.9 to 9.5	2	5	
			11.7 to 12.3	3	2	
University of Queensland	0.163	0.143	--	--	1	Nappe flow NA3. Loud noise observed at first drop only in absence of nappe ventilation.

Notes: d_t: tailwater depth; I: number of wavelengths in the nappe; Nb of Exp.: number of experimental data; W: channel (or crest) width.

tion of splitters (0.6 m high) along the dam crest. Other methods to prevent nappe instabilities include modifications of the crest profile or a roughening of the crest (e.g. addition of stones to the spillway crest).

Practically most nappe flow situations on stepped spillways are characterised by short falling nappes. If the ventilation is adequate, the risk of nappe oscillation and associated noise is small. Lack of nappe aeration, sometimes observed in rectangular channels, may be remedied by small sidewall offsets to ventilate the nappe cavity.

Nappe ventilation

When artificial ventilation is necessary, the required air flow rate may be estimated as:

$$\frac{q_{air}}{q_w} = 0.19 * \left(\frac{h - d_p}{d_b} \right)^{0.95} \tag{3.11}$$

where q_{air} is the nappe aeration. Equation 3.11 derives from the work of Levin (1968: 28-37) on bottom outlet aeration and nappe ventilation (Table 3.3).

Table 3.3. Nappe ventilation.

Reference (1)	Nappe ventilation (guidelines) (2)	Comments (3)
Levin (1968, pp. 28-37)	$\frac{q_{air}}{q_w} = 0.19 * \left(\frac{h - d_p}{d_b} \right)^{0.95}$ $\frac{q_{air}}{q_w} = 0.21 * \left(\frac{h - d_p}{d_b} \right)^{1.03}$	3 < Fr < 10 where Fr is a Froude defined in terms of nappe thickness. Bottom outlet aeration and nappe ventilation 13 < Fr < 15. Bottom outlet aeration.
Bos (1976)	$\frac{q_{air}}{q_w} = 0.1 * \left(\frac{d_t}{d} \right)^{-1.5}$	Based upon the data of Howe et al. (1955). Developed for sharp-crested weirs.

Notes: d_t: tailwater depth; d: upstream water depth above crest.

For a wide stepped spillway (Pedlar River Dam; Figure 2.2; Table 2.6), Schuyler (1909) described a ancient ventilation system: each step was ventilated by a 0.15 m diameter pipe open at each sidewall end and three open ends extending to the vertical face of the step. At the Pedlar River spillway, the maximum discharge implied a maximum critical depth over step height d_c/h of about 0.8 to 2.4 (i.e skimming flow regime). The ventilation system was designed probably for flow rates smaller than the maximum discharge.

3.3 HYDRAULIC CHARACTERISTICS OF NAPPE FLOWS WITH HYDRAULIC JUMPS

3.3.1 *Basic flow properties*

Nappe flows with hydraulic jump are characterised by the occurrence of critical flow conditions at each step edge. With horizontal steps, the flow properties may be analysed as a succession of drop structures (e.g. Fig. 3.6). Considering a horizontal step, the flow conditions near the end of the step change from subcritical to critical next the brink of the step[2]. At the brink the pressure distribution departs from the hydrostatic pressure distribution because of the fluid acceleration and the flow depth d_b equals:

$$d_b = 0.715 * d_c \tag{3.12}$$

where d_c is the critical flow depth (Rouse, 1936) (Fig. 3.7). Downstream of the brink, the nappe trajectory may be computed using trajectory equations, potential flow calculations, complex numerical methods or approximate methods (e.g. Montes, 1992).

The flow properties at nappe impingement and downstream of nappe impact may be derived from momentum considerations (App. 3). In practice the following correlations are recommended:

[2]The critical depth d_c is measured about $3*d_c$ upstream of the brink where the pressure distribution is hydrostatic (e.g. Henderson, 1966).

$$\frac{d_1}{h} = 0.54 * \left(\frac{d_c}{h}\right)^{1.275} \tag{3.13}$$

$$\frac{d_2}{h} = 1.66 * \left(\frac{d_c}{h}\right)^{0.81} \tag{3.14}$$

$$\frac{d_p}{h} = \left(\frac{d_c}{h}\right)^{0.66} \tag{3.15}$$

$$\frac{L_d}{h} = 4.30 * \left(\frac{d_c}{h}\right)^{0.81} \tag{3.16}$$

$$\frac{d_i}{h} = 0.687 * \left(\frac{d_c}{h}\right)^{1.483} \tag{3.17}$$

$$\tan\theta = 0.838 * \left(\frac{d_c}{h}\right)^{-0.586} \tag{3.18}$$

where d_p is the height of water in the pool behind the overfalling jet, d_1 is the supercritical flow depth downstream of nappe impact, d_2 is the subcritical depth downstream of the hydraulic jump, L_d is the distance from the drop wall to the position of the depth d_1, d_i is the nappe thickness at impingement, θ is the angle of the nappe with the horizontal at impact (Fig. 3.7; App. 3). Equations 3.12 to 3.18 were validated with numerous experimental data for aerated nappes (App. 3).

Downstream of nappe impact, the roller length of fully-developed hydraulic jump may be estimated as:

$$\frac{L_r}{d_1} = 8 * \left(\left(\frac{d_c}{d_1}\right)^{3/2} - 1.5\right) \tag{3.19}$$

where L_r is the roller length (HAGER et al., 1990)[3].

The above calculations (Eqs 3.12 to 3.19) were developed for ventilated nappes and for flat horizontal steps. They are not valid for inclined steps and pooled steps.

[3]Equation 3.19 was developed for horizontal rectangular channels with upstream and downstream hydrostatic pressure distributions. The assumption of upstream hydrostatic pressure distribution is improper if the jump is located close to the nappe impact.

Figure 3.6. Gabion drop structure near Pisino, Croatia (1974) (courtesy of Officine Maccaferri). Note the downstream counterweir and the concrete linings protecting both the drop crest and counterweir.

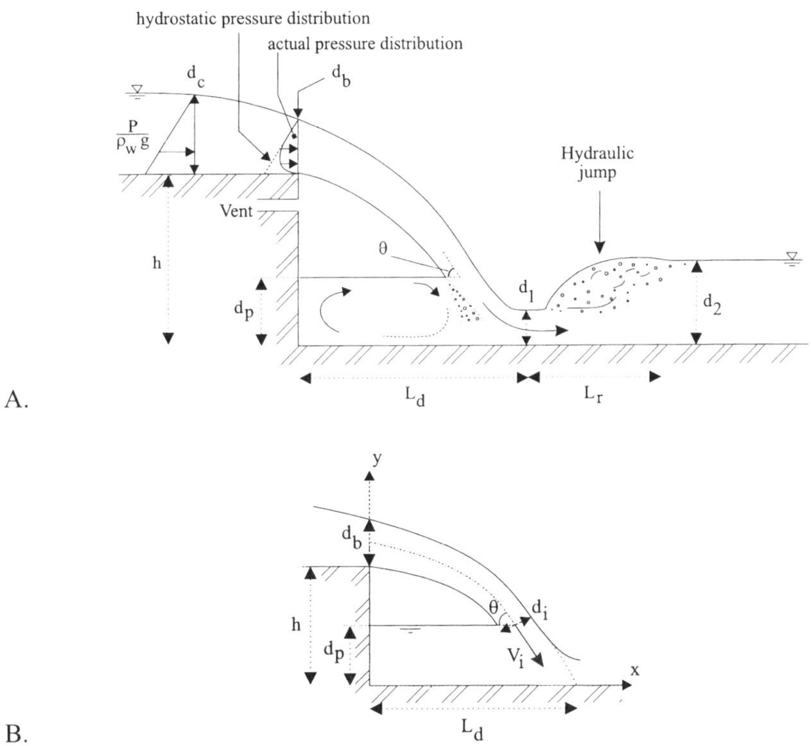

Figure 3.7. Flow at a drop structure. A. Flow at a drop structure. B. Detail of the nappe geometry.

Application to nappe flows with hydraulic jump
Along a stepped chute, a nappe flow with hydraulic jump is characterised by critical flow conditions taking place at the step brink. Equations 3.13 to 3.19 provide the main flow parameters for the sub-regime NA1 and they were successfully compared with experimental data (e.g. Peyras et al., 1991, 1992). Peyras et al. indicated that these equations can be applied also with reasonable accuracy to nappe flows with partially developed jumps (sub-regime NA2).

3.3.2 *Energy dissipation*

In nappe flow with hydraulic jump, the head loss at any intermediary step equals exactly the step height. The energy dissipation occurs by jet breakup and jet mixing, and with the formation of a hydraulic jump on the step. The total head loss along the chute ΔH equals the difference between the maximum head available H_{max} and the residual head at the downstream end of the channel (Fig. 3.8A). The residual head in the supercritical flow (i.e. H_1) equals:

$$\frac{H_{res}}{d_c} = \frac{H_1}{d_c} = \frac{d_1}{d_c} + \frac{1}{2} * \left(\frac{d_c}{d_1}\right)^2 \tag{3.20}$$

assuming a quasi-hydrostatic pressure distribution at the downstream end of the channel, an assumption that is not strictly correct. In dimensionless form, the rate of energy dissipation becomes:

$$\frac{\Delta H}{H_{max}} = 1 - \left(\frac{\frac{d_1}{d_c} + \frac{1}{2} * \left(\frac{d_c}{d_1}\right)^2}{\frac{3}{2} + \frac{H_{dam}}{d_c}}\right) \qquad \text{Un-gated chute} \tag{3.21a}$$

$$\frac{\Delta H}{H_{max}} = 1 - \left(\frac{\frac{d_1}{d_c} + \frac{1}{2} * \left(\frac{d_c}{d_1}\right)^2}{\frac{H_{dam} + H_o}{d_c}}\right) \qquad \text{Gated chute} \tag{3.21b}$$

where H_{dam} is the dam crest head above downstream toe and H_o is the reservoir free-surface elevation above the chute crest[4]. The residual energy $H_{res} = H_{max} -$

[4]The relationship between the maximum head available and dam height is $H_{max} = H_{dam} + 1.5*dc$ for an un-gated channel, and $H_{max} = H_{dam} + H_o$ for a gated waterway (Fig. 3.8A).

ΔH is to be dissipated at the toe of the chute in a downstream dissipation basin.

Combining Equations 3.13 and 3.19, the total energy loss becomes:

$$\frac{\Delta H}{H_{max}} = 1 - \left(\frac{0.54 * \left(\frac{d_c}{h}\right)^{0.275} + \frac{3.43}{2} * \left(\frac{d_c}{h}\right)^{-0.55}}{\frac{3}{2} + \frac{H_{dam}}{d_c}} \right) \quad \text{Un-gated chute} \quad (3.22a)$$

$$\frac{\Delta H}{H_{max}} = 1 - \left(\frac{0.54 * \left(\frac{d_c}{h}\right)^{0.275} + \frac{3.43}{2} * \left(\frac{d_c}{h}\right)^{-0.55}}{\frac{H_{dam} + H_o}{d_c}} \right) \quad \text{Gated chute} \quad (3.22b)$$

For ungated structures, the dimensionless head loss (Eq. 3.22a) is plotted as a function of the dimensionless flow rate (d_c/h) and the number of steps in Figure 3.8. The results are compared with drop structure data (Moore, 1943; Rand, 1955; Stephenson, 1979a; Rajaratnam & Chamani, 1995) and stepped chute data (Horner, 1969; Pinheiro & Fael, 2000). Note that the data of Horner were obtained down a 22.8° slope chute with flat horizontal steps. The flow conditions were a nappe flow regime without hydraulic jump (NA3).

Figure 3.8 indicates that most of the flow energy is dissipated on the stepped chutes for large dams (i.e. large number of steps). For a given dam height and stepped geometry, the rate of energy dissipation decreases with increasing discharge. Figure 3.8B demonstrates a good agreement between laboratory data and calculations (Eq. 3.22a).

Discussion

Although Equations 3.21 and 3.22 were obtained for nappe flows with fully developed hydraulic jumps (sub-regime NA1), it is believed that they are applicable to nappe flow with partially-developed hydraulic jumps (sub-regime NA2). In Figure 3.8B, Equation 3.22 for 10 steps compares favourably with the data of Pinheiro & Fael (2000), obtained with partially-developed humps. Peyras et al. (1991) performed experiments for nappe flows with fully and partially-developed hydraulic jumps. The rate of energy dissipation of nappe flows with partially-developed hydraulic jumps (sub-regime NA2) was within 10% of the values obtained for nappe flows with fully developed hydraulic jump for similar flow conditions.

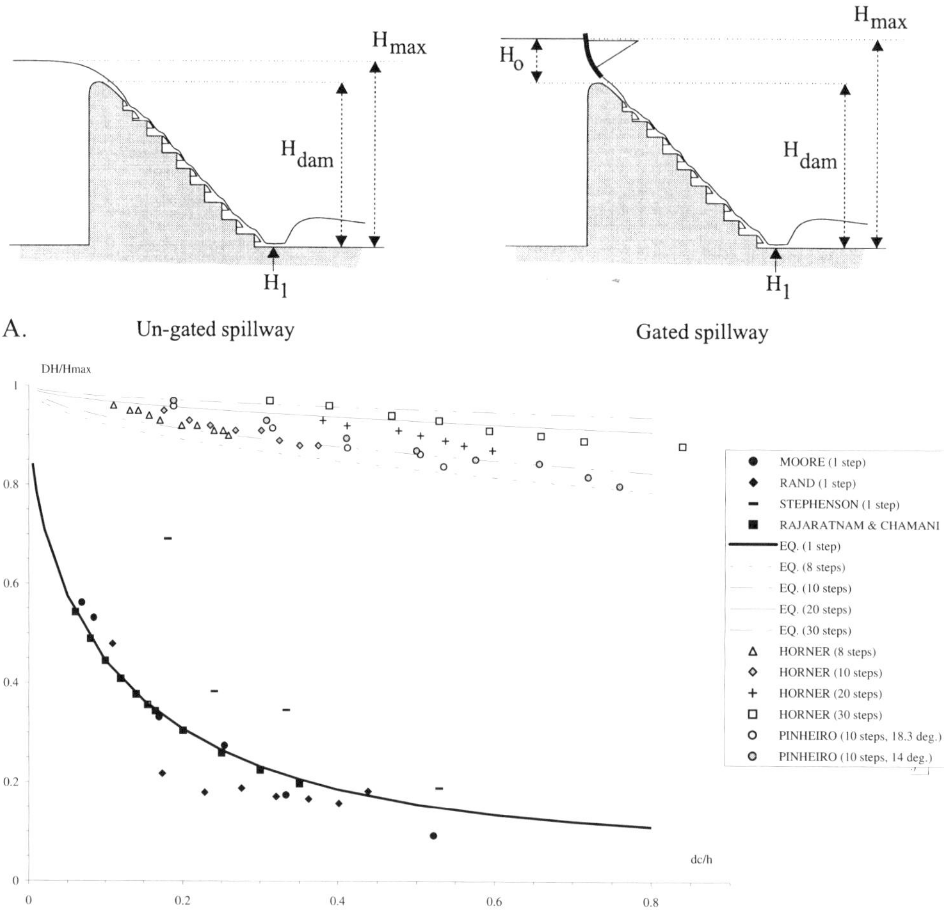

B.

Figure 3.8. Energy dissipation in nappe flow regime. A. Definition sketch. B. Comparison between Equation 3.22a and experimental data (Moore, 1943; Rand, 1955; Horner; 1969; Stephenson, 1979a; Rajaratnam & Chamani, 1995; Pinheiro & Fael, 2000).

3.4 HYDRAULIC CHARACTERISTICS OF NAPPE FLOWS WITHOUT HYDRAULIC JUMPS

In a nappe flow without hydraulic jump, the flow is basically supercritical at any point. Critical flow conditions are not observed at the step brink. The nappe trajectory may be estimated using Equations 3.1 to 3.8, and some researchers[5] give details of the brink flow characteristics and of the jet shape for supercritical flows. However the basic flow properties have not been studied, except by Horner (1969) and at the University of Queensland (e.g. Chanson & Toombes, 1997a; Toombes & Chanson, 2000).

[5]e.g. Rouse (1943), Rajaratnam & Muralidhar (1968), Hager (1983), Marchi (1993).

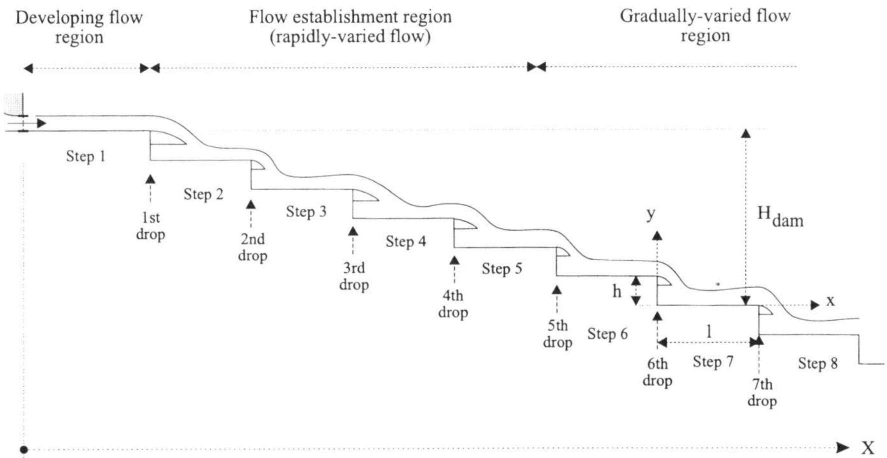

Figure 3.9. Sketch of longitudinal free-surface profile for a nappe flow without hydraulic jump (sub-regime NA3). Gated inflow.

The properties of nappe flow without hydraulic jump (NA3) cannot be predicted with analytical calculations. The best source of information derives from experimental investigations (Horner, 1969; the University of Queensland; Table 1.1). Experiments at the University of Queensland were conducted down a 24 m long 3.4° slope chute (h = 0.143 m, l = 2.4 m) with supercritical inflows (Fr = 2 to 10) (Fig. 3.9). Horner conducted experiments with slopes ranging from 22.8 to 40°, horizontal and inclined upwards steps (h= 0.03 to 0.45 m, δ = 0, +5, 10, 15, 20°) and with an uncontrolled horizontal crest. Thus, this section (3.4) has focused on horizontal step results.

3.4.1 *Longitudinal free-surface profiles*

For all investigations of nappe flow without hydraulic jump (sub-regime NA3), the inflow was two-dimensional until it reached the first drop and little aeration was observed. At the first drop, the deflected nappe is a free-falling jet and the downstream flow is highly turbulent. The flow on the first few steps is rapidly varied and this region is called a flow establishment region. It is characterised by three-dimensional flow patterns: e.g. shock waves and sidewall standing waves (Fig. 3.10; Chanson & Toombes, 1998). Significant changes from step to step are observed. For example, shock waves are observed on Steps 2 and sometimes 4, but the cross-waves are barely seen on Step 3. The rate of energy dissipation is significant: e.g. $\Delta H/H_{max} \approx$ 60 to 65% at the first three drops (University of Queensland). Horner (1969) observed that the "transitory zone" was about 4 to 5 steps long. At the University of Queensland, the rapidly-varied flow region (flow establishment region) extended along 3 to 4 steps (i.e. Steps 2, 3 and 4; Fig. 3.9). Downstream of the flow establishment region, the flow becomes gradually-varied, with continued disparity of the flow

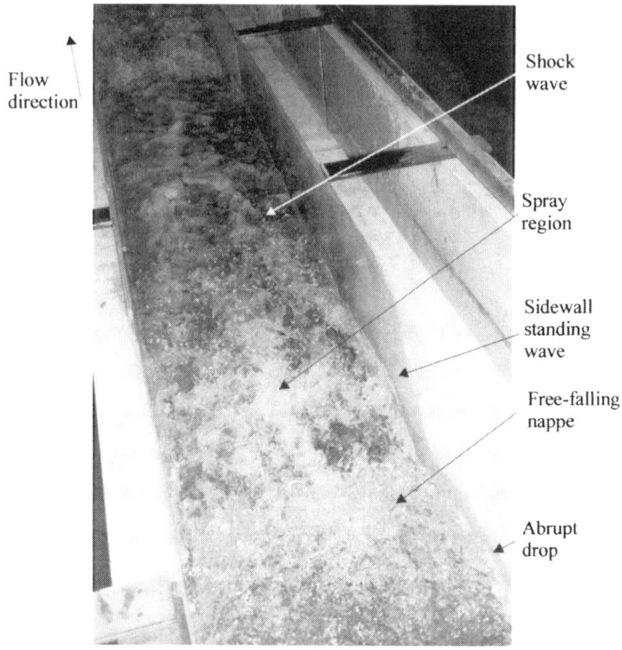

A.

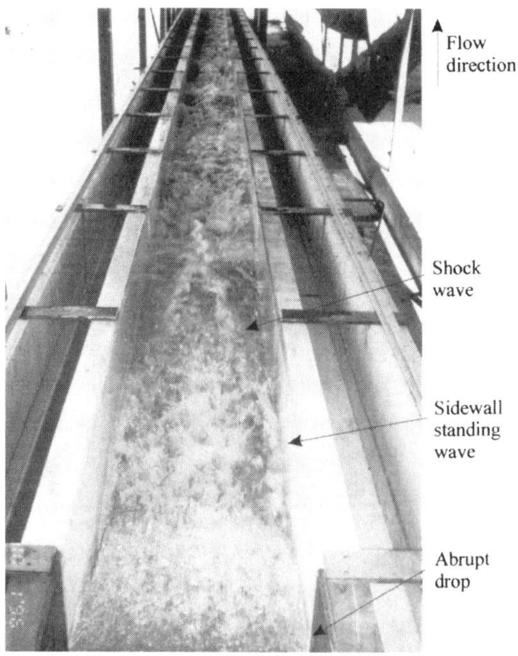

B.

Figure 3.10. Photograph of three-dimensional free-surface patterns at the first drop. A. High-speed photograph (q_w = 0.83 m^2/s, α = 3.4°, W = 0.5 m, h = 0.143 m) (Chanson & Toombes, 1998). B. Photograph of shockwaves (q_w = 0.13 m^2/s, α = 3.4°, W = 0.5 m, h = 0.143 m) (courtesy of Dr. L. Toombes).

properties from step to step (Fig. 3.9). Horner called this region the "uniform zone". The flow is supercritical at any point in the gradually-varied flow region.

Discussion
With inclined upwards steps, Horner (1969) reported some flow conditions with supercritical nappe flow (NA3) in the first part of the chute followed by a region of nappe flow with hydraulic jump (NA1 or NA2). He called these "transition category".

3.4.2 *Gradually-varied flow properties*

In the gradually-varied flow region, the flow patterns are nearly identical from one step to another. The flow is a succession of free-falling nappes and supercritical open channel flows on each and every step. At the upstream end of each step, the flow is characterised by a free-falling nappe, an air cavity and a pool of recirculating fluid (Figs 3.2C and 3.11). Air bubble entrainment occurs at both the upper and lower jet interfaces, with additional air being entrained by a plunging jet mechanism at the intersection of the lower nappe with the recirculating pool. At the lower nappe, significant interfacial air entrainment is observed and it may be modelled as:

$$C = \frac{1}{2} * \left(1 - \operatorname{erf}\left(\frac{y - Y_{50}}{2 * \sqrt{\frac{D_t}{V} * x}}\right)\right) \qquad \text{Lower nappe} \quad (1.5)$$

where y is the vertical co-ordinate, Y_{50} is the location where C = 50%, D_t is an average air bubble diffusivity, V is the jet velocity, x is the horizontal distance measured from the upstream step edge (Section 1.4.3). At the upper nappe, the air entrainment process is more complex. The air-water interface is affected by turbulent eddies acting next to the free-surface and by a rapid pressure redistribution downstream of the step brink. A modified solution of the Diffusion Equation is:

$$C = \frac{1}{2} * \left(1 - \operatorname{erf}\left(\frac{Y_{50} - y}{2 * \sqrt{\frac{D_t}{V} * x * \left(1 + k * \frac{Y_{50} - y}{x}\right)}}\right)\right) \qquad \text{Upper nappe} \quad (1.6)$$

Equations 1.5 and 1.6 are compared with air concentration data in Figure 3.12. The results show a reasonable agreement between theory and data.

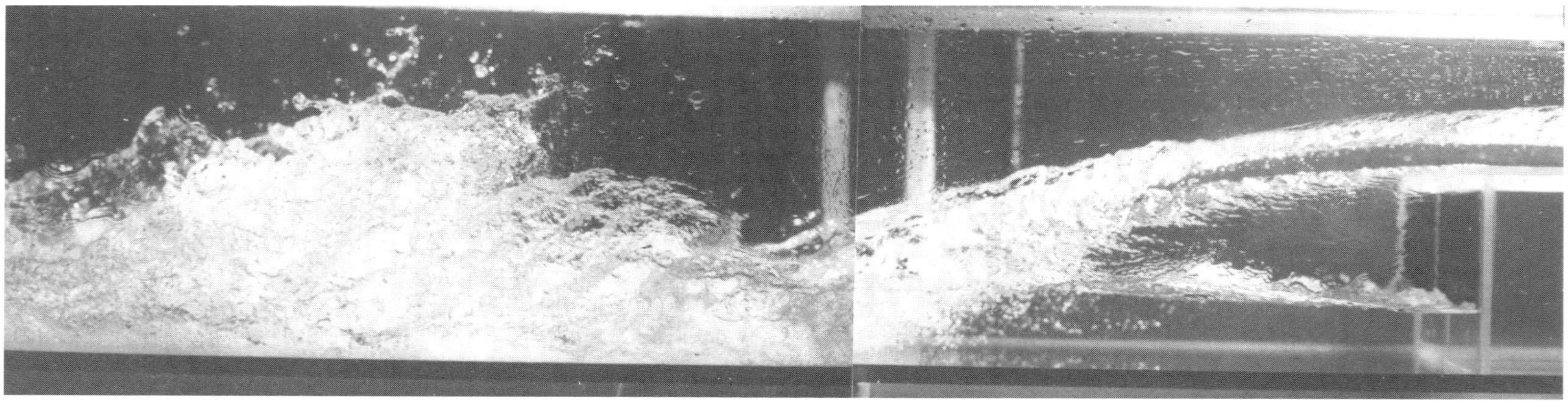

Figure 3.11. Side view of a nappe flow with hydraulic jump (NA3) (courtesy of Dr. L. Toombes). University of Queensland: $q_w = 0.112$ n = 3.2°, W = 0.35 m, h = 0.143 m, 1st drop.

Immediately downstream of nappe impact, a considerable spray is generated (Horner, 1969; University of Queensland). Visual observations at the University of Queensland showed some large droplets reaching heights of up to 0.5 m above the invert. A strong flow aeration is observed as illustrated by the air concentration profiles: e.g. at x = 0.15, 0.25 and 0.5 m in Figure 3.12A.

Further downstream the supercritical flow is decelerated and the aerated flow pattern is similar to that observed in self-aerated flows and in the flow downstream of an aeration device (e.g. Chanson, 1989b). The distribution of air concentration may be described by an air bubble diffusion model:

$$C = 1 - \tanh^2\left(K' - \frac{y}{2 * D' * Y_{90}}\right) \tag{1.7}$$

where tanh is the hyperbolic tangent function and y is the distance normal to the invert, Y_{90} is the distance where C = 90%, D' is a dimensionless turbulent diffusivity and K' is an integration constant (Section 1.4.3; App. 6). Equation 1.7 was tested successfully in parts of the spray region and in the supercritical flow region (Fig. 3.12), although it was developed originally and validated for smooth-invert chute flow.

Figure 3.12 presents also the dimensionless distributions of bubble count rates $F_{ab} * d_c / V_c$[6], where F_{ab} is the bubble frequency[7] and V_c is the critical flow velocity. For a given flow velocity and void fraction, the bubble count rate F_{ab} is inversely proportional to the mean bubble size, and directly proportional to the air-water specific interface area (e.g. Chanson, 1997a). The relationship between the bubble frequency and air content exhibits a characteristic parabolic shape which is best fitted by:

$$\frac{F_{ab}}{(F_{ab})_{max}} = 4 * C * (1 - C) \tag{3.23}$$

where the maximum bubble frequency $(F_{ab})_{max}$ is always observed for C = 50% (Fig. 3.12B). The parabolic shape (i.e. Eq. 3.23) was observed first in open channel flows (Chanson, 1997b) and later in high-velocity water jets (Brattberg et al., 1998) and hydraulic jumps (Chanson & Brattberg, 2000). The result suggests a similarity of air-water flow patterns between the four flow situations.

3.4.3 *Energy dissipation*

Although many researchers present the rate of energy dissipation $\Delta H/H_{max}$, practising engineers require information on the residual head to assess the need for

[6]Data recorded with a single-tip conductivity probe (0.35mm diameter inner electrode).
[7]bubble count rate defined as the number of bubbles impacting a probe per second.

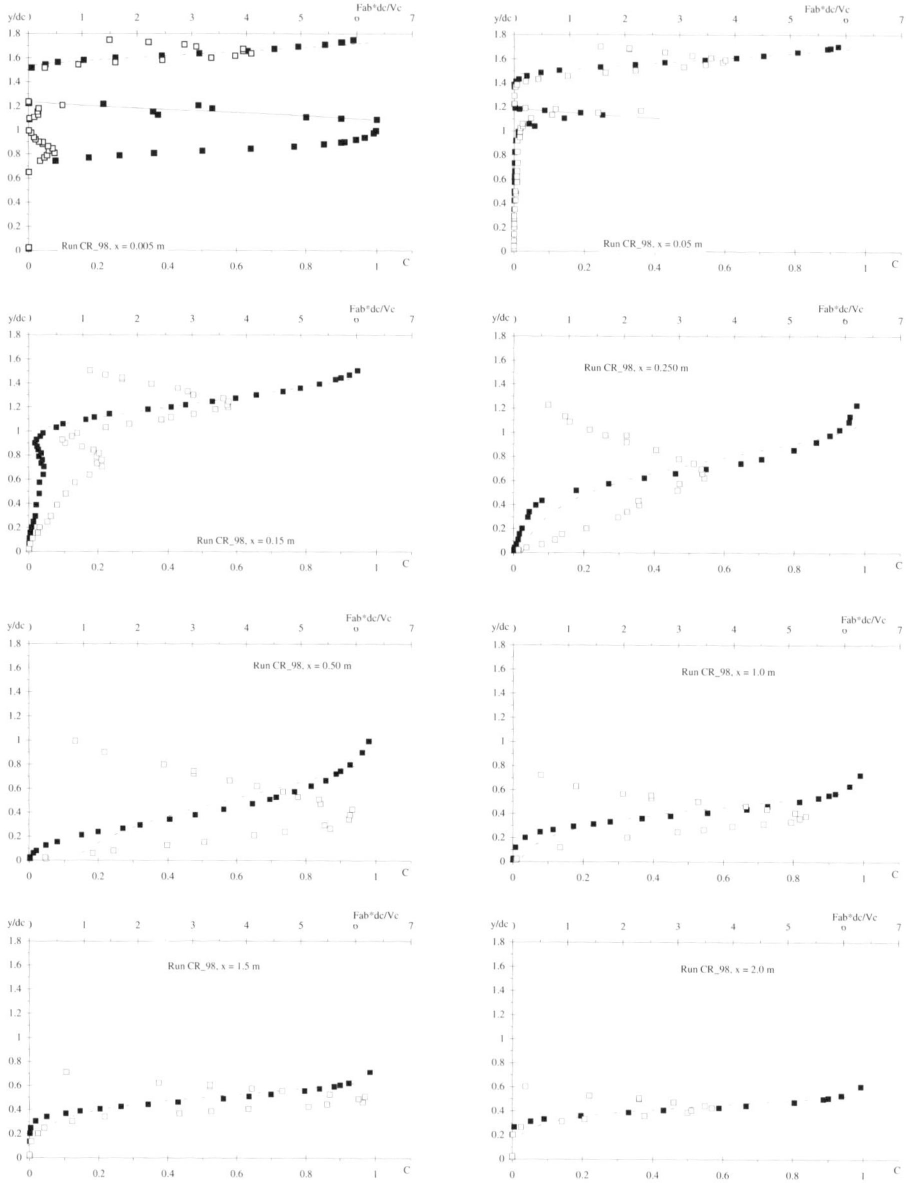

Figure 3.12. Air-water flow properties in the gradually-varied flow region (sub-regime NA3). University of Queensland: $\alpha = 3.4°$, W = 0.5 m, h = 0.143 m, supercritical inflow, centreline data. A. Dimensionless air concentration and bubble frequency distributions (q_w = 0.11 m²/s, h = 0.143 m, Step No. 9) in the free-falling nappe (x = 0.005 and 0.05 m), spray zone (x = 0.15, 0.25 and 0.5 m) and supercritical flow (x=1, 1.5 and 2 m). Legend: Black symbols = void fraction data; White symbols = dimensionless bubble count rate $F_{ab}*d_c/V_c$; Dotted/dashed line = void fraction theory.

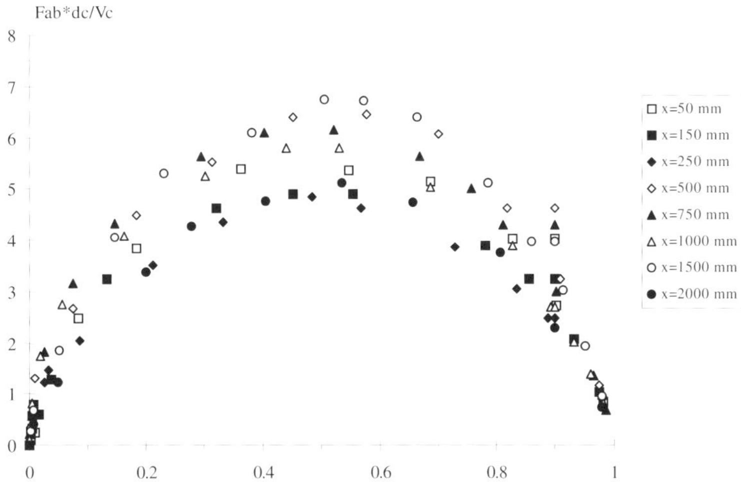

Figure 3.12. Air-water flow properties in the gradually-varied flow region (sub-regime NA3). University of Queensland: $\alpha = 3.4°$, W = 0.5 m, h = 0.143 m, supercritical inflow, centreline data. B. Dimensionless relationship between air concentration and air bubble frequency ($q_w = 0.15$ m²/s, h = 0.143 m, Step No. 9).

a downstream stilling structure. Figure 3.13A shows the dimensionless residual head H_{res}/d_c as a function of the dimensionless drop height H_{dam}/d_c where H_{dam} is the total drop in bed elevation (Fig. 3.8A). The total head was calculated based upon clear-water depth data (Eq. 1.10) for the data of University of Queensland (UQ), where d was deduced from air concentration distributions measured in the supercritical flow immediately upstream of the step brink where the streamlines were parallel to the invert and the pressure distribution was hydrostatic.

Figure 3.13A presents a comparison between flat smooth-invert chute at the University of Queensland and Nihon University ($\alpha = 4°$ and $5.7°$ respectively), nappe flow without hydraulic jump NA3 (Horner, 1969; Pinheiro & Fael, 2000; University of Queensland) and skimming flow ($\alpha = 5.7°$). The invert slope α and step height are given in the legend. The results highlight a greater energy dissipation in stepped chute flow (regime NA3) than on smooth-invert chute. Although the data were obtained with different slopes and step height, Figure 3.13A suggests smaller residual heads in skimming flow (SK1) than for nappe flows (NA3) with identical dam height, channel width and discharge. For nappe flow without hydraulic jump, the experimental results are best correlated by:

$$\frac{H_{res}}{d_c} = 6.0 * \left(\frac{H_{dam}}{d_c}\right)^{-0.30} \qquad 2 < \frac{H_{dam}}{d_c} < 20 \tag{3.24a}$$

$$\frac{H_{res}}{d_c} = 3.34 * \left(\frac{H_{dam}}{d_c}\right)^{-0.027} \qquad 30 < \frac{H_{dam}}{d_c} < 75 \tag{3.24b}$$

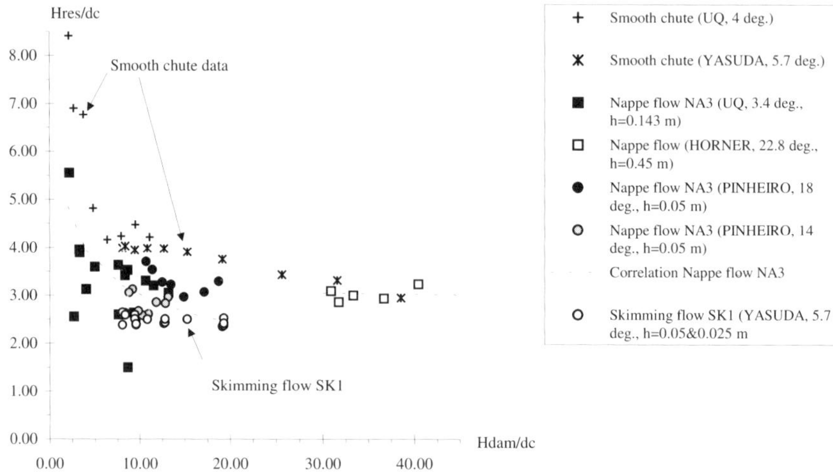

Figure 3.13. Energy dissipation in a nappe flow sub-regime NA3. A. Dimensionless residual head H_{res}/d_c as a function of the relative drop height H_{dam}/d_c. Comparison between experimental data (Horner, 1969; Pinheiro & Fael, 2000; University of Queensland), smooth chute data (Chanson, 1997b; Yasuda & Ohtsu, 1999) and skimming flow data (Yasuda & Ohtsu, 1999).

The rate of energy dissipation may also be expressed in terms of the friction slope $S_f = -\partial H/\partial s$, where s is the curvi-linear coordinate measured along the pseudo-bottom formed by the step edges. The friction slope is basically the slope of the total head line (T.H.L.). A modified form of the friction slope is introduced:

$$\frac{S_f}{\frac{1}{8} * Fr^2} \tag{3.25}$$

where Fr is the local Froude Number. For a wide channel, the Energy Equation (i.e. Backwater Equation) yields:

$$\frac{S_f}{\frac{1}{8} * Fr^2} = \frac{\tau_0}{1/8 * \rho_w * U_w^2} \tag{3.26}$$

where τ_0 is an average boundary shear stress and U_w is the flow velocity (Eq. 1.12). On a smooth chute, Equation 3.26 is the Darcy-Weisbach Friction Factor f characterising the friction loss. In nappe flow, the modified friction slope (Eq. 3.25) characterises the mean energy loss caused by jet disintegration, nappe impact and friction loss downstream while it characterises the form drag loss in skimming flow (Rajaratnam, 1990; Chanson et al., 2000).

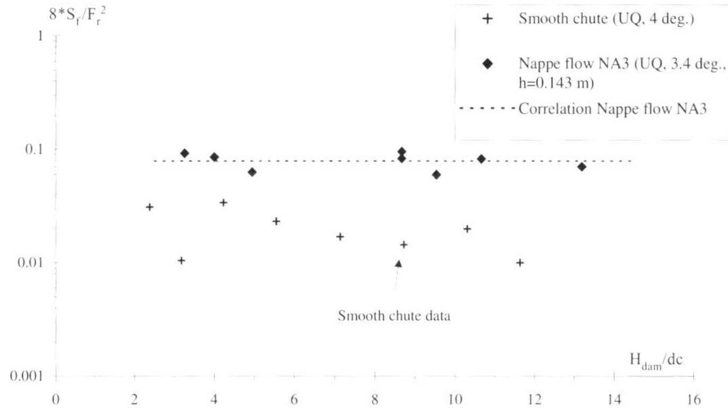

Figure 3.13. Energy dissipation in a nappe flow sub-regime NA3. B. Modified friction slope $8*s_f/Fr^2$ as a function of the dimensionless drop height H_{dam}/d_c. Comparison between nappe flow NA3 data (University of Queensland) and smooth-invert data (Chanson, 1997a).

Figure 3.13B presents experimental data of the modified friction slope $8*S_f/Fr^2$ for nappe flow NA3 and smooth chute data (Chanson, 1997b). The data are correlated by:

$$\frac{S_f}{\frac{1}{8}*Fr^2} = 0.079 \qquad 2.5 < H_{dam}/d_c < 14.5 \text{ (Data: UQ)} \qquad (3.27)$$

While the rate of energy dissipation $\Delta H/H_{max}$ characterises the overall head loss in the flow establishment zone and in the gradually-varied flow, the modified friction slope reflects upon the rate of energy dissipation per unit length in the gradually-varied flow region. For a prismatic channel, S_f tends to the pseudo-invert slope $S_o = \sin\alpha$. The experimental results show higher dissipation rates on stepped cascades with nappe flow than in smooth chute flows (Fig. 3.13B).

3.4.4 *Discussion: a prototype experience*

A nappe flow situation was observed on the Gold Creek Dam cascade (Fig. 3.14). The spillway consisted of a broad-crest (55 m wide, 61 m long), followed by the remains of the first step[8] and eleven identical steps (h = 1.5 m; l = 4 m; horizontal steps). On 2 May 1996, flood waters spilled over the spillway following a 200 mm rainfall in twenty-four hours[9].

Visual observations of the spillway operation indicate that the water flowed down the chute as a succession of free-falling nappes and supercritical flows (nappe flow NA3); the flow was highly aerated (i.e. "white waters") (Fig. 3.14); free-surface aeration took place downstream of the second drop (Fig.

[8]Erased in 1975 to enlarge the discharge capacity.
[9]The discharge was estimated as 27 m^3/s (Chanson & Whitmore, 1998).

A.

B.

Figure 3.14. Nappe flow above a prototype spillway: Gold Creek Dam (Australia, 1890), h = 1.5 m, l = 4 m, W = 55 m. Stepped spillway in operation in 1996 ($Q_w = 27\ m^3/s$). A. View from downstream. At the chute toe, the flow turns to the right (i.e. left of the photograph); there is no dissipation structure. Note the cross-waves intersecting at the step edges and on the free-falling nappes. B. View from downstream of the "cockscomb" waves, characteristic of shock waves intersecting next to the step edges.

3.14A); at the end of the chute, the water surface had a foam appearance and the distinction between the air-water flow and the spray was impossible (Fig. 3.14B). Figure 3.14B shows free-surface discontinuities that are characteristic of shock wave intersection next to the step edges. Viewed from downstream (e.g. Fig. 3.14B), the free-surface of the upper nappes exhibits a series of lozenges encompassed in between the shock waves. The angle of the shock waves with the flow direction was about 20 to 30° at the downstream end of the spillway, suggesting a flow Froude Number of about 3 to 4 (i.e. d ~ 0.12 to 0.14 m) (Chanson & Toombes, 1997b).

3.5 DESIGN OF CHUTES WITH NAPPE FLOW REGIME

A number of researchers proposed design criterion for stepped cascades operating with nappe flows. A review is presented in Table 3.4, including pooled-step chutes. At the beginning of the 20th century, Schoklitsch (1937) recommended a design such as:

$$\frac{h}{l} = \left(3 + 4.3 * \left(\frac{d_c}{h} \right)^{0.81} \right)^{-1} \tag{3.28}$$

Recently, a number of dams have been built in South Africa with stepped spillways. From this experience, Stephenson (1991) suggested that the most suitable conditions for nappe flow situations were a nappe flow regime with fully developed hydraulic jump (sub-regime NA1). Stephenson's (1991) recommendations imply relatively large steps and flat slopes. This situation is not often practical; but it may apply to relatively flat spillways, natural streams, creeks, river training and storm waterways. For steep channels or small step heights, a skimming flow regime is more desirable and will achieve larger rate of energy dissipation (Chapter 6). A related form of nappe flow design is observed in natural streams: i.e. the step-pool bed form in mountain rivers (Fig. 3.15).

Sidewall height

Design engineers must size the sidewall height to prevent spray overflow. In practice, the characteristic air-water depth Y_{90} may be used as a design parameter. For nappe flow with hydraulic jump, Equations 3.1 to 3.8, and 3.14 and 3.19 may predict the upper nappe trajectory and the hydraulic jump location. Downstream of the hydraulic jump, backwater calculations may be conducted in the subcritical flow region.

Table 3.4. Design criterion for channel with nappe flow.

Reference (1)	Geometry (2)	Tailwater level (3)	Comments (4)
Flat steps			
Binnie (1913)	$h/l < 1/3$ $h < 0.305$ m (=1 ft)	N/A	Horizontal steps at upstream end.
Schoklitsch (1937)	$\frac{h}{l} = \frac{h}{L_d + 3 * h}$	N/A	Horizontal steps.
	$\frac{h}{l} = \left(3 + 4.3 * \left(\frac{d_c}{h}\right)^{0.81}\right)^{-1}$	N/A	Practical correlation (after Chanson, 1995a).
Schleicher (1942)	$h = 0.5$ m	N/A	For river training. Criterion for nappe and skimming (SK1) flows.
Stephenson (1991)	$h/l < 0.20$ $d_c/h < 1/3$	N/A	Horizontal steps.
Pooled steps			
Binnie (1913)	$h/l = 1/10$	$d_t = h$	Pooled steps.
Schoklitsch (1937)	$h/l = 1/3$	$d_t > 2$ m	Pooled steps

Note: d_t: tailwater depth (i.e. pool height).

Figure 3.15. Step pools in mountain stream. Nishizawa-keikoku River near Mitomi village, Japan (photograph taken in November 1998).

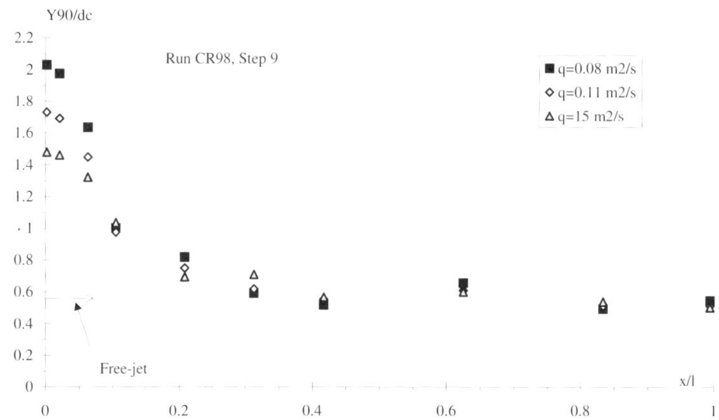

A.

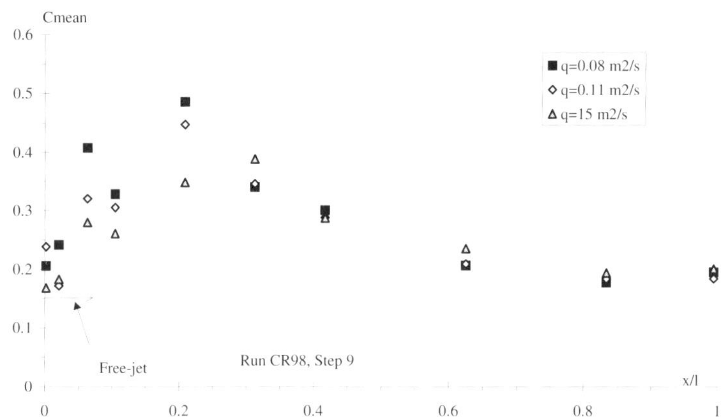

B.

Figure 3.16. Longitudinal variations of the dimensionless air-water depth Y_{90}/d_c and mean air concentration C_{mean} on a step in the gradually-varied flow region. UQ, h = 0.143 m, centreline data. A. Longitudinal variations of the dimensionless air-water depth Y_{90}/d_c. B. Longitudinal variations of the mean air concentration.

For nappe flow without hydraulic jump (NA3), experimental results are presented in Figure 3.16. Note that higher sidewalls are required in the flow establishment region. Chanson & Toombes (1997b) observed sidewall air-water depth Y_{90}/d_c up to 1.1 to 1.2 on the upper steps.

For a stepped channel over an embankment structure, the risk of erosion of the non-overflow section must be taken into account and the training wall height may be estimated based upon the Y_{98} characteristic depth. A similar criterion may be used to prevent the developing spray to generate fog or ice in the surroundings in winter.

CHAPTER 4

The transition flow regime

4.1 INTRODUCTION

For a given stepped chute geometry, a range of flow rates gives an intermediary flow regime between nappe flows at low discharges and skimming flows at large flow rates. This is known as the *transition flow regime* (Section 1.3). Some researchers reported transitory flow conditions between nappe and skimming flow: e.g. Elviro & Mateos (1995), Chanson (1996). Ohtsu & Yasuda (1997) were probably the first to introduce the concept of a "transition flow" regime, although they did not elaborate on its flow properties. It was suggested that the transition between nappe and skimming flow is characterised by strong hydrodynamic fluctuations and that these flow conditions should be avoided (Chanson, 1995a). Up to this time, the flow properties in the transition flow regime cannot be predicted by theoretical considerations. Little information is available, relying mainly upon flow visualisations at Nihon University (NH) and air-water flow measurements at the University of Queensland (UQ) (Table 4.1).

Visually a transition flow does not exhibit the "quasi-smooth" free-surface appearance of skimming flow nor the succession of free-jets observed in nappe flow. The dominant flow features are nappe stagnation on each horizontal step face associated with a downstream spray. These are observed at each and every step downstream of the inception point of free-surface aeration (Figs 4.1 & 4.2). It is the purpose of this section to present experimental observations, to propose first conclusions and to highlight the specific features of the transition flow regime.

4.2 FLOW PATTERNS

Longitudinal flow pattern and flow establishment region

Flow visualisations and measurements indicate that the upstream flow is rapidly varied. Further downstream, the flow becomes gradually-varied. On a flat chute (UQ, $\alpha = 3.4^{\circ}$), the upstream flow pattern shows some analogy with the flow establishment region in a nappe flow without hydraulic jump NA3 (Fig.

Table 4.1. Detailed experimental investigations of transition flows.

Reference (1)	α degrees (2)	h m (3)	q_w m^2/s (4)	d_c/h (5)	Remarks (6)
University of Queensland	3.4	0.071	0.08	1.22	L = 24 m. W = 0.5 m. Supercritical inflow: d_o = 0.03 m. Flat horizontal steps.
University of Queensland	21.8	0.10	0.046 to 0.10	0.6 to 0.98	L = 2.7. W = 1 m. Broad-crest with smooth inflow sidewall convergent (4.8:1 contraction). Low upstream turbulence. Flat horizontal steps. Series of 9 experiments.

3.9). At the first drop, the flow was a deflected nappe associated with downstream shock wave propagation. On Steps 2, 3 and 4, the flow was very chaotic and significant changes in flow properties were observed from one step to the next one. For example, shock waves observed on Step 2 were unseen on Step 3 and occurred sometimes on Step 4. The rate of energy dissipation was significant: i.e. $\Delta H/H_{max} \approx$ 30 to 40% at the first three drops. Further downstream (i.e. Step 5 onwards), the flow became gradually-varied with slow variations of the flow properties from step to step. The flow features were neither those of nappe flow nor skimming flow. The flow did not exhibit the quasi-homogeneous appearance observed in skimming flow, and on each step, significant longitudinal variations of flow properties were observed (Figs 4.1 & 4.2).

On steep slopes (NH, UQ α=22°), the upstream flow was smooth and transparent. In the first few steps, the free-surface was undular in phase with the stepped geometry. Some aeration in the step corners was observed immediately

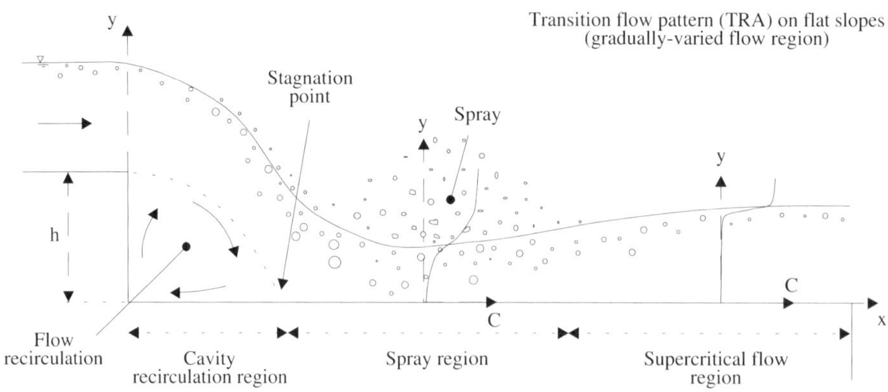

Figure 4.1. Transition flow regime: air-water flow patterns in the gradually-varied flow region. A. Flat chute (e.g. UQ, α = 3.4°).

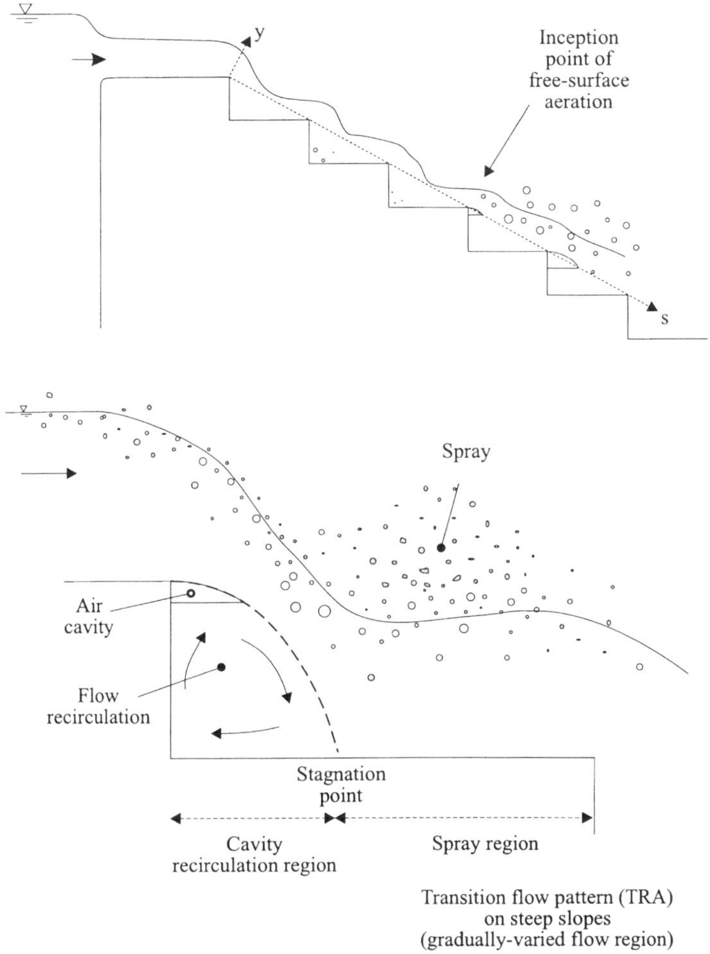

Figure 4.1. Transition flow regime: air-water flow patterns in the gradually-varied flow region. B. Steep chute (e.g. UQ, $\alpha = 22°$).

upstream of the inception point of air entrainment. Downstream of the inception point, significant splashing occurred. At the lowest flow rates (e.g. $d_c/h = 0.65$), air cavities existed beneath the nappes and the air cavity shapes alternated from step to step. For example, a small air cavity could be observed followed by a larger nappe cavity at the downstream step, then a smaller one and so on.

Downstream gradually-varied flow region

Further downstream, the air-water flow becomes gradually-varied with slow variations of the flow properties between adjacent steps. For all slopes, the dominant flow features are the impact of the lower nappe streamline on the step horizontal face, the stagnation point, and the downstream spray. On flat rectangular channels, the cavity beneath the nappe may be filled with recirculating water, while an air cavity is observed on steeper slopes. To an observer standing on the bank, the succession of stagnation points and downstream splashing gives a chaotic appearance to the flow. Splashing may be observed to reach heights[1] in excess of five to seven times the step height.

On a flat slope (UQ, $\alpha = 3.4°$), the flow at each step is characterised by a pool of recirculating waters, and significant spray and water deflection immediately downstream of the stagnation point (Fig. 4.1). Air concentration measurements suggest that the recirculation region is basically non-aerated. In the spray region, the depth-averaged mean air content C_{mean} may reach up to 40%. Downstream, the supercritical flow is decelerated up to the downstream step edge (Fig. 4.1). Shock waves are observed in the supercritical flow region. These are somehow similar to those observed in supercritical flows downstream of a single drop (e.g. Chanson & Toombes, 1998).

On steep slopes (NH, UQ α=22°), the flow pattern is dominated by the lower nappe impact and stagnation point, and the spray and water droplets ejection downstream. There is no decelerated supercritical flow zone because the step length is too short (Fig. 4.1). For a range of particular flow conditions (UQ α=22°, 0.6 to $0.65 \le d_c/h \le 0.8$), a deflecting flow was observed sometimes[2]. Liquid fractions (i.e. (1-C)) greater than 10% were measured at distances up to $1.6*d_c$.

4.3 AIR-WATER FLOW PROPERTIES

Experimental observations suggest different flow behaviours between steep and flat chutes (Figs 4.1 & 4.2). The former is typical of dam and weir spillways while the latter geometry is representative of storm waterways and irrigation structures.

4.3.1 *Air-water flow properties on a flat stepped chute*

In the transition flow regime, the gradually-varied flow (GVF) pattern exhibits significant longitudinal variations along one step (Fig. 4.3). Visually the flow appears very chaotic, with numerous droplet ejections (e.g. Fig. 4.2) and a substantial spray downstream of the stagnation point. In parts of the spray region and in the supercritical flow region, the air concentration distributions may be estimated as:

[1]Measured normal to the chute invert.

[2]Such a deflecting jet flow was seen few steps downstream of the inception point: e.g. four steps downstream of the inception point of air entrainment in Figure 4.5.

A.

B.

Figure 4.2. Photographs of the transition flow regime. A. 19° slope chute at Nihon University ($h = 0.04$ m, $d_c/h = 1.05$). Sideview in the gradually-varied flow region. B. 3.4° slope chute at the University of Queensland ($h = 0.071$ m, $q_w = 0.08$ m^2/s, $d_c/h = 1.22$). Looking upstream at Step No. 9.

C.

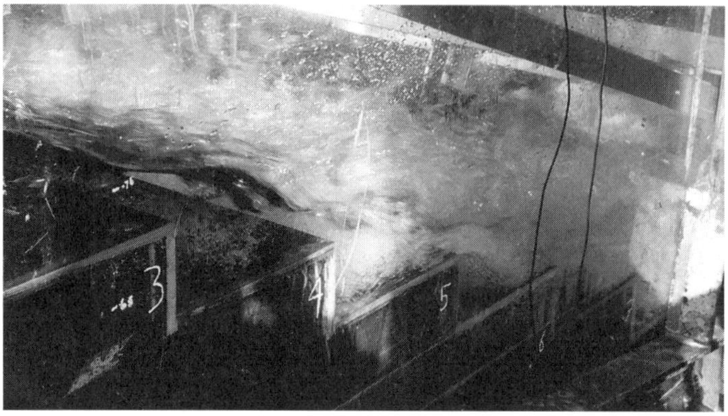

D.

Figure 4.2. Photographs of the transition flow regime. C. 22° slope chute at the University of Queensland (h = 0.1 m, q_w = 0.103 m²/s, d_c/h = 1.0). Three-quarter view from upstream, with inception of air entrainment at 4th Step. D. 22° slope chute at the University of Queensland (h = 0.1 m, q_w = 0.066 m²/s, d_c/h = 0.77). View from upstream, with a probe support in background.

$$C = 1 - \tanh^2\left(K' - \frac{y}{2 * D' * Y_{90}}\right) \tag{1.7}$$

where tanh is the hyperbolic tangent function and y is the distance normal to the invert, Y_{90} is the distance where C = 90%. The dimensionless turbulent diffusivity D' and the integration constant K' are functions of the mean air content C_{mean} only (Section 1.4.3). Equation 1.7 is compared with experimental data in Figure 4.3, which includes dimensionless distributions of bubble count rates $F_{ab} * d_c / V_c$ ([3]), where F_{ab} is the bubble frequency, d_c is the critical depth

[3]Data recorded with a single-tip conductivity probe (0.35 mm diameter inner electrode). The bubble count rate, F_{ab}, is the defined as the number of bubbles impacting the probe sensor per second (units: Hertz).

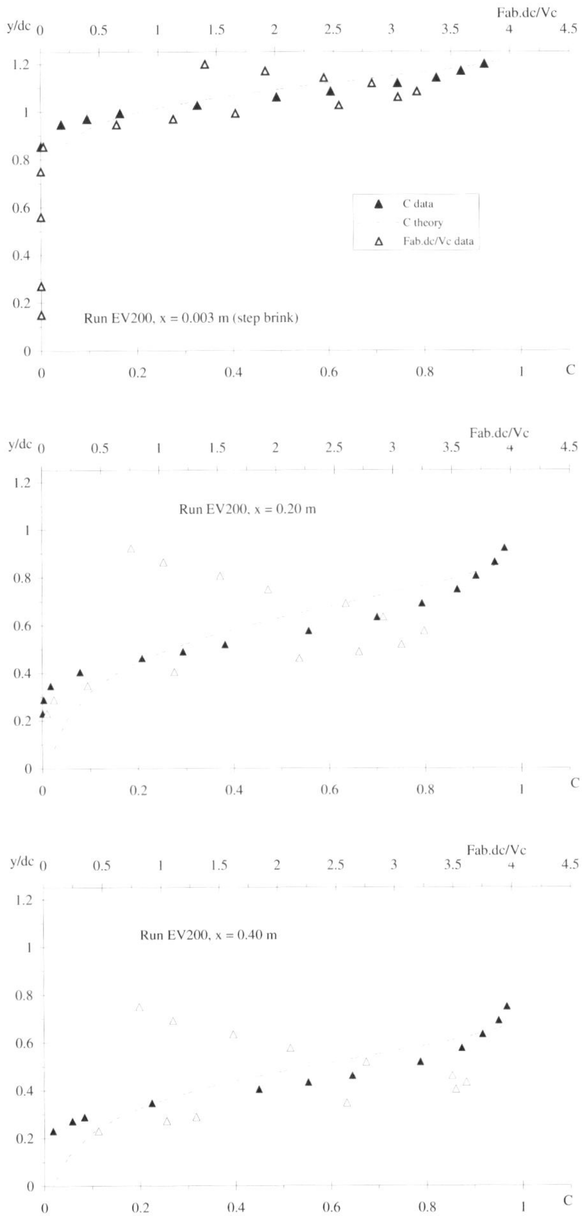

Figure 4.3. Air-water flow properties in the gradually-varied flow region (Transition flow TRA). University of Queensland: $\alpha = 3.4°$, supercritical inflow, centreline data. A. Dimensionless air concentration and bubble frequency distributions ($q_w = 0.08$ m²/s, h = 0.07 m, Step No. 16) in the recirculation zone (x = 0), impact zone (x = 0.2 and 0.4 m) and supercritical flow zone (x = 0.8 and 1 m). Legend: Black symbols = void fraction data; White symbols = dimensionless bubble count rate $F_{ab}*d_c/V_c$; Dashed line = void fraction theory.

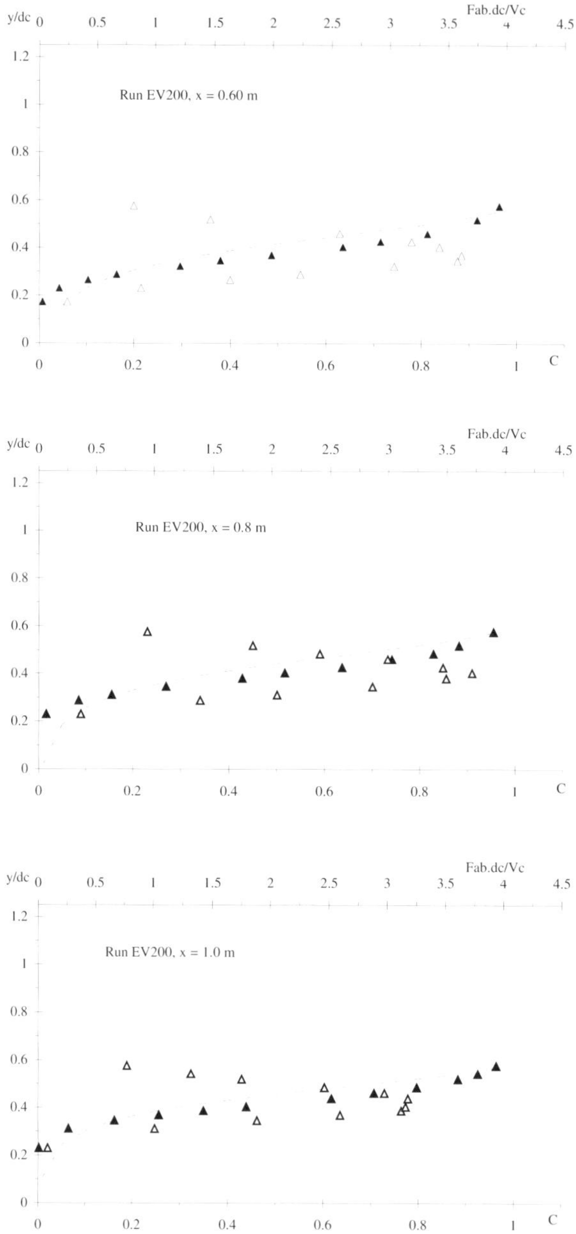

Figure 4.3. Legend see previous page.

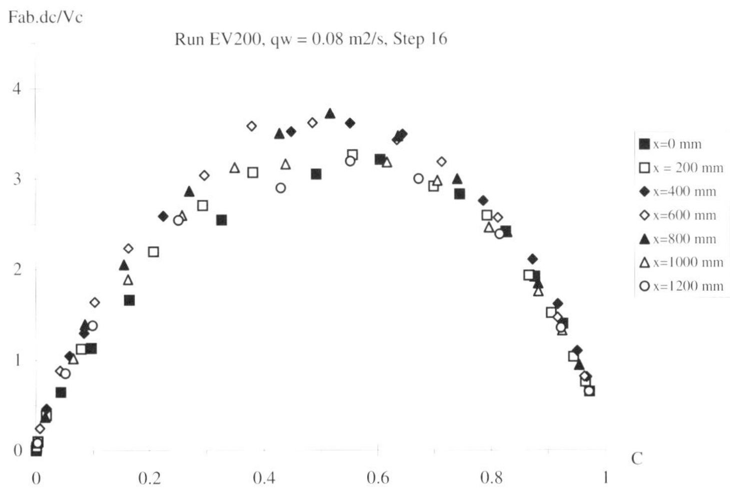

Figure 4.3. Air-water flow properties in the gradually-varied flow region (Transition Flow TRA). University of Queensland: $\alpha = 3.4°$, supercritical inflow, centreline data. B. Dimensionless relationship between air concentration and air bubble frequency ($q_w = 0.08$ m²/s, h = 0.07 m, Step No. 16).

and V_c is the critical flow velocity. For a given flow velocity and void fraction, the bubble count rate F_{ab} is inversely proportional to the mean bubble size, and directly proportional to the air-water specific interface area (e.g. Chanson, 1997c). The relationship between the bubble frequency and air content exhibits a characteristic parabolic shape which is best fitted by:

$$\frac{F_{ab}}{(F_{ab})_{max}} = 4 * C * (1 - C) \tag{3.23}$$

where the maximum bubble frequency $(F_{ab})_{max}$ is always observed for C = 50% (Figure 4.3).

4.3.2 *Air-water flow properties on a steep chute*

On steep chutes (NH, UQ), the inception point of air entrainment[4] is clearly defined (e.g. Fig. 4.2). Downstream, typical air-water flow properties are shown in Figures 4.4 and 4.5 where the dimensionless bubble count rate and void fraction are plotted as functions of the dimensionless distance y/d_c, y being measured normal to the pseudo-bottom formed by the step edges. All the data were recorded at the step edges. The dimensionless relationship between the void fraction C and the bubble frequency is also presented.

Figure 4.4 highlights a rapid free-surface aeration downstream of the incep-

[4]That is, the inception of free-surface aeration.

tion point. The mean air content C_{mean} and maximum bubble count rate $(F_{ab})_{max}$ exhibit a monotonic increase in the downstream flow direction. The relationship between void fraction and bubble count rate follows nearly a parabolic law (Eq. 3.23; Fig. 4.4B).

Figure 4.5 presents a particular flow rate in the lower range of the transition flow regime. Significant changes in air-water flow characteristics were observed between one step edge and the next one. At the fourth step downstream of the inception point, a deflecting jet was observed. Large amounts of splashing were measured up to $1.5*d_c$ and some spray overtopped the 1.25 m high sidewalls. The nappe re-attached the main flow at the next downstream step edge.

4.4 DESIGN FLOW PROPERTIES

Figure 4.6 summarises longitudinal variations of depth-averaged flow properties for flat and steep slopes: e.g. mean air concentration, maximum bubble count rate. The results may provide useful information for design purposes.

Figure 4.6A presents longitudinal variations of the mean air content and maximum bubble count rate along a step on a flat slope (UQ α=3.4°). The mean air concentration is maximum in the spray region and minimum at the downstream end of each step. The fluctuations in dimensionless maximum bubble frequency $(Fab)_{max}*d_c/V_c$ are relatively small. A maximum is observed downstream of the spray region.

On a steep chute (UQ α=22°), rapid free-surface aeration is observed downstream of the inception point of air entrainment (Fig. 4.6B). Observed values of mean air concentration are in excess of equilibrium values acknowledged in skimming flows (Section 5.3.1). In Figure 4.6B, the horizontal axis is the dimensionless longitudinal distance s/d_c, where s is measured from the first brink (i.e. at the downstream end of the broad crest). Maximum bubble count rate data indicate a rapid increase downstream of the inception point, and the data show no obvious asymptotical trend. Mean flow velocity data suggest a rapid increase in flow velocity between the crest (i.e. $U_w/V_c = 1$) up to the inception point (where $U_w/V_c \sim 2.7$ to 3), followed by a nearly constant dimensionless velocity. For nine experiments, the experimental observations suggest an asymptotical value of $U_w/V_c \sim 2.7$ at the end of the chute for $\alpha = 22°$, h = 0.1 m, $H_{dam} = 0.7$ m and $0.6 < d_c/h < 1$.

4.5 DISCUSSION: INSTABILITIES IN TRANSITION FLOWS

Although two prototype accidents and one failure were experienced with transition flow conditions (Chapter 8), some engineers refuse to consider the risks of accidents associated with hydrodynamic instabilities, even after large cracks

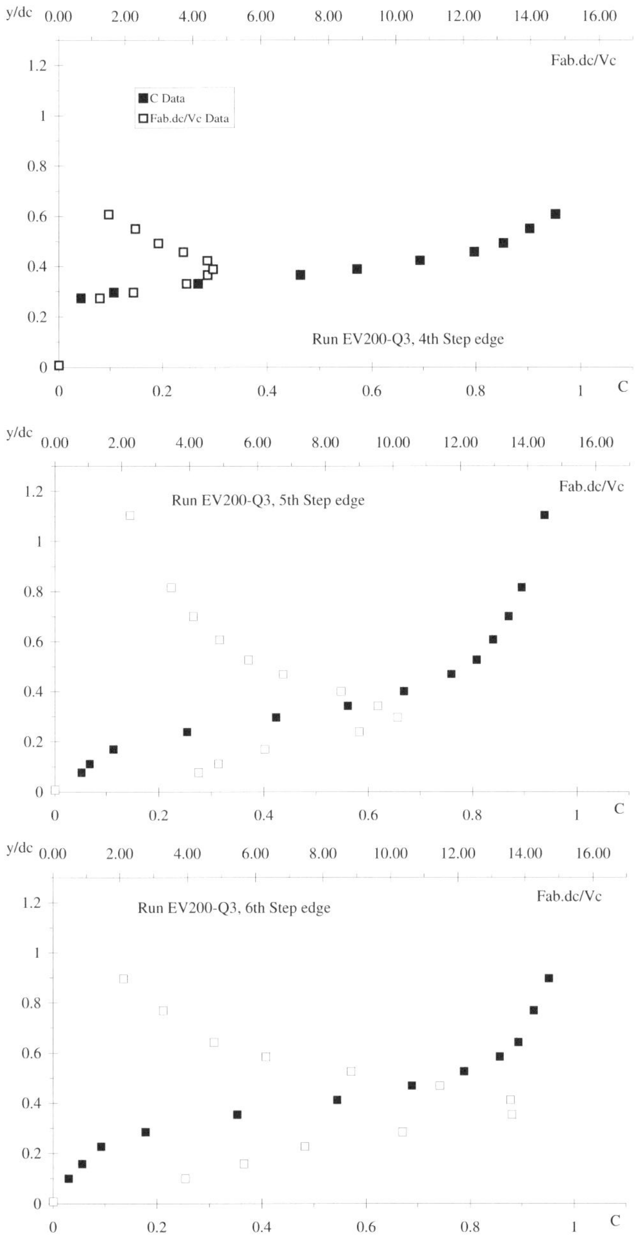

Figure 4.4. Air-water flow properties in a transition flow (TRA). University of Queensland: α = 22°, centreline data measured at step edge. A. Dimensionless air concentration and bubble frequency distributions (d_c/h = 0.87, h = 0.10 m).
1st Step edge (immediately) downstream of the inception point of air entrainment; 2nd Step edge d/s of inception point; 3rd Step edge d/s of inception point; 5th Step edge d/s of inception point.
Legend: Black symbols = void fraction data; White symbols = dimensionless bubble count rate. $F_{ab}*d_c/V_c$; Dashed line = void fraction theory.

A.

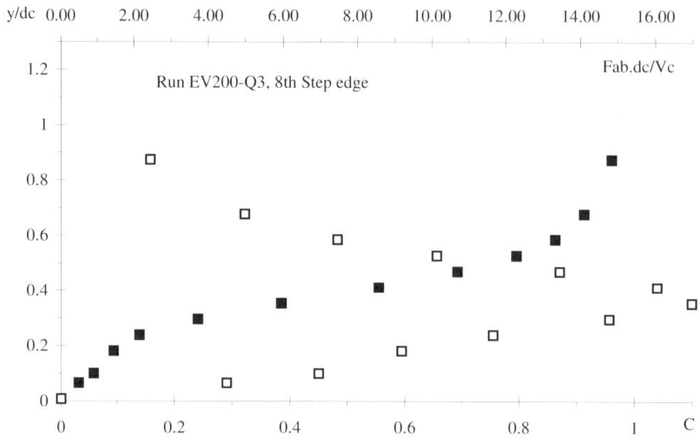

B.

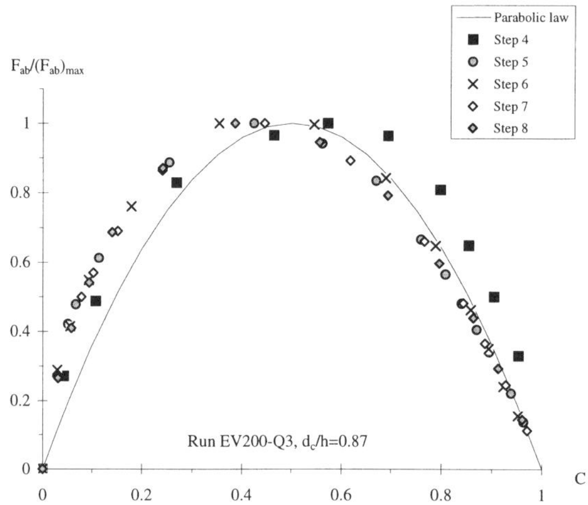

Figure 4.4. Air-water flow properties in a transition flow (TRA). University of Queensland: α = 22°, centreline data measured at step edge. A. Legend see previous page. B. Dimensionless relationship between air concentration and air bubble frequency (d_c/h = 0.87, h = 0.1 m).

were observed underneath two prototype spillway chutes (Chapter 8). The above results have demonstrated that the transition flow is characterised by a chaotic flow behaviour associated with rapid variations of the flow properties on each step. It is believed that these unsteady fluctuations are associated with unsteady pressure fluctuations on the step faces and fluid-structure interactions may take place.

On a steep slope (UQ α=22°), a range of flow conditions was associated with deflecting jet flows (e.g. Fig. 4.5). For such flow conditions, the chute sidewall height must be sized to at least $Y_{90} \sim 1.6*d_c$ or even larger than $1.4*Y_{90} = 2.2*d_c$ if splashing is not acceptable[5].

[5]For example, with a road next to the spillway chute and high risks of frost and icy conditions.

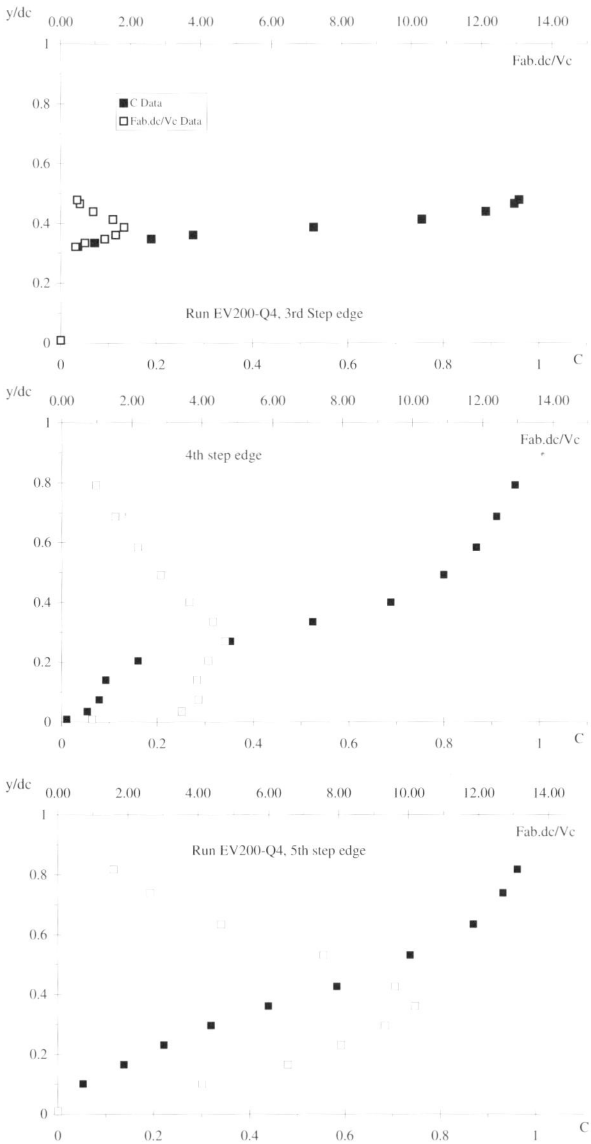

Figure 4.5. Air-water flow properties in a transition flow (TRA). University of Queensland: α = 22°, centreline data measured at step edge. A. Dimensionless air concentration and bubble frequency distributions (d_c/h = 0.77, h = 0.10 m).
1st Step edge (immediately) downstream of the inception point of air entrainment; 2nd Step edge d/s of inception point; 3rd Step edge d/s of inception point; 4th Step edge d/s of inception point; 5th Step edge d/s of inception point; 6th Step edge d/s of inception point.
Legend: Black symbols = void fraction data; White symbols = dimensionless bubble count rate. $F_{ab}*d_c/V_c$; Dashed line = void fraction theory.

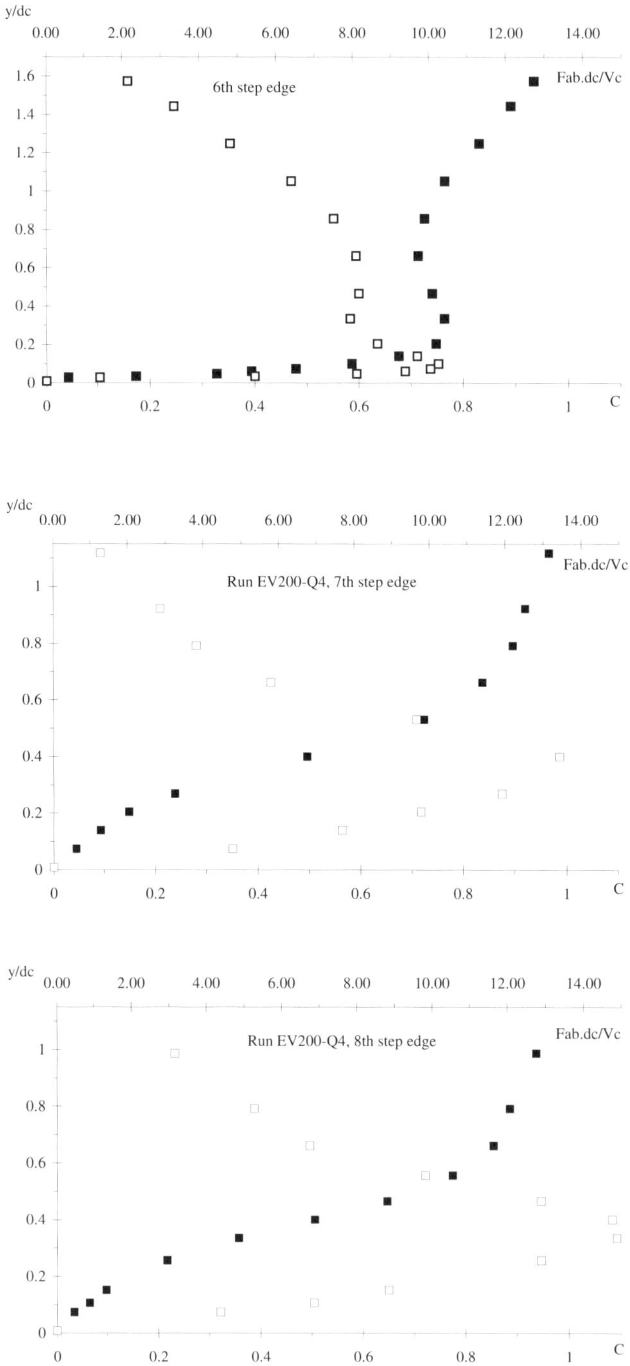

Figure 4.5. Air-water flow properties in a transition flow (TRA). University of Queensland: $\alpha = 22^{\circ}$, centreline data measured at step edge. A. Legend see previous page.

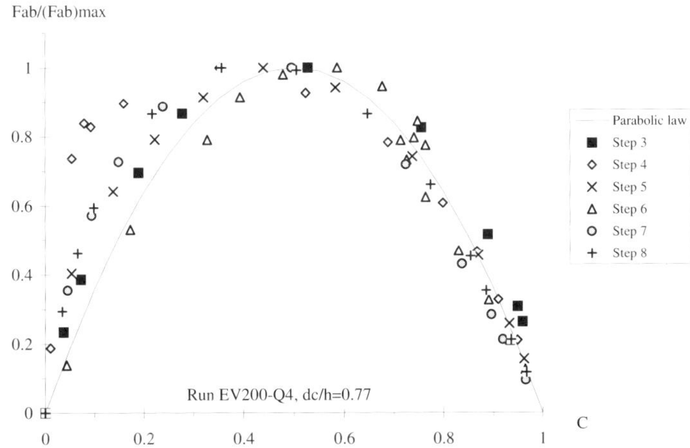

Figure 4.5. Air-water flow properties in a transition flow (TRA). University of Queensland: $\alpha = 22°$, centreline data measured at step edge. B. Dimensionless relationship between air concentration and air bubble frequency ($d_c/h = 0.77$, $h = 0.1$ m).

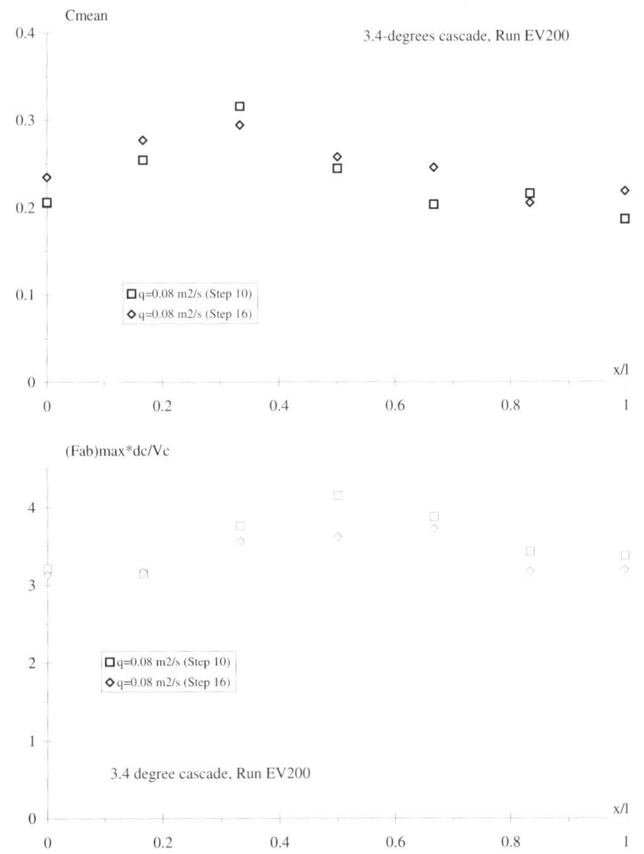

Figure 4.6. Longitudinal variations of the air-water flow properties in a transition flow (TRA): experimental observations. A. University of Queensland: $\alpha = 3.4°$, $q_w = 0.08$ m^2/s, h – 0.07 m, centreline data. Top: mean air concentration. Middle: maximum dimensionless bubble count rate. Bottom: dimensionless air-water flow depth Y_{90}/d_c.

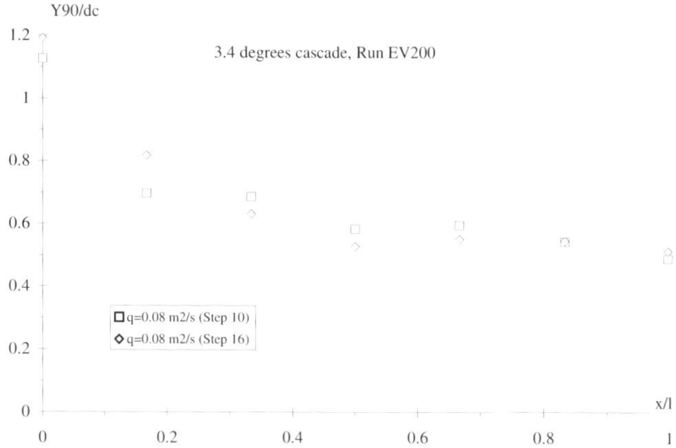

A.

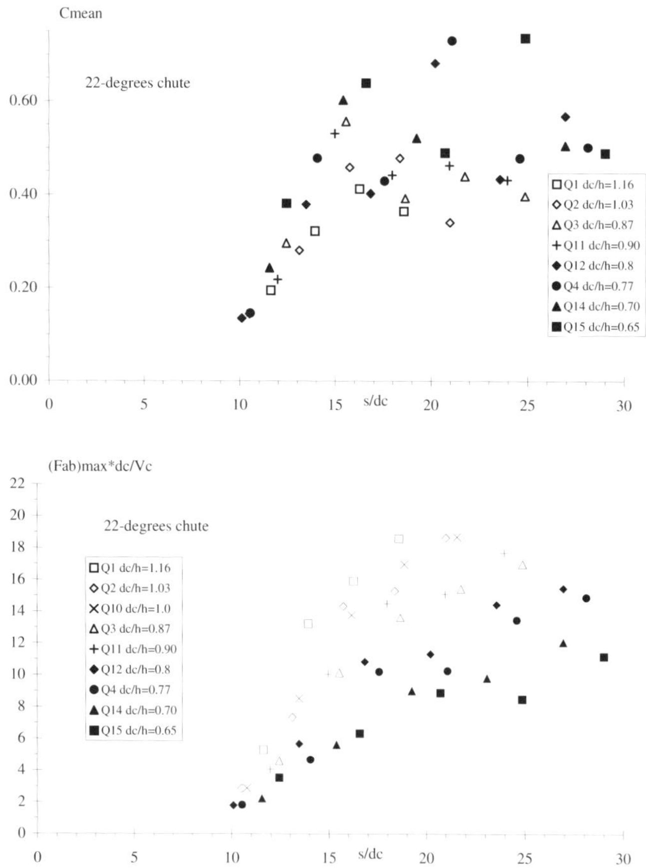

B.

Figure 4.6. Longitudinal variations of the air-water flow properties in a transition flow (TRA): experimental observations. A. Legend see previous page. B. University of Queensland: $\alpha = 22°$, h = 0.10 m, data measured at step edge. Top: mean air concentration. Middle: maximum dimensionless bubble count rate. Bottom: dimensionless mean flow velocity.

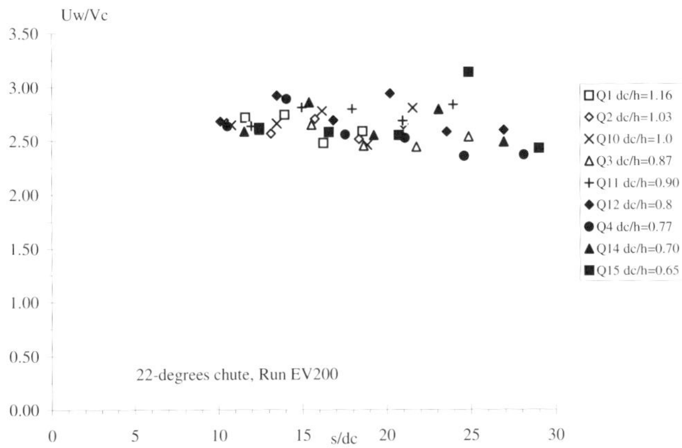

Figure 4.6. Longitudinal variations of the air-water flow properties in a transition flow (TRA): experimental observations. B. Legend see previous page.

Practically, a stepped spillway must be designed to avoid the transition flow regime, unless a rigorous hydraulic and structural modelling is conducted. If the spillway is designed to operate as a skimming flow regime at design discharges, transition flow conditions are unavoidable with an uncontrolled crest. It is recommended that the transition flow regime occur at low discharges and that physical modelling be conducted for a range of flow rates characterised by transition flow (TRA).

CHAPTER 5

Hydraulics of skimming flows

5.1 PRESENTATION

For large discharges, the waters flow down a stepped channel as a coherent stream "skimming" over the steps. The external edges of the steps form a pseudo-bottom over which the flow skims. Beneath the pseudo-bottom, recirculating vortices develop and recirculation is maintained through the transmission of shear stress from the main stream (Fig. 5.1). Small-scale vorticity is also generated at the corner of the steps. Most of the flow energy is dissipated to maintain the circulation of the vortices.

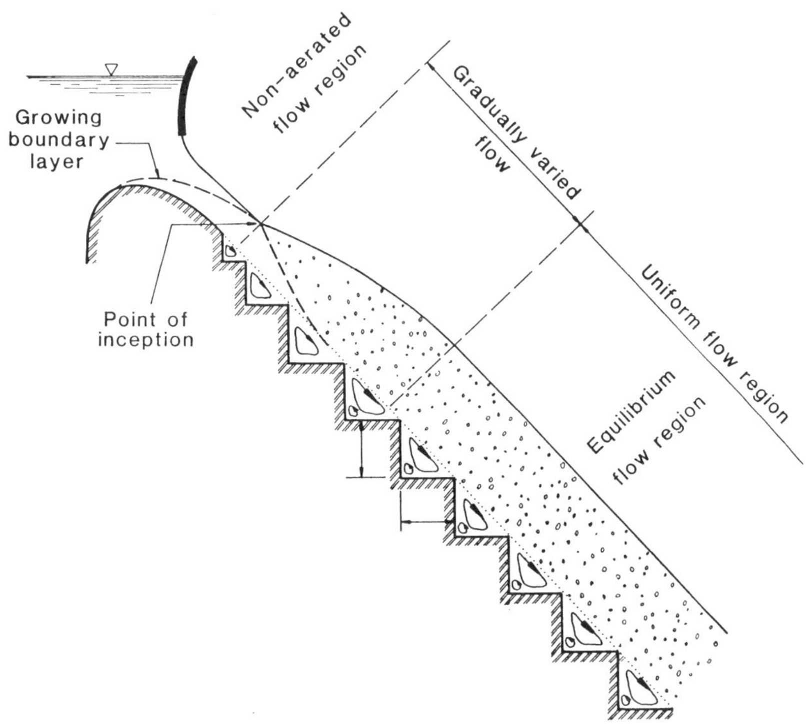

Figure 5.1. Flow regions above a stepped chute with skimming flow regime.

For an observer standing on the bank, a skimming flow exhibits an appearance similar to self-aerated flows down a smooth-invert chute (Fig. 5.2). At the upstream end, the free-surface is smooth and glassy. Next to the boundary however, turbulence is generated and a boundary layer develops (Fig. 5.1). When the outer edge of the boundary layer reaches the free surface, the turbulence initiates natural free surface aeration (Figs 5.1 & 5.2). The location of the start of air entrainment is called the point of inception. Downstream, the flow is fully-developed and rapid free-surface aeration is observed. Far downstream the flow will reach uniform equilibrium and for a given discharge any measure of flow depth, air concentration and velocity distributions will not vary along the chute. This region is defined as the uniform equilibrium flow region. The hydraulic characteristics of skimming flows are presented in this section. It must be understood that the following analysis applies to *steady flows* in *rectangular prismatic channels with flat horizontal steps*.

Figure 5.2. Photographs of skimming flows. A. Cantref dam (Wales, 1892) (courtesy of Mr. L. Stuart Davies, Welsh Water). h = 0.3048 m, $\alpha = 34°$, W = 9.14 m, design flow rate: 157 m^3/s, $Q_w = 11\ m^3/s$. Photograph taken in March 1997.

B.

C.

Figure 5.2. Photographs of skimming flows. B. Trigomil Dam spillway, Mexico (courtesy of Drs. J.L. Sanchez-bribiesca and F. Gonzalez-Villareal). $\alpha = 51.3°$, h = 0.3 m (unprotected RCC), W = 75 m. Q_w = 1,017 m^3/s in January 1992. C. Skimming flow at Nihon University. $\alpha = 19°$, h = 0.04 m, W = 0.4 m, d_c/h = 1.85. Photograph taken in November 1998.

Figure 5.2. Photographs of skimming flows. D. Skimming flow at Cedex Hydraulics Laboratory (courtesy of Mr. H. Bandler). $\alpha = 53.1°$, $H_{dam} = 3.8$ m, W = 0.8 m.

Basic flow patterns

Experimental observations suggest a number of different recirculation flow patterns beneath the skimming stream depending upon the invert slope (Fig. 5.3). For flat channels ($\alpha < 15$ to 25°), the cavity of recirculating fluid is oblong and relatively thin (Fig. 5.3A). Large stable recirculating eddies cannot develop and the vortices do not fill the entire cavity between the edges of adjacent steps. The free-surface is not parallel to the pseudo-bottom formed by the step edges but it exhibits an "undular" pattern in phase with the step geometry. The vortex generation associated with each mixing layer is disturbed by the next step and might interfere with that of the adjacent steps (Fig. 5.4). The flow depth will control in part the vertical extent of the mixing layer and the vortex interference region. The main flow parameters become the distance between two adjacent step edges and the depth of flow[1].

For very flat slopes ($\alpha < 12$ to 15°), the flow is characterised by the impact of the mixing on the next step, a three-dimensional unstable recirculation in the wake and some skin friction drag on the step downstream of the wake

[1]This reasoning is confirmed by flow resistance data showing good correlation between the friction factor and the relative roughness, the latter being proportional to the edge spacing to flow depth ratio (Sub-section 5.3.3).

impact (Fig. 5.3A). The flow pattern is called a *wake-step interference* sub-regime (SK1). For larger slopes (12 to 15° < α < 15 to 25°), the wake interferes with the next mixing layer and there is little skin-friction drag component (Fig. 5.3B). This pattern is called a *wake-wake interference* sub-regime (SK2).

On steep slopes (α > 25 to 30°), a stable recirculation is observed in the cavities between the edges of adjacent steps as shown by photographs (e.g. Peyras et al., 1991; Diez-Cascon et al., 1991; Figs 5.3C & 5.16). The flow pattern is called a *recirculating cavity flow* sub-regime (SK3). The recirculating flow appears asymmetrical with most vorticity activity occurring in the downstream part of the cavity. At irregular intervals, there are strong fluid ejections and inflows (Section 5.3.3). For an observer standing on the bank, the aerated free-surface has the same appearance as a self-aerated flow down a smooth-invert spillway.

With flows past two-dimensional cavity, researchers observed a mechanism of stable recirculation for a range of cavity dimensions (Table 5.1; Fig. 5.5). Stable re-circulatory flow patterns[2] were observed for groove height to length ratio larger than 0.4 to 0.45. For a horizontal step, the cavity height to length ratio equals cosα*sinα and a value of 0.4 would correspond to α = 27°. The results (Table 5.1) would imply stable recirculation for invert slopes larger than 27° on stepped chutes.

Crest designs and deflecting jet flow

The shape of the crest profile is important to achieve a proper flow behaviour on the upstream steps. Model and prototype observations showed the possibility of jet deflection at the first steps if the upstream steps are too high. Figure 5.6 shows a deflecting nappe "bypassing" the first nine steps and impacting on a downstream apron. An analytical condition[3] for jet deflection is:

[2]Photographic evidences of stable flow recirculation include Townes & Sabersky (1966) and Furuya et al. (1976).

[3]Developed by Chanson (1996) based upon a cavity filling model. If the first step cavity is filled with water, the pressure forces of the vertical and horizontal step faces on the flow counterbalance the hydrostatic pressure force of the main stream flow, and the waters flow (i.e. skim) over the recirculating fluid without water deflection. If the cavity below the jet is not filled, the water leaving the step edge will drop as a free-fall jet and hit the horizontal face of the next step, with deflecting jets of waters which then might "by-pass" (i.e. skip over) the following step(s).

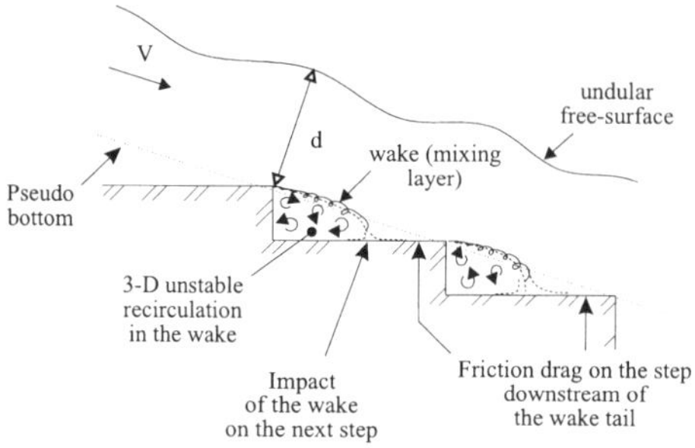

A. (SK1) Unstable recirculation with wake-step interference

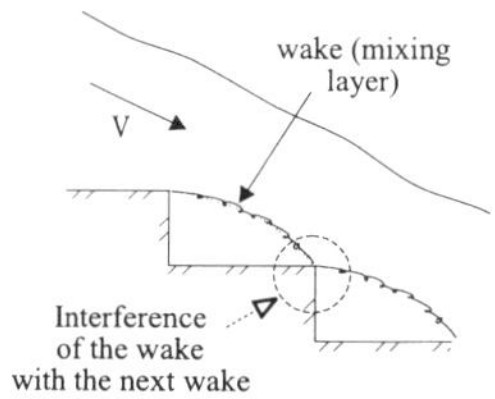

B. (SK2) Unstable recirculation with wake-wake interaction

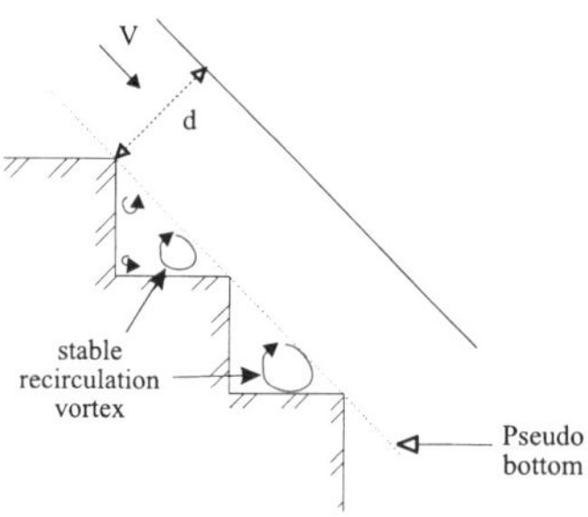

C. (SK3) Stable recirculation

Figure 5.3. Flow patterns in the cavity between adjacent steps. A. Wake-step interference sub-regime (SK1) (flat slopes). B. Wake-wake interference sub-regime (SK2) (slope about 20 to 25º). C. Recirculating cavity flow sub-regime (SK3) (steep slopes).

$$\frac{d_c}{h} < \frac{Fr_b^{2/3} * \sqrt{1+\frac{1}{Fr_b^2}}}{\sqrt{1+2*Fr_b^2*\left(1+\frac{1}{Fr_b^2}\right)^{3/2}*\left(1-\frac{\cos\alpha_b}{\sqrt{1+\frac{1}{Fr_b^2}}}\right)}} \tag{5.1}$$

Table 5.1. Observations of stable re-circulatory flow motion in two-dimensional cavity.

Reference (1)	Condition for stable recirculation (2)	Remarks (3)
<u>Water flow experiments</u> Adachi (1964)	$\frac{L_s - l_s}{k_s} \leq 2.4$	Rectangular wooden bars.
Townes & Sabersky (1966)	$\frac{V_* * k_s}{\nu} \leq 150$	Square grooves (i.e. L_S-l_S = k_S).
Knight & Macdonald (1979)	$\frac{L_s - l_s}{k_s} \leq 2.5$	Two-dimensional strip roughness in open channel. Square strips.
<u>Wind tunnel experiments</u> Maull & East (1963)	$\frac{L_s - l_s}{k_s} \leq 2.5$	Flow past a two-dimensional rectangular cavity.
Kistler & Tan (1967)	$\frac{L_s - l_s}{k_s} \leq 2.2$	Flow past a two-dimensional rectangular cavity.
Furuya et al. (1976)	$\frac{L_s - l_s}{k_s} \leq 2$	Two-dimensional circular strip roughness.[a]

Notes: V_*: shear velocity; (L_s - l_s): groove or cavity length measured parallel to the flow direction (Fig. 5.5); [a]: (L_s - l_s) is measured between the outer edges of the circular rods (Fig. 5.5).

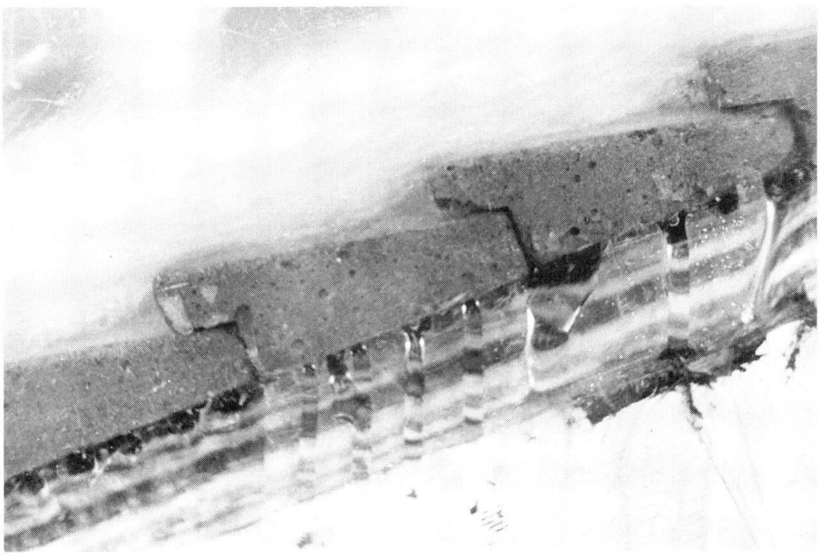

Figure 5.4. Skimming flow with wake-step interference sub-regime SK1 past overlapping concrete blocks (courtesy of Dr. Baker, University of Salford). Flow from the right to the left (q_w = 0.17 m²/s, W = 0.6 m) α = 21.8°, h = 0.019 m, inclined downward steps: δ =. 8.3°.

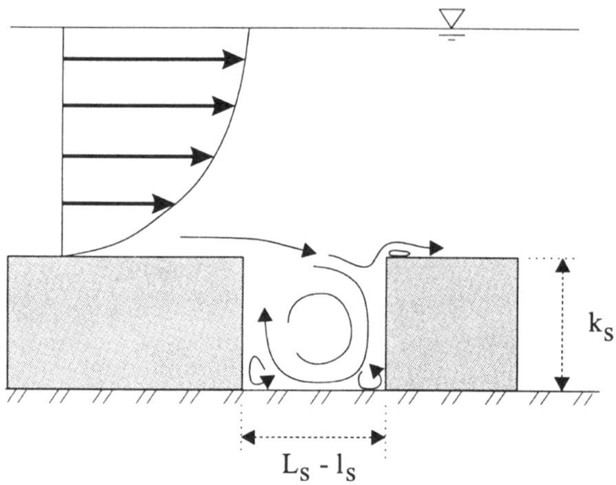

Figure 5.5. Example of stable flow recirculation observed in a two-dimensional cavity.

Figure 5.6. Deflecting jet at Goulburn Weir (1891, Australia). $Q_w = 5.21$ m³/s, h = 0.5 m, $d_c/h = 0.51$ on 30 January 2000. The concrete apron at the downstream end was added in 1986 when the original overflow gates were replaced by undershoot radial gates.

where $Fr_b = V_b/ Fr_b = V_b/\sqrt{g*d_b}$, d_b and V_b are the flow depth and velocity at the brink, and α_b is the angle between the invert and horizontal (Fig. 5.7B)[4]. Equation 5.1 was validated with large-size model observations (Chanson, 1996).

[4]The flow properties at the brink may be deduced from the Bernoulli Equation if the smooth crest is short enough to neglect the effect of the developing boundary layer. The angle α_b may be estimated as the local channel slope to a first approximation.

Figure 5.7 summarises the conditions for the onset of deflecting jets as a function of the brink Froude Number and jet angle. The results suggest that, for a given discharge, the risks of water deflection may be reduced by decreasing the first step height h, by placing the first step as upstream as possible and possibly by changing the profile curvature near the first step. These simple conclusions are in agreement with the experimental observations of BaCaRa (1991) and other researchers (e.g. Fig. 1.1). Figure 5.7 provides a design chart which could be used to assess the risks of jet deflection at the first steps when a skimming flow regime is designed.

In practice, various shapes of spillway crest have been used in prototypes: e.g. broad-crest (Gold Creek, Aus.), circular crest (Le Pont, Fra.), ogee crest (Trigomil, Mex.). For concrete dams, the ogee crest design is common: e.g. WES profile at Monksville Dam, Creager profile at M'Bali Dam. With RCC dams, a wide crest is preferable to allow truck traffic to supply concrete and for the roller compacting process. In Queensland (Australia), some stepped diversion weirs were designed with provision for an inflatable rubber dam. These designs have a relatively long flat crest inclined upwards to avoid neg-

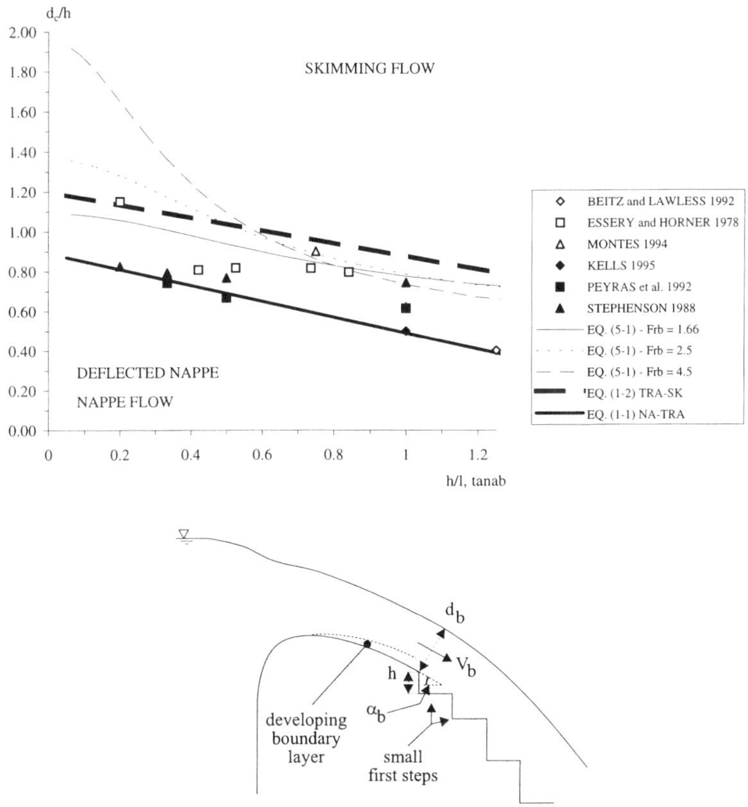

Figure 5.7. Onset of jet deflections: comparison with the onset of nappe and skimming flow regimes (Section 1.3).

ative pressures on the crest. Most gabion stepped weirs have a flat crest followed directly by the first step.

5.2 DEVELOPING FLOW REGION AND INCEPTION OF AIR ENTRAINMENT

5.2.1 *Developing flow properties*

On a chute spillway, the flow is accelerated by the gravity force component in the flow direction. For an ideal fluid flow, the free-stream velocity may be deduced from the Bernoulli Equation:

$$V_{max} = \sqrt{2*g*(H_{max} - d*\cos\alpha)} \qquad \text{Ideal fluid flow (5.2)}$$

where H_{max} is the upstream total head, α is the channel slope and d is the flow depth.

In practice, however, friction losses occur. At the upstream end, a turbulent boundary layer is generated by bottom friction and it develops in the flow direction. The location where its outer edge reaches the free surface is called the inception point of air entrainment[5]. Downstream, the turbulence next to the free-surface becomes large enough to initiate natural free surface aeration (Fig. 5.2). The characteristics of the inception point are L_I and d_I: L_I is the distance from the start of the growth of the boundary layer and d_I is the depth of flow at the point of inception.

In the developing flow region, the flow consists of a turbulent boundary layer next to the invert and an ideal-fluid flow region above. In the boundary layer, the velocity distribution follows closely a power law:

$$\frac{V}{V_{max}} = \left(\frac{y}{\delta_{BL}}\right)^{1/N} \qquad 0 < y/\delta < 1 \text{ (5.3)}$$

where δ_{BL} is the boundary layer thickness, y is the distance normal to the pseudo-bottom and V_{max} is the ideal-fluid velocity (Eq. 5.2). Careful LDA measurements showed that $N \approx 5$ on a stepped chute upstream of the inception point (Fig. 5.8). In Figure 5.8, Equation 5.3 is compared also with experimental results obtained in a smooth turbulent boundary layer past a flat plate (e.g. Schlichting, 1979). The data of Ohtsu & Yasuda (1997) show further that the turbulent velocity fluctuations in the boundary layer are large reflecting the strong turbulence generated by the steps.

[5]The inception point is defined as the inception point of free-surface aeration.

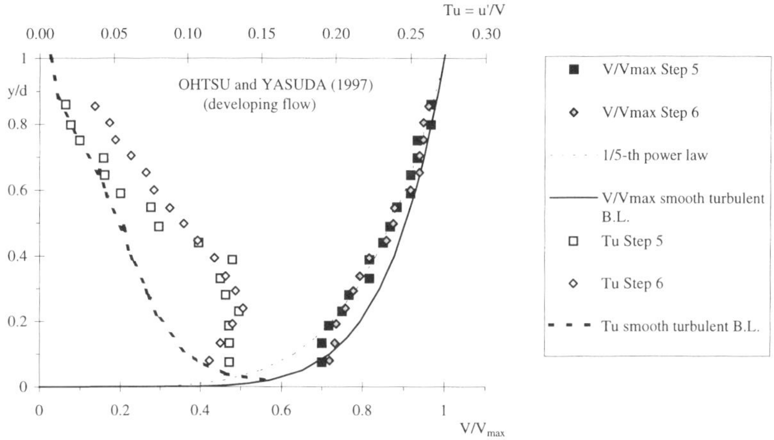

Figure 5.8. Velocity and streamwise velocity fluctuation distributions in the developing flow region (Ohtsu & Yasuda, 1997). $\alpha = 19^o$, h = 0.05 m, $q_w = 0.089\ m^2/s$. Measurements at step edges upstream of the inception point. Comparison with Equation 5.3 and smooth turbulent boundary layer data.

The relationship between the water depth, boundary layer thickness and free-stream velocity derives from the continuity equation:

$$q_w = V_{max} * \left(d - \frac{\delta_{BL}}{N+1} \right) \tag{5.4}$$

where d is the flow depth measured normal to the pseudo-bottom.

5.2.2 *Point of inception*

The position of the inception point is primarily a function of the discharge and the step roughness:

$$\frac{L_I}{h * \cos\alpha} = f_1(F_*, \alpha) \tag{5.5}$$

$$\frac{d_I}{h * \cos\alpha} = f_2(F_*, \alpha) \tag{5.6}$$

where F_* is a Froude Number defined in terms of the roughness height: $F_* = q_w / \sqrt{g * \sin\alpha * (h * \cos\alpha)^3}$, $h * \cos\alpha$ is the step roughness height and α is the channel slope (Chanson, 1995a). A statistical data analysis indicates that the flow properties are best correlated by:

$$\frac{L_I}{h^* \cos\alpha} = 9.719 * (\sin\alpha)^{0.0796} * (F^*)^{0.713} \tag{5.7}$$

$$\frac{d_I}{h^* \cos\alpha} = \frac{0.4034}{(\sin\alpha)^{0.04}} * (F^*)^{0.592} \tag{5.8}$$

It yields:

$$\frac{d_I}{L_I} = 0.06106 * (\sin\alpha)^{0.133} * \left(\frac{L_I}{h^* \cos\alpha}\right)^{-0.17} \tag{5.9}$$

Model and prototype data (Table 1.1) are compared with Equations 5.7 and 5.8 in Figure 5.9 where $k_s = h^*\cos\alpha$. Note the good agreement between the statistical correlations of Chanson (1994d) and prototype data at Trigomil Dam. The latter is shown in operation in Figure 5.2 where the inception point of free-surface aeration is clearly visible. The results (Fig. 5.9) indicate that the rate of boundary layer growth on stepped chutes is approximately 2.8 times larger than that on smooth channels. For a given channel geometry, Equation 5.7 implies that the location of the inception point moves downstream with increasing discharges.

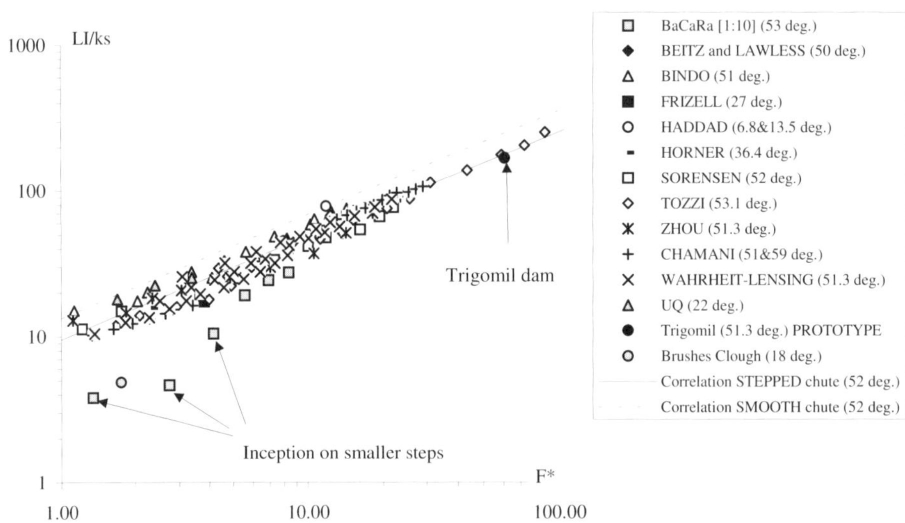

Figure 5.9. Characteristics of the inception point of air entrainment. A. L_I/k_s as a function of F_*. Model data: BaCaRa (1991), Beitz & Lawless (1992), Bindo et al. (1993), Frizell & Mefford (1991), Haddad (1998), Horner (1969), Sorensen (1985), Tozzi (1992), University of Queensland, Wahrheit-Lensing (1996), Zhou (1996). Prototype data: Trigomil Dam (Sanchez-Bribiesca and Gonzalez-Villareal, 1996), Brushes Clough Dam (Baker, 1994).

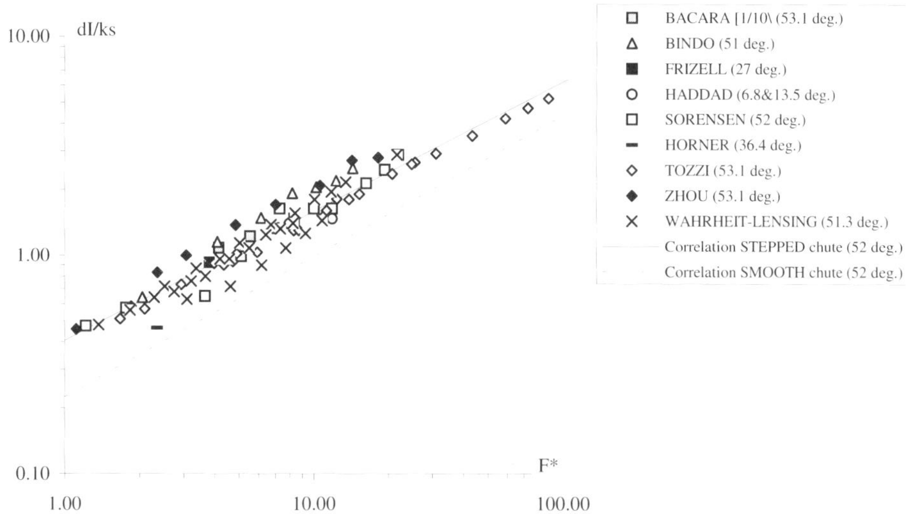

Figure 5.9. Characteristics of the inception point of air entrainment. B. d_I/k_s as a function of F_*. Model data: BaCaRa (1991), Bindo et al. (1993), Frizell and Mefford (1991), Haddad (1998), Horner (1969), Sorensen (1985), Tozzi (1992), Wahrheit-Lensing (1996), Zhou (1996).

Figure 5.10. Effects of the discharge on the location of the inception of air entrainment (courtesy of Mr. Royet, Cemagref). Flow from the right to the left: $\alpha = 51.3°$, h = 0.06 m. A. $q_w = 0.057$ m²/s.

B.

C.

Figure 5.10. Effects of the discharge on the location of the inception of air entrainment (courtesy of Mr. Royet, Cemagref). Flow from the right to the left: $\alpha = 51.3°$, h = 0.06 m. B. $q_w = 0.079$ m^2/s. C. $q_w = 0.158$ m^2/s.

D.

Figure 5.10. Effects of the discharge on the location of the inception of air entrainment (courtesy of Mr. Royet, Cemagref). Flow from the right to the left: $\alpha = 51.3°$, h = 0.06 m. D. $q_w = 0.266$ m^2/s.

In Figure 5.10, several photographs show a stepped spillway model (α = 51.3°, h = 0.06 m) with discharges between 0.06 and 0.27 m^2/s. For the smallest discharge, the point of inception is located near the top of the channel. For the largest discharge, the water is transparent all along the channel, indicating the absence of free-surface aeration: the outer edge of the turbulent boundary layer does not reach the free-surface before the downstream end of the channel.

Discussion

Figure 5.9 includes data obtained with various types of crest profiles (ogee, broad-crest). The results suggests that the type of crest profile has little influence on the characteristics of the inception point, provided that no deflecting jet is observed. Note that Equations 5.7 and 5.8 are based upon experimental data obtained with channel slopes ranging from 20 to 55°. Great care must be taken when applying these equations outside of that range.

5.2.3 *Rapidly-varied flow next to the inception point*

Visual observations and detailed point measurements indicate that the flow properties are rapidly varied next to, and immediately downstream, of the inception point of air entrainment. Side view observations suggest that some air is entrapped in the step cavity(ies) upstream of the visual location of free-surface aeration (Fig. 5.11). Immediately upstream, the flow is extremely turbulent and the free-surface appears to be subjected to a flapping mechanism. At irregular time intervals, a water jet impinges onto the step horizontal face and air is trapped in the step cavity[6]. Few instants later, a rapid unsteady flow bulking is observed downstream (Fig. 5.11). The process is extremely rapid: i.e. it occurs typically in less than 0.1 to 0.2 seconds.

Precise LDA velocity measurements by Ohtsu & Yasuda (1997) indicate that the turbulent velocity fluctuations are large immediately upstream of the inception point, with dimensionless streamwise fluctuations u'/v of about 15 to 18% and normal velocity fluctuations v'/v of about 2 to 4%, where u' and v' are the root mean square of longitudinal and lateral components of turbulent velocity respectively (Fig. 5.12). In Figure 5.12, observed values of $v' \approx 0.14$ m/s next to the free-surface are large enough to initiate air bubble entrainment (Eqs 1.1 & 1.2; Section 1.4.1).

Immediately downstream of the inception point, time-averaged air concentration data show a sudden aeration. For example, an increase in mean air concentration $\Delta C_{mean} = 0.25$ along $\Delta s = 6.5*d_c$ down a 30° slope for $d_c/h = 5.2$

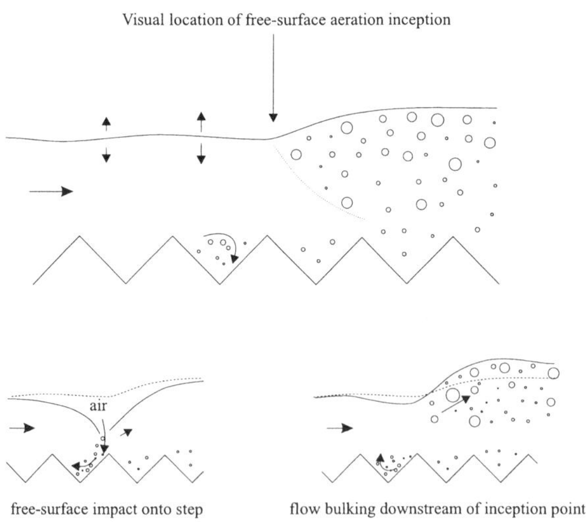

Figure 5.11. Rapidly-varied flow region at the inception point of free-surface aeration. A. Flow mechanisms next to the inception point.

[6]For example, Horner (1969), Chamani (2000), Matos (2000), University of Queensland ($\alpha = 22°$).

B.

C.

Figure 5.11. Rapidly-varied flow region at the inception point of free-surface aeration. B. University of Queensland experiment: $\alpha = 22^{\circ}$, h = 0.10 m, W = 1 m, d_c/h = 1.23. Flow from top left to bottom right. Note the step cavity aeration upstream of free-surface aeration location. C. Laboratory experiment: $\alpha = 50^{\circ}$, h = 0.02 m, W = 0.5 m, d_c/h = 8.0. Flow from top left to bottom right. Note the step cavity aeration upstream of the free-surface aeration location and the significant splashing downstream of the inception point.

(Boes, 1999); an increase in mean air concentration $\Delta C_{mean} = 0.55$ in 18 step heights down a 53° slope for $d_c/h \leq 2$ (Matos, 2000); an increase in mean air concentration $\Delta C_{mean} = 0.32$ in 2 step heights down a 22° slope for $d_c/h = 1.2$ (University of Queensland).

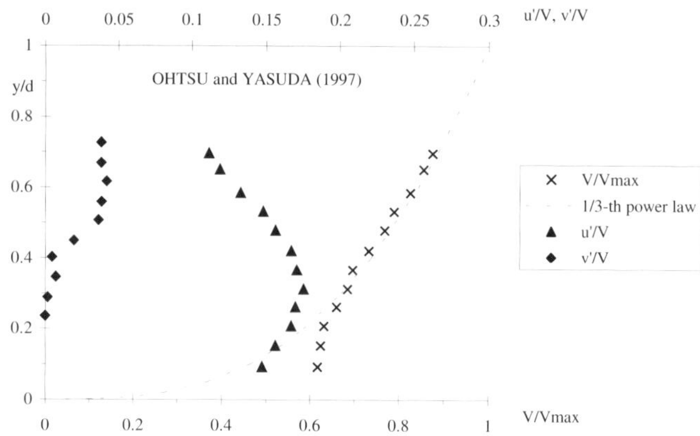

Figure 5.12. Velocity and velocity fluctuation distributions at the point of inception (Ohtsu & Yasuda, 1997). $\alpha = 19°$, h = 0.05 m, $q_w = 0.089$ m²/s. Measurements at the step edge (Step 7).

5.3 FULLY-DEVELOPED FLOW CHARACTERISTICS

5.3.1 *Air entrainment*

Downstream of the inception point, the flow is fully-developed and substantial air entrainment if observed (Figs 5.1 & 5.2). Turbulence acting next to the free-surface is large enough to overcome both surface tension for the entrainment of air bubbles and buoyancy to carry downwards the bubbles (Section 1.4.1). The diffusion of air bubbles may be approximated by a simple diffusion model:

$$C = 1 - \tanh^2\left(K' - \frac{y}{2 * D' * Y_{90}}\right) \tag{1.7}$$

where C is the void fraction (or air concentration), tanh is the hyperbolic tangent function, y is the distance normal to the pseudo-invert formed by the step edges, Y_{90} is the distance where C = 90%, D' is a dimensionless turbulent diffusivity and K' is an integration constant (App. 6). D' and K' are functions of the mean air content C_{mean} only (Table 5.2), where C_{mean} is the depth-averaged air content defined in terms of Y_{90}:

$$C_{mean} = \int_0^{Y_{90}} C * dy \tag{5.10}$$

Equation 1.7 was tested successfully with stepped chute data obtained in models and prototype (Fig. 5.13) although it was originally developed for smooth-invert chute flows (Chanson, 1995b, 1997a). Figure 5.13C and D presents further dimensionless distributions of bubble count rate downstream of the inception point of air entrainment.

Far downstream, at uniform equilibrium, the mean air concentration C_{mean} tends to a constant C_e function of the channel slope α only (Chanson, 1995a). For slopes less than 50°, the equilibrium mean air concentration may be approximated by: $C_e = 0.9*\sin\alpha$. Uniform equilibrium air content data for smooth chutes[7] are shown in Figure 5.14. The data are compared with measured mean air contents on stepped chutes. Stepped cascade results are close to and below the corresponding uniform equilibrium mean air concentration suggesting that the upper limit C_e, observed in smooth chutes, may also apply to stepped spillways (Fig. 5.14).

Discussion

In mountain rivers and rockfill channels, visual and photographic observations show a large amount of free-surface aeration (i.e. "white waters")[8]. Knauss (1979) analysed the data of Hartung & Scheuerlein (1970), obtained with great natural roughness and steep slopes. He indicated that the quantity of entrained air on rockfill channels was estimated as:

$$C_{mean} = 1.44*\sin\alpha - 0.08 \tag{5.11}$$

This result has a similar form as the uniform equilibrium mean air content approximation: $C_e = 0.9*\sin\alpha$ (Fig. 5.14).

In summary, experimental observations (Figs 5.13 & 5.14) show a comparable rate of air entrainment for smooth and rough channels, and stepped cas-

Table 5.2. Relationship between C_{mean}, D' and K' (after Chanson, 1997a)

C_{mean} (1)	D' (2)	K' (3)
0.01	0.007312	68.70445
0.05	0.036562	14.0029
0.10	0.073124	7.16516
0.15	0.109704	4.88517
0.20	0.146489	3.74068
0.30	0.223191	2.567688
0.40	0.3111	1.93465
0.50	0.423441	1.508251
0.60	0.587217	1.178924
0.70	0.878462	0.896627

[7]Data: Straub & Anderson (1958), Aivazyan (1986). Re-analysis: Chanson (1993a,1997a).
[8]e.g. Judd & Peterson (1969), Bathurst (1978) and Thorne & Zevenbergen (1985).

cades, validating the analogy between aerated flows in smooth and stepped chutes (Chanson, 1993c).

5.3.2 *Velocity distribution*

In fully-developed air-water flows, the velocity distribution may be approximated by a power law:

$$\frac{V}{V_{90}} = \left(\frac{y}{Y_{90}}\right)^{1/N} \tag{5.12}$$

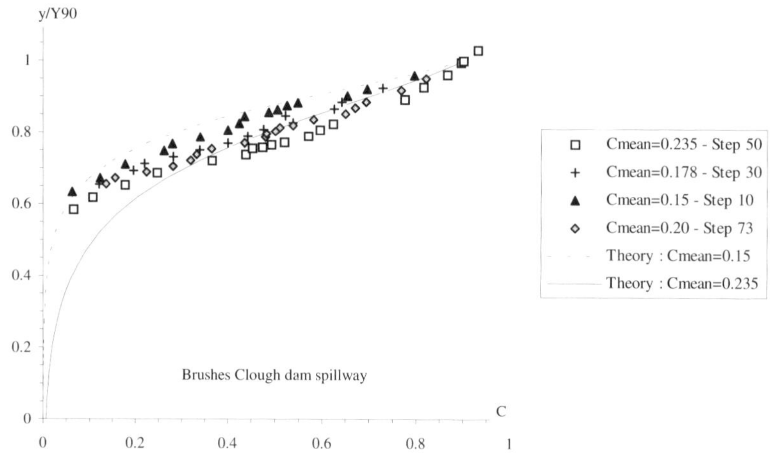

A.

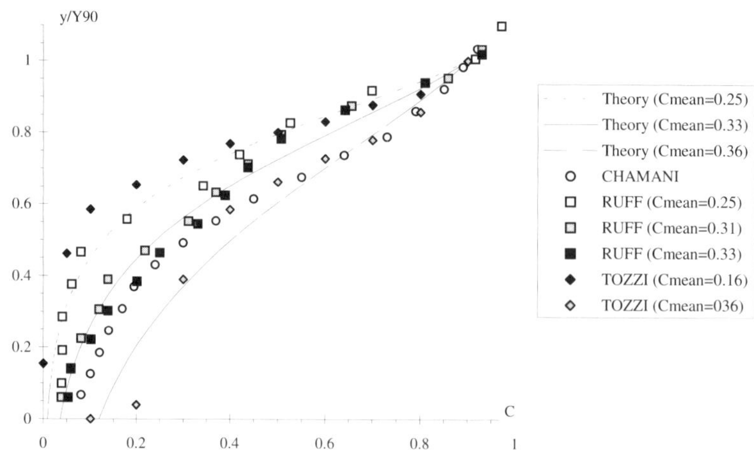

B.

Figure 5.13. Void fraction distributions in stepped chute flows: comparison with Equation 1.7. A. Prototype data (Baker, 1994). B. Model data (Ruff & Frizell, 1994; Tozzi et al., 1998; Chamani & Rajaratnam, 1999a).

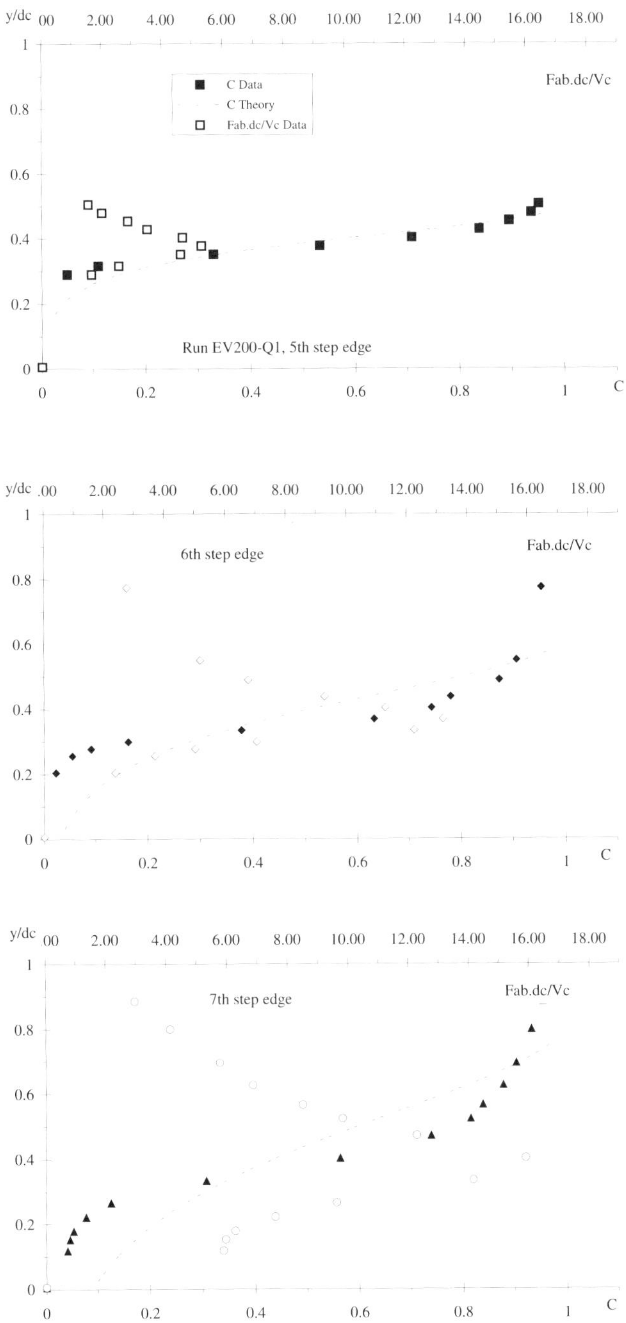

Figure 5.13. Void fraction distributions in stepped chute flows: comparison with Equation 1.7. C. Dimensionless air concentration and bubble count rate distributions. Model data $\alpha = 22°$, h = 0.1 m, l = 0.25 m, W = 1 m, $q_w = 0.124$ m²/s (University of Queensland). Legend: Black symbols = void fraction data; White symbols = bubble count rate data $F_{ab}*d_c/V_c$; Dashed line = theory (Eq. 1.7).

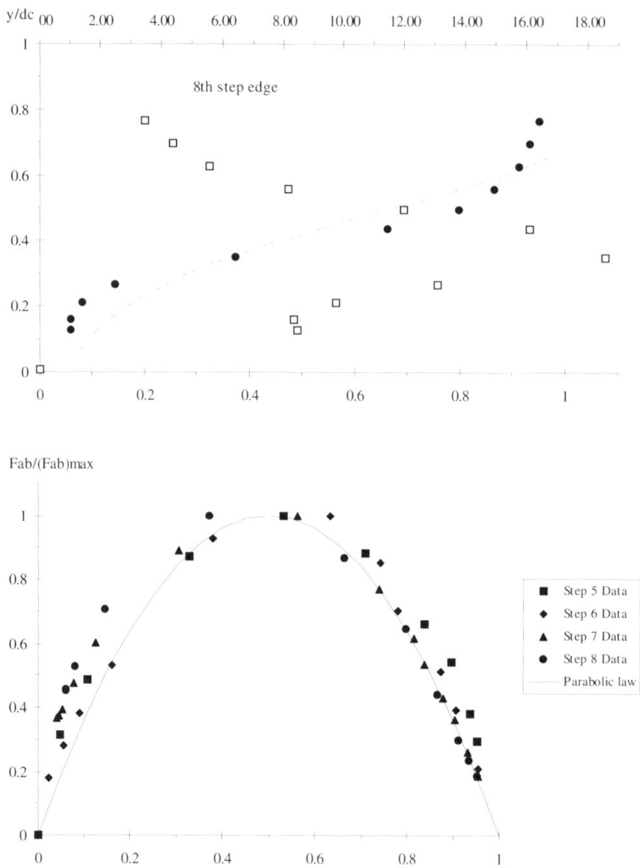

Figure 5.13. Void fraction distributions in stepped chute flows: comparison with Equation 1.7. D. Dimensionless relationship between air concentration and bubble count rate. Model data $\alpha = 22°$, h = 0.1 m, l = 0.25 m, W = 1 m, $q_w = 0.124$ m²/s (University of Queensland).

where V_{90} is the characteristic velocity at $y = Y_{90}$. Air-water velocity measurements by Matos (2000) showed that the exponent of the velocity distribution is about N = 3.7 (Fig. 5.15). The result is consistent with the earlier works of Frizell (1992) and Tozzi (1992).

In uniform non-aerated flows above a smooth chute, Chen (1990) derived a theoretical relationship between the velocity distribution exponent N and the Darcy Friction Factor f as:

$$N = K^* \sqrt{\frac{8}{f}} \tag{5.13}$$

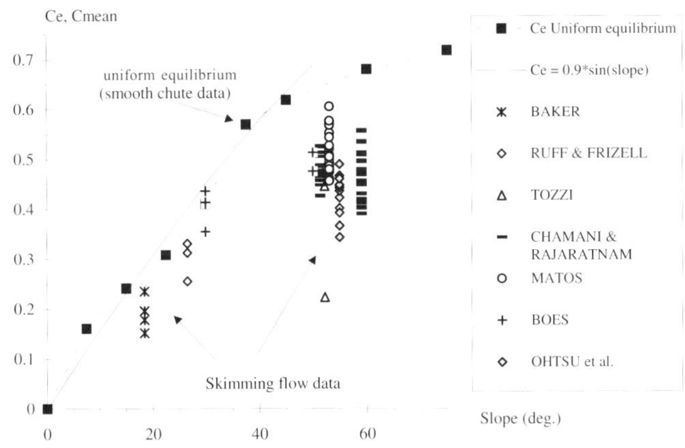

Figure 5.14. Uniform equilibrium mean air concentration: comparison between gradually-varied stepped flow data and smooth chute results at uniform equilibrium. Prototype data: Baker (1994). Laboratory data: Ruff & Frizell (1994), Tozzi et al. (1998), Chamani & Rajaratnam (1999a), Ohtsu et al. (2000), Matos (2000), Boes (2000).

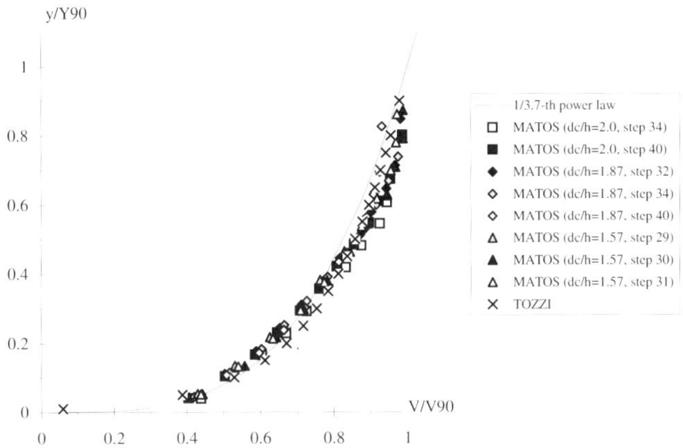

Figure 5.15. Dimensionless velocity distributions in air-water flows down stepped chutes. Laboratory data: Tozzi (1992), Matos (2000).

where K is the Von Karman Constant ($\varkappa = 0.4$). For N = 3.7, Equation 5.13 predicts f = 0.094. Such a value is of the same order of magnitude as the experimental values presented in Section 5.3.3.

Discussion

Bathurst (1988) and Ferro & Baiamonte (1994) measured velocity distributions in rockfill channel flows. The velocity distributions exhibited a flat S-shape which can be fitted by Equation 5.12 with an exponent N slightly above unity.

Practically the characteristic velocity V_{90} may be deduced from the continuity equation for the water phase as:

$$\frac{V_{90}}{U_w} = \frac{1}{1 - C_{mean}} * \left(\int_0^1 (1 - C) * y'^{1/N} \, dy' \right)^{-1} \tag{5.14}$$

where C is the local air concentration (Eq. 1.7), y' = y/Y_{90} and C_{mean} is the depth-average mean air content (Chanson, 1989b).

5.3.3 *Flow resistance and cavity flow*

While the flow resistance on smooth-invert chutes is primarily skin friction, skimming flows over stepped chute are characterised by significant form losses. The water skims over the step edges with formation of recirculating vortices between the main stream and the step corners (Fig. 5.16). *Form drag* is predominant. Kazemipour & Apelt (1983) stressed that to try to account for the form losses with a Gauckler-Manning or Darcy-Weisbach formula is unsatisfactory. Nevertheless it is still common practice including for skimming flows and this study is no exception[9].

In *uniform equilibrium flows*, the momentum principle states that the boundary friction force equals "exactly" the gravity force component in the flow direction: i.e. the "weight-of-water" component acting parallel to the pseudo-bottom formed by the step edges. It yields:

$$\tau_o * P_w = \rho_w * g * A_w * \sin\alpha \qquad \text{Uniform equilibrium} \tag{5.15}$$

where τ_o is the average shear stress between the skimming flow and recirculating fluid underneath, P_w is the wetted perimeter, ρ_w is the water density, g is the gravity constant, A_w is the water flow cross-section area and α is the mean bed inclination angle. In dimensionless terms, and for a wide channel, Equation 5.15 becomes:

$$f_e = \frac{8 * \tau_o}{\rho_w * U_w^{\,2}} = \frac{8 * g * \left(\int_{y=0}^{y=Y_{90}} (1 - C) * dy \right) * \sin\alpha}{U_w^{\,2}} \qquad \text{Uniform equilibrium} \tag{5.16}$$

where f_e is the Darcy Friction Factor for air-water flow, C is the local void fraction, y is measured normal to the pseudo-invert formed by the step edges, Y_{90} is the distance normal to the pseudo-bottom where C = 90% and U_w is the water flow velocity (Eq. 1.12). The dimensionless shear stress is analogous to

[9]This approach facilitates further comparison between smooth and stepped chute data.

A. 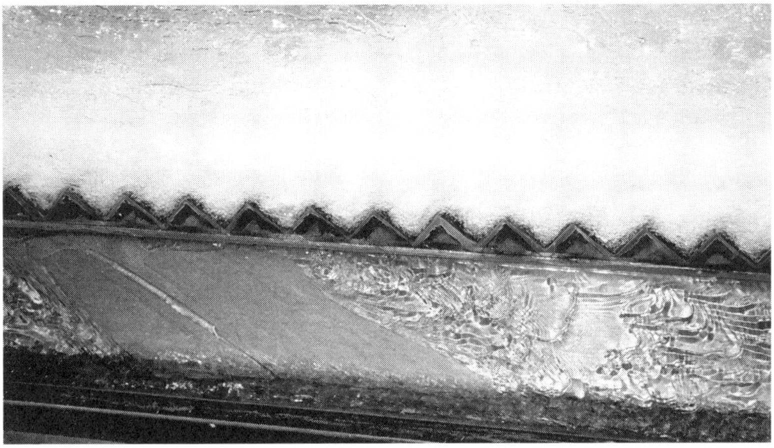

B. 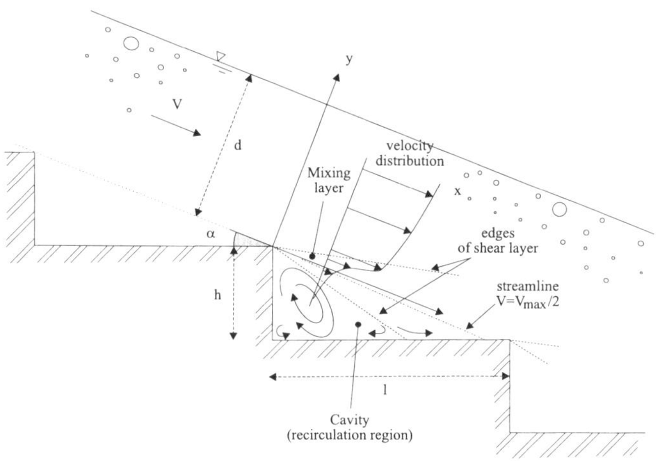

Figure 5.16. Skimming flow. A. Cavity recirculation and flow aeration. Flow from the left to the right. $\alpha = 50°$, h = 0.02 m, W = 0.5 m, $d_c/h = 8.0$. B. Definition sketch.

the Darcy Friction Factor[10] but it characterises the form drag loss.

In *gradually-varied flows*, the average shear stress between the skimming flow and the cavity recirculation "must" be deduced from the friction slope S_f[11]. For a wide channel the energy equation yields:

$$f_e = \frac{8 * \tau_o}{\rho_w * U_w^2} = \frac{8 * g * \left(\int_{y=0}^{y=Y_{90}} (1-C) * dy \right) * S_f}{U_w^2} \quad \text{Gradually-varied flow (5.17)}$$

[10]For a smooth invert chute, the boundary shear stress is defined as: $\tau_o = f_e * \rho_w * U_w^2/8$ (e.g. Chanson, 1999a, p. 74).

[11]The friction slope is the slope of the total head line (Henderson, 1966; Chanson, 1999a).

where the friction slope equals $S_f = -\partial H/\partial s$, H is the mean total head and s is the curvilinear coordinate along the flow direction. It must be understood that free-surface aeration is always substantial in prototype skimming flows and its effects must be taken into account as shown in Equations 5.15 to 5.17.

Experimental results

Several researchers studied skimming flows past large roughness elements in pipes and open channels (see review in Chanson, 1995a). Experimental results indicated that the classical flow resistance calculations must be modified to take into account the shape of the roughness elements. Dynamic similarity of skimming flows is, however, complex because of the role of the steps in enhancing turbulent dissipation and free-surface aeration.

For uniform equilibrium flow down a prismatic rectangular channel with horizontal steps, the dominant flow feature is the momentum exchange between the free-stream and the cavity flow within the steps (Fig. 5.16). A complete dimensional analysis yields:

$$F_1\left(\rho_w, \mu_w, \sigma, g, U_w, d, C_{mean}, h, l, \alpha, W, k_s'\right) = 0 \tag{5.18}$$

where d is the clear-water depth (Eq. 1.10), U_w is the water flow velocity (at normal flow conditions), C_{mean} is the mean air concentration, h, l and α are the step height, step length and channel slope, W is the channel width, k_s' is the skin roughness height, g is the gravity acceleration, μ_w and ρ_w are the dynamic viscosity and density of water respectively and σ is the surface tension. For flat horizontal steps, the additional relationship holds: $\alpha = \tan^{-1}(h/l)$, and the cavity dimensions are its length h/sinθ, depth h*cosθ and breath W[12]. For horizontal steps, Equation 5.18 may be rewritten in dimensionless term as:

$$F_2\left(\frac{U_w}{\sqrt{g*d}};\rho_w\frac{U_w*d}{\mu_w};\frac{g*\mu_w^4}{\rho_w*\sigma^3};C_{mean};\frac{d}{h};\frac{W}{h};\alpha;\frac{k_s'}{h}\right) = 0 \tag{5.19}$$

From left to right, the dimensionless terms are Froude, Reynolds and Morton numbers, the amount of entrained and the last four terms characterise the cavity shape and skin friction on the cavity walls.

Equation 5.19 illustrates that a Froude similitude does not describe the complexity of stepped spillway flows. BaCaRa (1991,1997) described a systematic laboratory investigation of the M'Bali Dam spillway with model scales of 1:10, 1:21.3, 1:25 and 1:42.7. For the scales 1:25 and 1:42.7, the flow resistance was improperly reproduced with a Froude similitude. The Proceedings of the International Symposium on Scale Effects in Modelling Hydraulic Structures

[12]In the general case of inclined steps, at least four parameters are required to defined the cavity geometry.

(IAHR, Esslingen, Germany, 1984, edited by H. Kobus) suggested that air entrainment is poorly reproduced on small-size scale models. For air-water flow studies, the writer recommended that "model studies of air-water flows should be always investigated on near full-scale apparatus: 0.1 < scale ratio < 10 (or 1:10 < scale < 10:1), and preferably 0.2 < scale ratio < 5" (Chanson, 1997a: 236).

In the present section, flow resistance in skimming flows has been re-analysed based upon over 40 model studies and 4 prototype investigations with channel slopes ranging from 5.7 up to 55° (Table 1.1)[13]. The analysis is based upon Equations 5.16 and 5.17 for uniform equilibrium and gradually-varied flows respectively. The raw analysis (over 660 data) shows that the flow resistance is larger on stepped chutes than on smooth channels, but steep stepped chute data show very little correlation (Fig. 5.17). Figure 5.17 regroups Darcy Friction Factor estimate f_i obtained neglecting free-surface aeration, and friction factor data f_e calculated using Equations 5.16 and 5.17. A conditional analysis was applied and only the following data were retained: prototype data, large-size model data (h > 0.020 m, Re > 1E+5), and well-documented model studies published since January 1997.

Basic results of the conditional analysis indicate different trends for flat and steep chutes (Fig. 5.18A & B). For flat chutes (regime SK1), prototype data[14] exhibit higher flow resistance than laboratory data:

$$\frac{1}{\sqrt{f}} = -1.224 - 1.245 * \mathrm{Ln}\left(\frac{h * \cos\alpha}{D_H}\right) \quad \text{Flat chute } (\alpha < 20°)\text{. Prototypes} \quad (5.20a)$$

$$\frac{1}{\sqrt{f}} = 2.430 - 0.2676 * \mathrm{Ln}\left(\frac{h * \cos\alpha}{D_H}\right) \quad \text{Flat chute } (\alpha < 20°)\text{. Models} \quad (5.20b)$$

with normalised correlation coefficients of 0.9525 (10 data) and 0.4872 (188 data) respectively and f is an average friction factor. By comparison, the Colebrook-White Formula for flat rough walls is:

$$\frac{1}{\sqrt{f}} = 1.14 - 4.605 * \mathrm{Ln}\left(\frac{k_s}{D_H}\right) \quad \text{Fully-rough turbulent} \quad (5.21)$$

The different trends between model and prototype data are presently not understood.

[13]This work derives from the study of Chanson et al., (2000).

[14]Obtained with horizontal steps (Dneiper hydroplant) and downward-inclined steps (Brushes Clough).

Table 5.3. Cavity ejection mechanisms: basic parameters.

Reference	Average ejection period ΔT s	Average ejection duration Δt s	Comments
(1)	(2)	(3)	(4)
Townes & Sabersky (1966)	$\Delta T = 0.1 * \frac{k_s}{V*}$	--	Monophase flow: U_W = 0.04 to 0.25 m/s. Square cavities: k_S = 0.0254 m.
Ethembabaoglu (1978)	$\Delta T = 0.91 * \frac{l_{cav}}{U_w}$	--	Water tunnel: U_W = 5.5 to 7.5 m/s. Rectangular cavities: k_S = 0.1 m, l_{cav}/k_S = 1.9 to 3.6.
Djenedi et al. (1994)	$\Delta T = 5.5 * \frac{\delta}{U_w}$	$\Delta t = \frac{^{\delta}BL}{U_w}$	Monophase flow: U_W = 0.4 m/s. Square cavities: k_S = 5 mm.
Tantiridge et al. (1994)	0.025	$\frac{\Delta t}{\Delta T} \sim \frac{2}{30}$	Fully-developed monophase flow (25.mm square channel): U_W = 0.43 m/s. Triangular cavity: k_S = 0.8 & 1.5 mm, $\alpha = 45°$.

Notes: k_s; cavity depth; l_{cav}: cavity length; U_w: free-stream velocity; V_*: shear velocity; δ_{BL}: boundary layer thickness.

For steep chutes (regime SK3), the friction factor data present no obvious correlation (see below) but they are distributed around two dominant values: f ≈ 0.17 and 0.30 (130 data) (Fig. 5.18C).

Discussion
Flow resistance in skimming flows is dominated by the recirculation process and by unsteady momentum exchanges between the main stream and cavity flows. The recirculating fluid will, at irregular time intervals, flow outward into the main flow and be replaced by fresh fluid. The ejection mechanism is sequential. Once one cavity outflow occurs, it induces a sequence of outflows at the downstream cavities (Fig. 5.19). Figure 5.19 shows four successive snap shots of sequential fluid ejections. The process was observed visually for slopes ranging from 5.7 to 55° but the sequential mechanism was possibly most effective on the steep chute geometries[15]. A similar pattern is observed with skimming flows past strip roughness[16].

The recirculation flow shows some asymmetry within the cavity, with most activity occurring in the downstream section. The observation is consistent with that of Tantirige et al. (1994). Occasionally, strong ejections of fluid away from the cavity into the skimming flow occur. It was proposed that the ejection process is caused by interactions between low-speed streaks and vorticity structures next to the pseudo-bottom formed by the step edges (Elavarasan et al., 1995). Both ejections and inflows occur only in the downstream

[15]Laboratory experiments: M'Bali stepped spillway model (Professor LejeunE), Nihon University, University of Queensland ($\alpha = 22°$).

[16]For example, Tantirige et al. (1994) with triangular strip roughness; Djenedi et al. (1994) and Elavarasan et al. (1995) with rectangular strip roughness (also called d-type roughness).

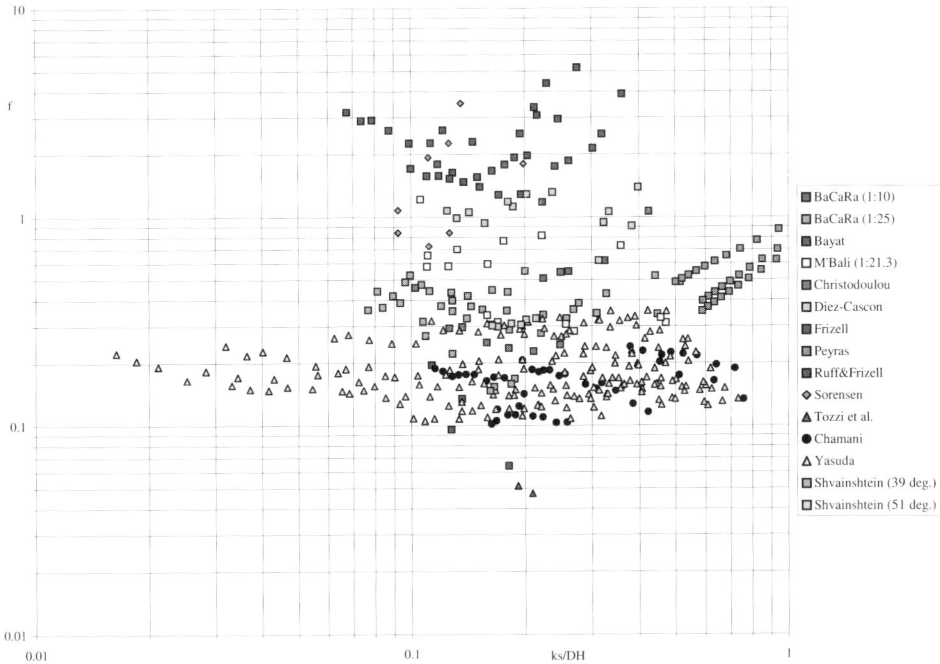

Figure 5.17. Darcy Friction Factor of skimming flows on steep stepped chute (α > 20°) {429 data}.

Laboratory data:

f_i (estimate neglecting air entrainment)	BaCaRa (1991), Bayat (1991), Bindo et al. (1993), Christodoulou (1993), Diez-Cascon et al. (1991), Frizell (1992), Peyras et al. (1991), Sorensen (1985), Yasuda & Ohtsu (1999), Shvainshtein (1999)
f_e (based upon air-water flow properties)	Ruff & Frizell (1994), Tozzi et al. (1998), Chamani & Rajaratnam (1999), Yasuda & Ohtsu (1999) (55° only)

region of the cavity[17]. The ejection phase is much shorter than the period between successive ejections (Table 5.3). The initiating mechanisms of the ejections resides within the fully-developed flow and not in the cavity flow itself.

For flat slopes (α < 20°), the flow resistance includes the form drag of the recirculating cavity plus skin friction on the downstream end of each step. The free-surface exhibits an "undular" pattern in phase with the step geometry (Section 5.1). Between adjacent steps, the cavity of recirculating fluid is affected by the interference of the mixing layer onto the downstream step.

For steep slopes (α > 20°), there is no skin friction between the main flow

[17]Stepped chute data: University of Queensland (α = 22°). Strip roughness data: Djenedi et al. (1994), Elavarasan et al. (1995).

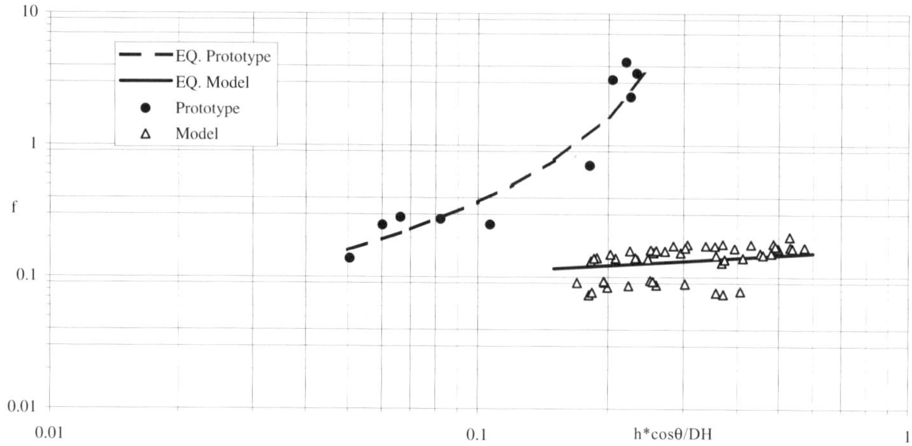

Figure 5.18. Flow resistance in skimming flow: conditional analysis. A. Flat stepped chute data ($\alpha < 20°$) {62 data}. Comparison with Equations 5.20a and 5.20b.

Prototype data:

f_i (estimate neglecting air entrainment)	Grinchuk et al. (1977)
f_e (based upon air-water flow properties)	Baker (1994)
Laboratory data:	
f_i (estimate neglecting air entrainment)	Yasuda & Ohtsu (1999)

and the step faces, and the form drag associated with the recirculation predominates. The cavity flow is three-dimensional as confirmed by visual observations. The flow resistance is dominated by the transfer of momentum between the cavity flow and the main stream. A simplified analytical model (App. 5) suggests that the pseudo-boundary shear stress may be expressed, in dimensionless form, as:

$$f_d = \frac{8 * \tau_o}{\varrho_w * U_w^2} = \frac{2}{\sqrt{\pi}} * \frac{1}{K} \tag{5.22}$$

where f_d is an equivalent Darcy Friction Factor estimate of the form drag, 1/K is the dimensionless rate of expansion of the shear layer (Fig. 5.16). In air-water mixing layers, Brattberg & Chanson (1998) observed K ~ 6 for V = 2 to 8 m/s. For K = 6, Equation 5.22 predicts $f_d \approx 0.2$ in steep chute skimming flows: i.e. close to the observed friction factor (Fig. 5.18B).

It is believed that the channel width affects the development and number of recirculating cells at each step. For steep chutes ($\alpha > 20°$), the data of Chamani & Rajaratnam (1999a) and Yasuda & Ohtsu (1999) suggest higher friction factors at low aspect ratios (i.e. $W/h \leq 10$) with identical flow conditions and

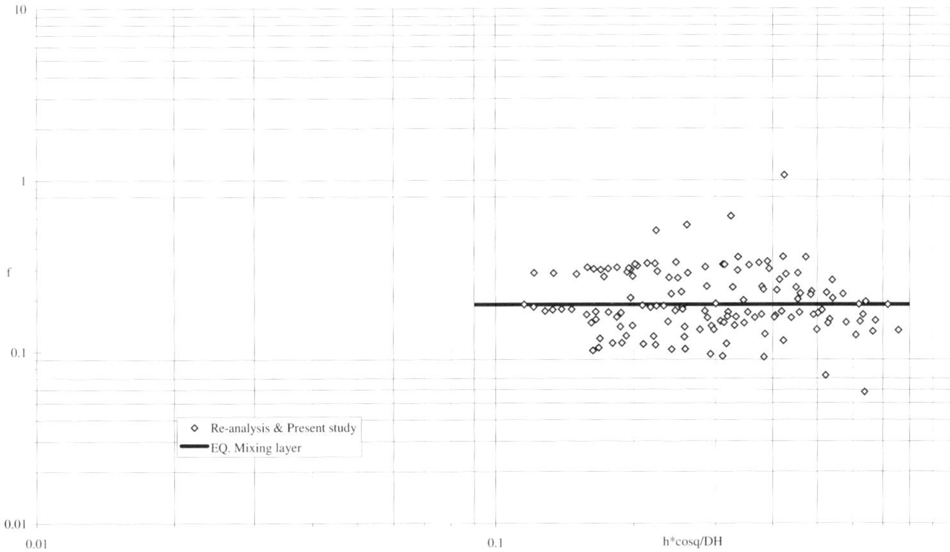

Figure 5.18. Flow resistance in skimming flow: conditional analysis. B. Steep stepped chute data (α > 20°) {140 data}. Comparison with Equation 5.22 (f_d = 0.2).

Laboratory data:

f_i (estimate neglecting air entrainment)	BaCaRa (1991), Yasuda & Ohtsu (1999), Shvainshtein (1999)
f_e (based upon air-water flow properties)	Chamani & Rajaratnam (1999), Yasuda & Ohtsu (1999) (55° only), Matos (2000)

geometry. Note that both studies used a constant channel breadth, 0.3 m and 0.4 m respectively, and the effect of the channel width was not specifically investigated. In a relevant study, Osaka & Mochizuki (1991) showed that the introduction of longitudinal thin ribs in square cavities induced a modification of the flow resistance.

Effects of air entrainment on the flow resistance

On smooth-invert chutes, the presence of air within turbulent boundary layers reduces the shear stress between the flow layers and hence the shear force (Jevdjevich & Levin, 1953; Wood, 1983; Chanson, 1994a). Chanson (1994a) proposed an estimate of the drag reduction:

$$\frac{f_e}{f} = 0.5 * \left(1 + \tanh\left(0.628 * \frac{0.514 - C_{mean}}{C_{mean} * (1 - C_{mean})}\right)\right) \quad \text{Smooth chute (5.23)}$$

where tanh is the hyperbolic tangent function, C_{mean} is the mean air concentration, f is the clear-water friction factor and f_e is the Darcy Friction Factor of

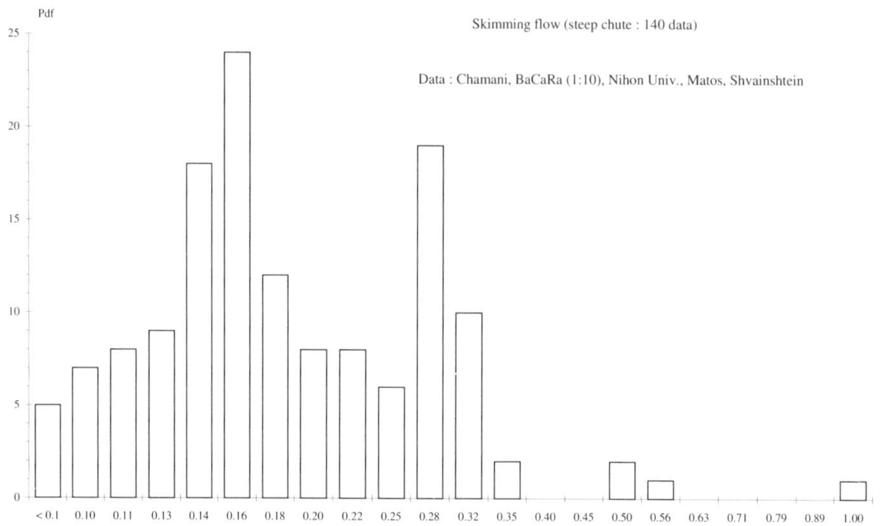

Figure 5.18. Flow resistance in skimming flow: conditional analysis. C. Probability distribution function of steep chute friction factor ($\alpha > 20^\circ$) {140 laboratory data}.

f_i (estimate neglecting air entrainment)	BaCaRa (1991), Yasuda & Ohtsu (1999), Shvainshtein (1999)
f_e (based upon air-water flow properties)	Chamani & Rajaratnam (1999), Yasuda & Ohtsu (1999) (55° only), Matos (2000)

air-water flow. Equation 5.23 was validated with model and prototype data (Fig. 5.20A). The study of Hartung & Scheuerlein (1970) on rockfilled channels showed a similar trend:

$$\frac{f_e}{f} = \frac{1}{\left(1 - 3.2 * \sqrt{f} * \mathrm{Log}_{10}\left(1 - C_{mean}\right)\right)^2} \qquad \text{Rockfill chute (5.24)}$$

On stepped chutes, some researchers[18] conducted detailed air concentration distribution measurements. The re-analysis of their data gives experimental values of the air-water flow friction factor showing f_e decreasing with increasing mean air content C_{mean}. The results are best correlated as:

$$\frac{f_e}{f} = 0.5 * \left(1 + \tanh\left(2.5 * \frac{0.5 - C_{mean}}{C_{mean} * \left(1 - C_{mean}\right)}\right)\right) \qquad \text{Stepped chute (5.25)}$$

[18]That is, Chamani & Rajaratnam (1999), Ohtsu & Yasuda (1999), Matos (2000).

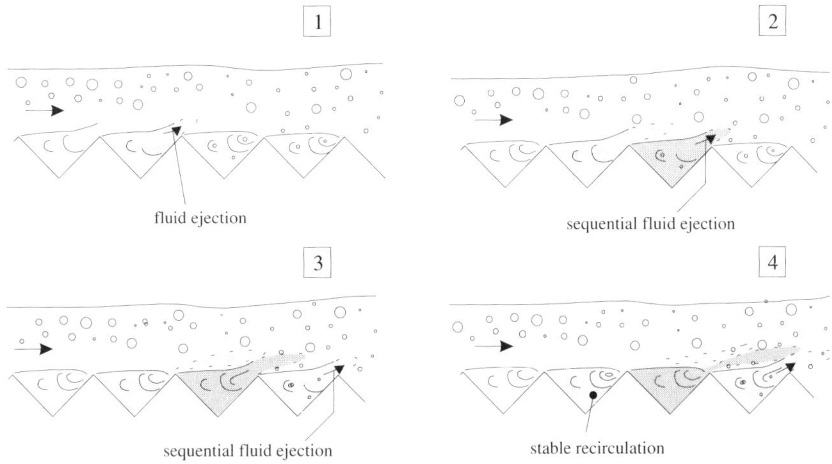

Figure 5.19. Sequential ejections of fluid cavity.

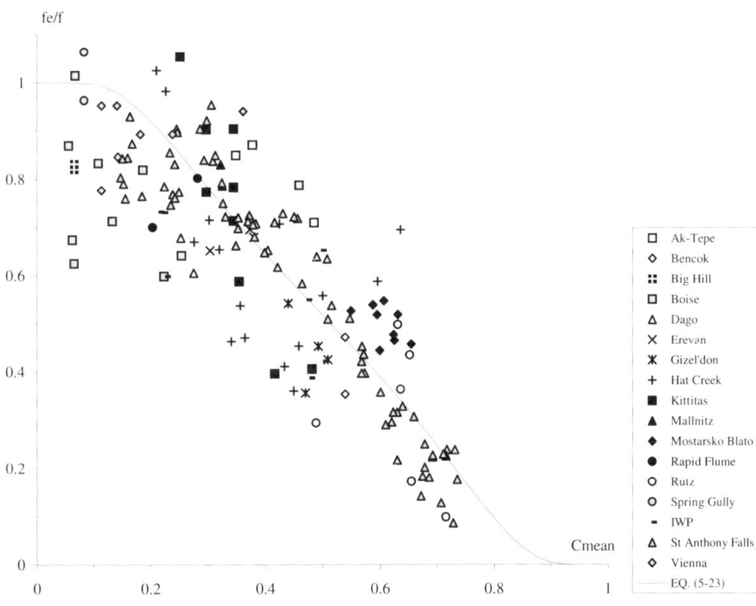

Figure 5.20. Drag reduction caused by free-surface aeration on spillway chutes. A. Drag reduction in smooth chutes. Comparison between Equations 5.23 and model and prototype smooth-invert data (after Chanson, 1994a).

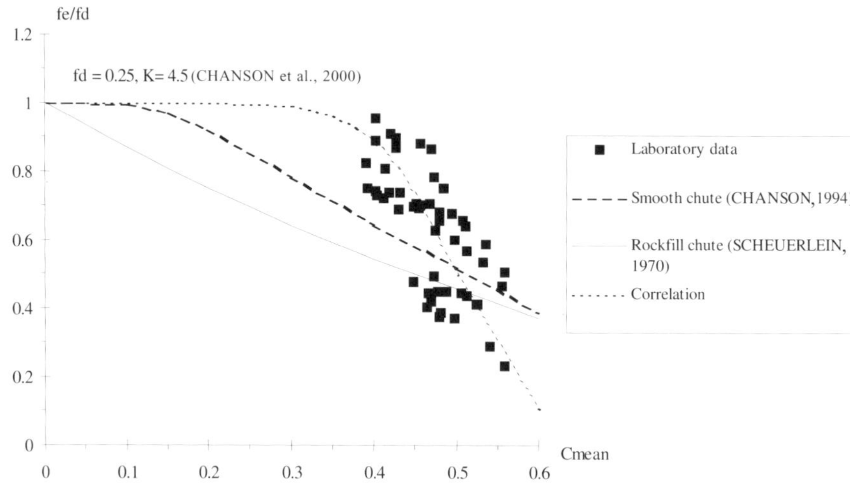

Figure 5.20. Drag reduction caused by free-surface aeration on spillway chutes. B. Drag reduction in stepped chutes. Comparison between Equations 5.23, 5.24 and 5.25, and experimental data. Laboratory data: Chamani & Rajaratnam (1999a), Ohtsu & Yasuda (1999), Matos (2000).

where the clear-water friction factor is calculated as:

$$f_d = \frac{2}{\sqrt{\pi}} * \frac{1}{K} \tag{5.22}$$

Equation 5.25 is compared with experimental data in Figure 5.20B assuming a mixing layer expansion rate: $1/K = 0.22$. Equation 5.25 was obtained for $0.38 < C_{mean} < 0.57$ and it should not be applied outside of that range. Nonetheless the results confirm Chanson's (1993c) assumption that a drag reduction process caused by air entrainment occurs on stepped spillways.

Mechanisms of drag reduction

Recent experimental results (Gyr, 1989; Bushnell & Hefner, 1990) showed that the main mechanisms of drag reduction include: 1. a reduction of the normal velocity fluctuation component (i.e. normal to the wall), 2. a modification of the turbulent bursting process, and 3. an increase of the sublayer thickness (see review in Chanson, 1997a). In smooth-invert chutes, drag reduction is caused by the interactions of entrained air bubbles with the sub-layer (Chanson, 1994a).

In stepped chutes, it is believed that drag reduction results from the interactions between the entrained air bubbles and the mixing layers (or wake)[19] downstream of each step edge (Fig. 5.16). Interactions between particle and turbu-

[19]Schlichting (1979) developed a complete analogy between wakes and shear layers (mixing layers). In the present book, the term mixing layer is used for both wake and shear layer.

lent structures in developing shear layers were visualised in dilute polymer solutions. Hibbert et al. (1982) and Riediger (1989) showed the existence of large-scale turbulent structures but a drastic reduction in number of small-scale eddies in the developing shear layers with polymer additives Separan™ AP-30 and AP-45.

Summary and design recommendations

Design engineers must take into account flow aeration effects to estimate the residual energy and to dimension stilling basins downstream of stepped chutes[20]. The residual head equals:

$$H_{res} = d * \cos\alpha + \frac{q_w^2}{2 * g * d^2} \tag{5.26}$$

where d is the clear-water depth (Eq. 1.10). In *uniform equilibrium flows*, the clear water depth equals:

$$\frac{d}{d_c} = \sqrt[3]{\frac{f_e}{8 * \sin\alpha}} \qquad \text{Uniform equilibrium (wide channel)} \quad (5.27)$$

where d_c is the critical depth and the Darcy Friction Factor f_e is estimated as:

$$\frac{f_e}{f_d} = 0.5 * \left(1 + \tanh\left(2.5 * \frac{0.5 - C_{mean}}{C_{mean} * (1 - C_{mean})}\right)\right) \tag{5.25}$$

$$f_d = \frac{2}{\sqrt{\pi}} * \frac{1}{K} \qquad \text{Steep chutes } (\alpha > 22^o) \quad (5.22)$$

with K = 4.5. Equations 5.22 and 5.25 characterise the form drag in air-water flows. In first approximation, the Darcy Friction Factor may be estimated as $f_e \approx 0.2$ for steep channels (Fig. 5.18). For flat chutes, Equation 5.20 must be used.

In *gradually-varied flows*, the clear-water depth must be deduced from the integration of the energy equation (see Section 5.3.4) in which the friction slope is estimated as:

$$S_f = \sqrt{\frac{f_e}{8} * \frac{q_w^2}{g * d^3}} \qquad \text{wide channel} \quad (5.28)$$

Alternately a simplified method may be used (Section 5.3.4; App. 7).

[20]The use of Equation 5.25 for stepped spillway design is conservative and increases the safety of the structure.

5.3.4 *Gradually-varied flow region*

Air entrainment calculations

In the gradually-varied flow region downstream of the inception point, experimental data show a gradual increase in mean air content along the chute (Fig. 5.21). Assuming a slow variation of the rate of air entrainment, a slow variation of the velocity with distance and a hydrostatic pressure distribution, the continuity equation for the air phase may be solved:

$$\frac{d}{ds} q_{air} = V_{ent} - C_{mean} * u_r * \cos\alpha \qquad (5.28)$$

where q_{air} is the quantity of entrained air within the flow, u_r is a bubble rise velocity and V_{ent} is the local entrainment velocity (Wood, 1985; Chanson, 1993a). The limit of Equation 5.28 is $C_{mean} = C_e$ in uniform equilibrium flows (Section 5.3.1, Fig. 5.14). After transformation, the continuity equation for the air phase yields in a prismatic channel:

$$\frac{d}{ds'} C_{mean} = (1 - C_{mean}) * \left(\frac{u_r * d_* * \cos\alpha}{q_w} * (C_e - C_{mean}) * (1 - C_{mean}) \right) \qquad (5.29)$$

where d_* is the flow depth at the reference location ($s = s_*$), $s' = (s - s_*)/d_*$, and s is the curvi-linear coordinate along the invert. For a channel of constant width and channel slope, the analytical solution of Equation 5.29 is:

$$\frac{1}{(1 - C_e)^2} * \mathrm{Ln}\left(\frac{1 - C_{mean}}{C_e - C_{mean}} \right) - \frac{1}{(1 - C_e) * (1 - C_{mean})} = k_o * s' + K_o \qquad (5.30)$$

where K_o and k_o are:

$$k_o = \frac{u_r * d_* * \cos\alpha}{q_w}$$

$$K_o = \frac{1}{1 - C_e} * \left(\frac{1}{1 - C_e} * \mathrm{Ln}\left(\frac{1 - C_*}{C_e - C_*} \right) - \frac{1}{1 - C_*} \right)$$

and C_* is the mean air concentration at the reference location ($s = s_*$).

Equation 5.29 allows the calculations of the average air concentration C_{mean} as a function of the distance along the chute independently of the velocity, roughness and flow depth. Results are compared with experimental data in Figure 5.21. The calculations were performed with $u_r = 0.4$ m/s and $C_* = 0.20$. The former value is characteristic of millimetric bubbles (App. 4) and it was used for the Aviemore Dam spillway modelling by Chanson (1993a). The experimental works of Boes (1999), Matos (2000) and at the University of Queensland (α=22°), suggest that $C_* = 0.20$ at the inception point is a reasonable approximation: i.e. to take into account the sudden flow aeration in the rapidly-varied flow immediately downstream of the inception point (Section 5.2.3).

Note that an alternative method relates the uniform equilibrium mean air content C_e to the friction slope S_f instead of the bed slope. The technique is more appropriate for flat stepped chutes downstream of high-head gates, but it introduces some coupling between the air entrainment calculations (Eq. 5.29) and the energy equation (Eq. 5.31).

Backwater calculations

In the gradually varied flow region, the differential form of the energy equation gives:

$$\frac{\partial H}{\partial s} = -S_f \tag{5.31}$$

where H is the mean total head, S_f is the friction slope (Eq. 5.28), and s is the curvilinear coordinate along the flow direction.

Several researchers, including the writer, have attempted unsuccessfully to apply the backwater equation (Eq. 5.31) based upon the Darcy-Weisbach Friction Factor without introducing "fudging" coefficients. It is believed that three basic assumptions of the backwater calculations are violated in skimming flows. These are: (H1) the flow is quasi one-dimensional, (H2) the flow must be gradually-varied and (H3) the flow resistance must be the same as for an uniform equilibrium flow (e.g. Chow, 1959: 217-218; Chanson, 1999a: 106-107). The form drag and associated cavity recirculation are very energetic processes and the flow properties in the mixing layer are rapidly varied (Fig. 5.16). A number of researchers showed drastic differences between data measured at the step edge and above the cavity for the same step, as well as from one step to the adjacent ones (e.g. Ohtsu & Yasuda, 1997; Matos et al., 1999). The concepts of gradual variation of the flow resistance and quasi-one-dimensional flow are inappropriate in a form drag-dominated skimming flow.

Some engineers compute the flow resistance using the Manning Formula: it is improper. For clear water flows, the Gauckler-Manning Coefficient $n_{Manning}$ and the friction factor f are related by $n_{Manning} = \sqrt{f/8*g}*(D_H/4)^{1/6}$. The use of the Manning Formula implies that the velocity distributions follow a 1/6th power law (Chen, 1990) and that the flow resistance is independent of the flow depth. This assumption is not realistic in extremely rough flows and in stepped channel flows (Section 5.3.2). The Gauckler-Manning Coefficient must be avoided for hydraulic calculations of stepped channels.

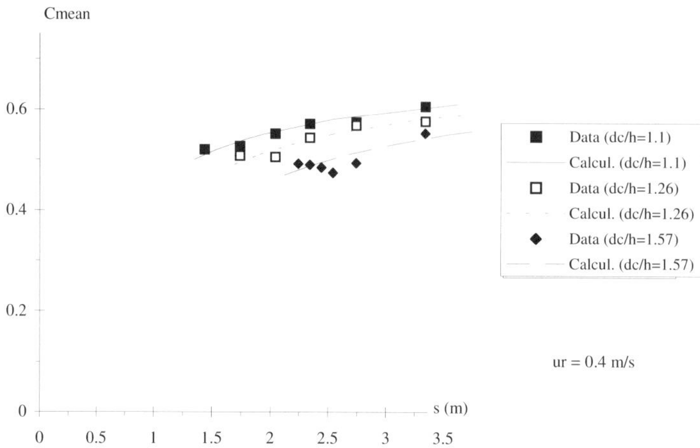

Figure 5.21. Self-aerated flows on stepped chute: comparison between computations Eq. 5.29 and experimental data. s: distance from the spillway crest. d_*: flow depth at the point of inception. $\alpha = 53.1°$, h = 0.08 m, $u_r = 0.40$ m/s (Matos, 2000).

Practical calculations

For steep chutes ($\alpha \sim 35$ to 55°), both the flow acceleration and boundary layer development affect the flow properties significantly. The complete flow calculations can be tedious and most backwater calculations are not suitable[21]. Complete calculations of developing flow and uniform equilibrium flow (App. 7) may be combined to provide a general trend which may be used for a preliminary design (Fig. 5.22).

Ideally, the maximum velocity at the downstream chute end is:

$$V_{max} = \sqrt{2 * g * (H_{max} - d * \cos\alpha)} \qquad \text{Ideal fluid flow} \quad (5.32)$$

where H_{max} is the upstream total head and d is the downstream flow depth (Fig. 5.22A). In practice the downstream flow velocity U_w is smaller than the theoretical velocity V_{max} because of friction losses. In Figure 5.22, the mean flow velocity is plotted as U_w/V_{max} versus H_{max}/d_c where V_{max} is the theoretical velocity (Eq. 5.32), H_{max} is the upstream total head and d_c is the critical depth. Both developing flow calculations and uniform equilibrium flow calculations are shown. Fitting curves must be plotted to connect these lines (Fig. 5.22B).

Experimental data based upon air concentration distribution measurements are compared with theoretical calculations in the developing flow region and in the uniform equilibrium flow region (App. 7). The curves are valid for skimming flow with slopes ranging from 35 to 55° (i.e. 1V:2.1H to 1V:0.7H).

[21]Because backwater calculations are valid only for fully-developed flows.

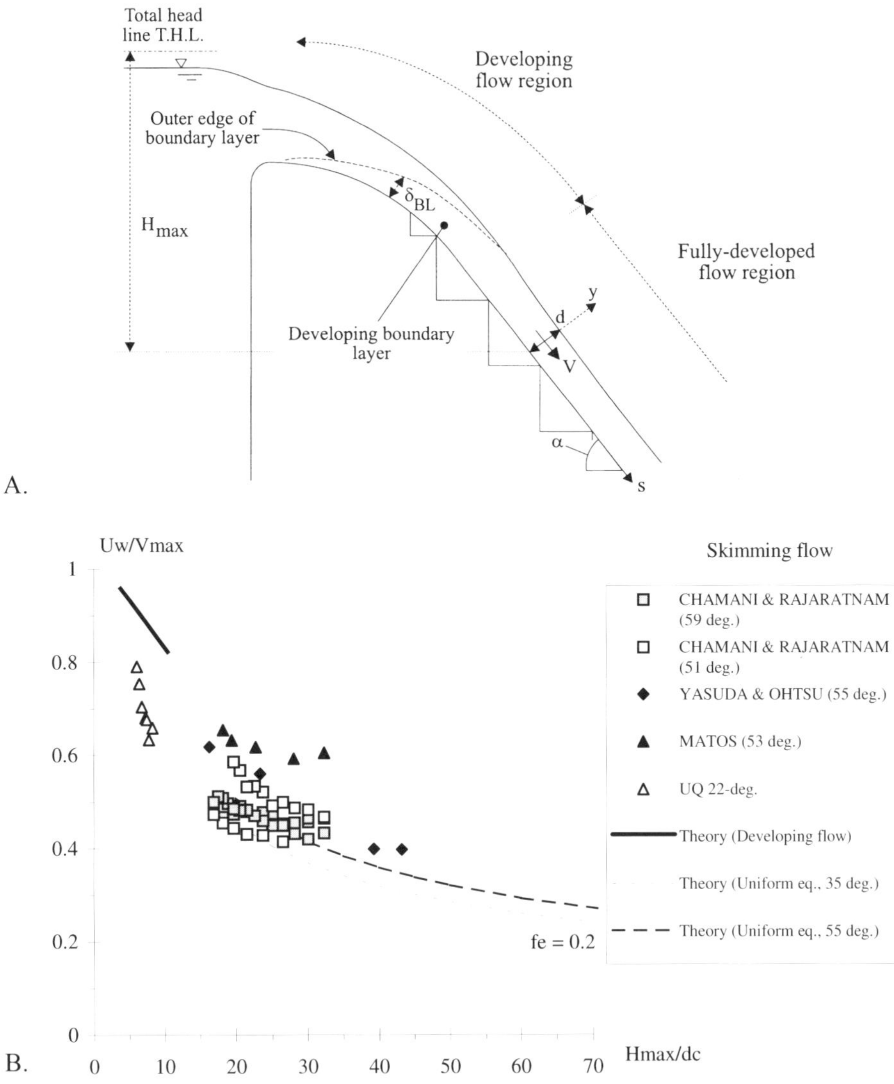

Figure 5.22. Residual flow velocity at the downstream end of the chute. A. Definition sketch. B. Data: Chamani & Rajaratnam (1999a); Yasuda & Ohtsu (1999); Matos (2000); University of Queensland (22°, h=0.1 m).

5.4 ENERGY DISSIPATION

At the downstream end of the chute, the residual head equals:

$$H_{res} = d * \cos\alpha + \frac{q_w^2}{2 * g * d^2} \tag{5.26}$$

where d is the clear-water depth (Eq. 1.10).

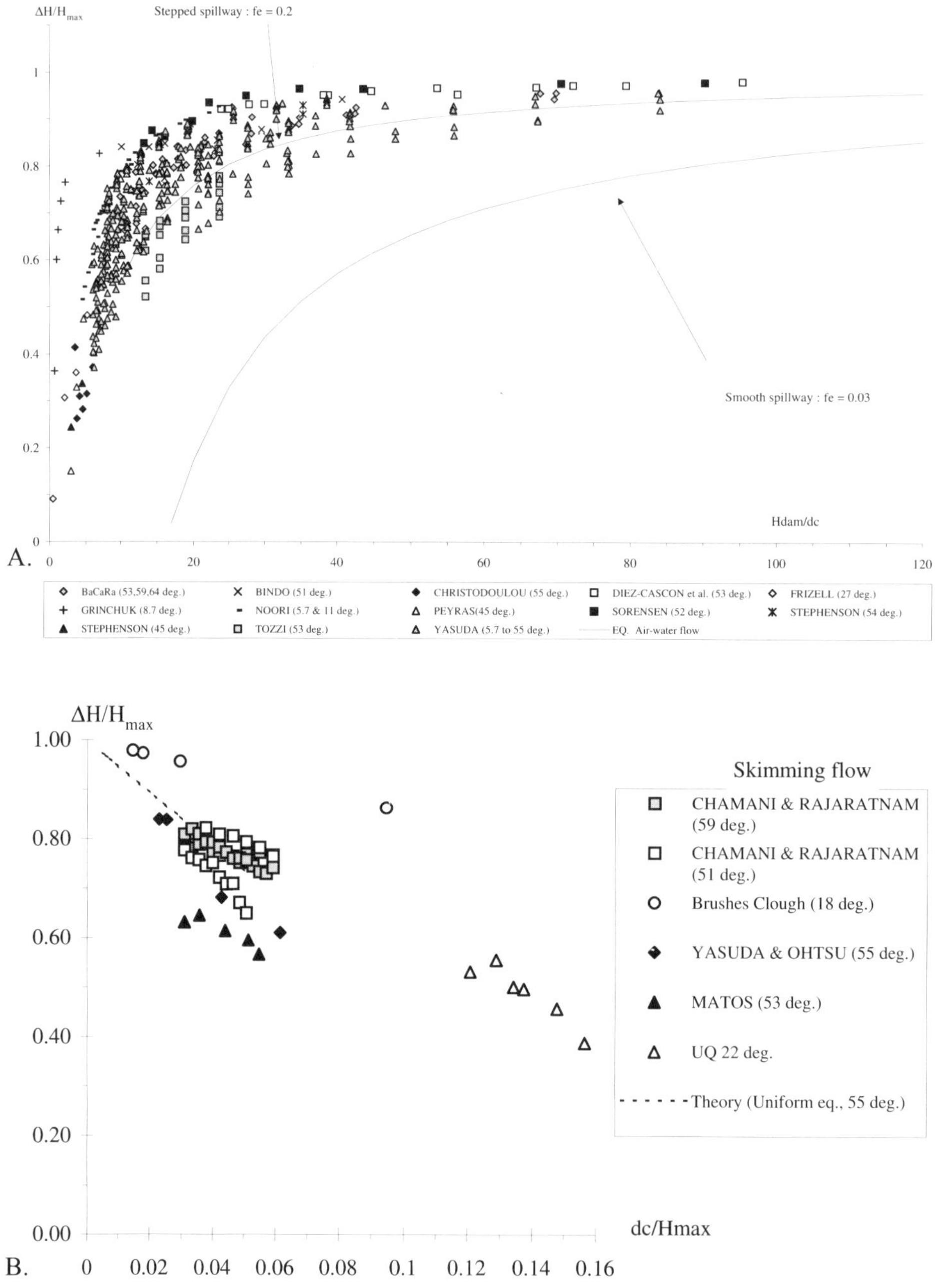

Figure 5.23. Rate of energy dissipation in skimming flow regime. A. Comparison between stepped and smooth-invert chutes. B. Rate of energy dissipation in skimming flows taking into account free-surface aeration. Data: Chamani & Rajaratnam (1999); Yasuda & Ohtsu (1999); Matos (2000); University of Queensland (22°, h=0.1 m).

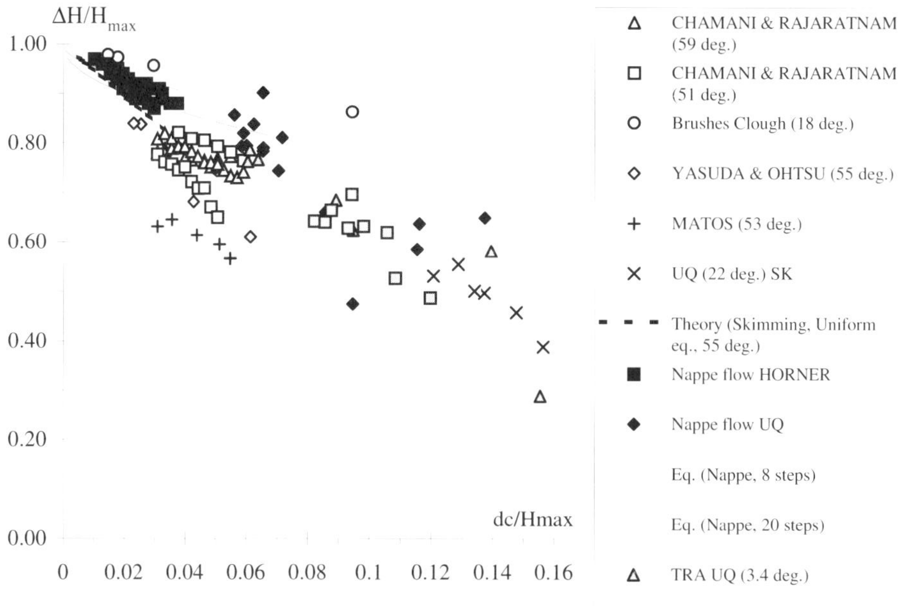

Figure 5.24. Comparison of energy dissipation in nappe flow, transition flow and skimming flow regimes.

For a very long chute in which the flow reaches *uniform equilibrium* at the downstream end, the total head loss ($\Delta H = H_{max} . H_{res}$) equals:

$$\frac{\Delta H}{H_{max}} = 1 - \frac{\left(\frac{f_e}{8 * \sin\alpha}\right)^{1/3} * \cos\alpha + \frac{1}{2} * \left(\frac{f_e}{8 * \sin\alpha}\right)^{-2/3}}{\frac{3}{2} + \frac{H_{dam}}{d_c}} \quad \text{Un-gated chute} \quad (5.33a)$$

$$\frac{\Delta H}{H_{max}} = 1 - \frac{\left(\frac{f_e}{8 * \sin\alpha}\right)^{1/3} * \cos\alpha + \frac{1}{2} * \left(\frac{f_e}{8 * \sin\alpha}\right)^{-2/3}}{\frac{H_{dam} + H_o}{d_c}} \quad \text{Gated chute} \quad (5.33b)$$

where H_{max} is the maximum head available, f_e is the friction factor of air-water flows, H_{dam} is the dam crest head above the downstream toe, H_o is the reservoir free-surface elevation above the chute crest. For an un-gated channel, the maximum head available and the weir height are related by: $H_{max} = H_{dam} + 1.5*d_c$. For a gated chute, the relationship is: $H_{max} = H_{dam} + H_o$. Equation 5.33 may be approximated for a very-high dam as:

$$\frac{\Delta H}{H_{max}} = 1 - \left(\left(\frac{f_e}{8 * \sin\alpha} \right)^{1/3} * \cos\alpha + \frac{1}{2} * \left(\frac{f_e}{8 * \sin\alpha} \right)^{-2/3} \right) * \frac{d_c}{H_{dam}} \tag{5.34}$$

Equations 5.32 and 5.33 show that the rate of energy dissipation increases with the height of the dam. In practice, however, it is more appropriate to consider the dimensionless residual head:

$$\frac{H_{res}}{d_c} = \left(\frac{f_e}{8 * \sin\alpha} \right)^{1/3} * \cos\alpha + \frac{1}{2} * \left(\frac{f_e}{8 * \sin\alpha} \right)^{-2/3} \tag{5.35}$$

Equation 5.35 implies that the residual head increases with $q_w^{2/3}$ for a given slope and friction factor.

The writer has computed the rate of energy dissipation (Eq. 5.33a) for a steep slope ($\alpha = 52°$) using two values of the friction factor: $f_e = 0.03$ and $f_e = 0.2$, that represent an average flow resistance on smooth spillways and on stepped chutes respectively (Section 5.3.3). The results (Fig. 5.23A) are compared with experimental data (Table 1.1), showing a larger energy dissipation on stepped channels for the same discharge and weir height. The agreement between data and Equation 5.33 is fair although there is a considerable scatter.

Detailed air concentration distribution measurements were re-analysed. The data provide accurate information on the "real" (clear-water) residual energy. The results are presented in Figure 5.23B, showing the rate of energy dissipation $\Delta H/H_{max}$ as a function of the dimensionless flow rate d_c/H_{max}. They highlight a significant reduction in energy dissipation with increasing discharge for a given dam height. They are compared with Equation 5.33a assuming $f_e = 0.2$. Despite careful and detailed measurements, the scatter of the data is considerable. It is felt that such a scatter might reflect the poor understanding of the form drag process and cavity flow mechanisms. Note that most data were obtained for steep chutes ($50 < \alpha < 60°$).

Comparison of energy dissipation between nappe, transition and skimming flows

Several researchers[22] suggested that higher energy dissipation is achieved in nappe flows than in skimming flow situations. Figure 5.24 compares the rate of energy dissipation in nappe flows (black symbols), transition flows (grey symbols) and skimming flows (white symbols and cross). Experimental data based upon air concentration distribution data are compared with calculations: i.e. Equation 5.33a in skimming flow assuming $f_e = 0.2$ and Eq. 3.22a in nappe flow for 8 and 20 steps. The results suggest that the largest rates of energy

[22]Ellis (1989), Peyras et al. (1991), Chanson (1994a), Chamani & Rajaratnam (1994).

dissipation are observed in nappe flows and the lowest in skimming flows. Yet such results tend to mask practical limitations: e.g. nappe flow situations require large step heights and flat slopes, while larger discharge capacities may be achieved in skimming flows. Note that skimming flow data cover a narrow range of slopes ($50 < \alpha < 60°$).

For skimming flow down short channels, the flow might not be fully developed at the downstream end for large flow rates. Analytical calculations (App. 7) demonstrate that the residual energy could be considerable. On short chutes, nappe flow situations will dissipate higher energy than skimming flow regime.

Discussion

Several important conclusions derive from the above results (Chapters 3, 4 and 5) which are widely recognised. These are:

1. Larger rates of energy dissipation are observed on stepped spillways than on smooth chutes.
2. For a given hydraulic structure (i.e. H_{dam} = constant), the rate of energy dissipation $\Delta H/H_{max}$ down the stepped chute decreases with increasing discharges. The result holds for both nappe and skimming flow regime.
3. The rate of energy dissipation increases with increasing dam heights for a constant discharge (i.e. q_w = constant).
4. For a stepped spillway, it is more appropriate to consider the residual energy rather than the rate of energy dissipation. Figure 5.22 provides a simple tool to estimate the residual flow velocity.
5. For small stepped weirs, the flow might not be fully developed at the toe of the spillway in a skimming flow regime. A nappe flow situation is preferable for greater energy dissipation.

The author wishes to emphasises that conclusion No. 2 is of primary importance when designing a stepped spillway! In the event of an exceptional flood leading to an overflow larger than the maximum design discharge, the large residual flow energy might induce substantial damage downstream of the spillway (see discussion in Chapter 8).

CHAPTER 6

Mass transfer on stepped cascades aeration, de-aeration, oxygenation, stripping, denitrification

6.1 INTRODUCTION

One of the most important water quality parameters in rivers and streams is the dissolved oxygen content (DOC). The oxygen concentration is a prime indicator of the quality of the water. Dams and weirs across a stream affect the air-water gas transfer dynamics. Deep and slow pools of water upstream of a dam reduce the gas transfer process and the natural re-aeration as compared to an open river. Some hydraulic structures (e.g. spillway, stilling basin) can enhance the oxygen transfer during water releases.

Aeration enhancement by macro-roughness is well-known in water treatment, and one form is the aeration cascade. In-stream cascades have been built along polluted and eutrophic streams (Table 6.1). For example, in Chicago, five re-aeration cascades were built recently to re-oxygenate the depleted waters of the Calumet waterway. In operation their aeration efficiency corrected

Figure 6.1. Flow aeration at an in-stream stepped cascade. Cunningham Weir, Dumaresq River, Australia in February 1998, during a low overflow (view from right bank).

Table 6.1. Re-aeration stepped cascades.

Cascade (1)	Ref. (2)	Purpose (3)	Characteristics (4)	Performances Prototype data (5)
Calumet waterway cascades, Chicago, USA	(1,5)	Five re-aeration cascades. Designed to re-oxygenate the depleted waters of the Calumet waterway.	Cascades with pooled steps: Station 1: 11.6 m^3/s, 4 steps (h = 1 m); Station 2: 1.59 m^3/s, 4 steps (h = 1 m); Station 3: 13.7 m^3/s, 3 steps (h = 1.5 m); Station 4: 13.7 m^3/s, 3 steps (h = 1.5 m); Station 5: 16.4 m^3/s, 4 steps (h = 1 m).	E_{15} = 0.95 (3-steps) to 1.0 (4-steps) for q_w = 0.021 m^2/s
Canyon Weir, USA	(2)	Re-aeration weir downstream of a hydropower station. Designed to re-oxygenate turbined waters.	Labyrinth weir (crest length: 118 m). Single drop (h = 1 m). Plunge pool depth: 1.9 m. Design flow conditions: 0.14 m^2/s.	E_{15} = 0.50 to 0.65 for Q_w = 3 to 14 m^3/s
Chatuge Weir, USA	(2,3)	Re-aeration weir downstream of Chatuge hydroproject in the Hiwasse River (North Carolina, USA). Designed to re-oxygenate turbined waters.	Hollow broad-crested weir. Single drop (h = 2.9 m). Plunge pool depth: 1.1 m. Design flow conditions: 1.2 m^2/s.	E_{15} = 0.63 to 0.73 for Q_w = 14 to 40 m^3/s
Petit-Saut re-aeration cascade, French Guyana	(7)	Re-aeration weir downstream of Petit-Saut Dam (French Guyana). Designed to re-oxygenate turbined waters and to remove methane.	Labyrinth weir. Two drops (h = 2 m each). Design flow conditions: 110 m^3/s.	Oxygen supersaturation achieved. E(methane) = 80% (initial methane level: > 10 g/m^3).
South Houlston Weir, USA	(2,4)	Re-aeration weir downstream of a hydropower station. Designed to re-oxygenate turbined waters.	Labyrinth weir (crest length: 640 m). Single drop (h = 2.3 m). Plunge pool depth: 0.91 to 1.37 m Design flow conditions: q_W = 0.185 m^2/s (0.11 m^2/s during turbine operation).	E_{15} = 0.55 to 0.70 for Q_w = 14 to 68 m^3/s
Montferland demonstration plant, the Netherlands	(6)	Nitrate removal from ground water by sulphur/limestone denitrification: aeration cascade to re-oxygenate depleted water at the end of the process.	3 weirs with 3 steps each (h = 0.7 m). Design flow conditions: 9.72 L/s.	

Ref.: (1) Cargill (1994); (2) Hauser & Morris (1995); (3) Hauser et al. (1992); (4) Rizk & Hauser (1993); (5) Robison (1994); (6) Hoek et al. (1992); (7) Gosse & Gregoire (1997).
Notes: E_{15}: aeration efficiency (in terms of DOC) at 15 Celsius.

to a temperature of 15 Celsius is nearly 95% (Robison, 1994). Similarly, stepped weirs are designed downstream of large dams to control the quality of water releases (e.g. nitrogen supersaturation effect). Despite the associated hydropower loss, a two-steps re-aeration cascade was added downstream of the Petit-Saut Dam in French Guyana to treat turbined waters which had unacceptably high methane content.

With large dams, another issue is downstream nitrogen supersaturation. In the Columbia and Snake rivers (USA), "gas bubble disease" caused by water supersaturated with nitrogen gas was a significant cause of fish mortality for Salmonids and Steelheads (Boyer, 1971; Smith, 1973).

Re-aeration cascades are used in water treatment for re-oxygenation, denitrification or VOC removals. The Montferland plant in the Netherlands was designed to remove nitrate from ground water by sulphur/limestone denitrification. It includes an aeration cascade to re-oxygenate depleted waters at the end of the process (Hoek et al., 1992). In the treatment of drinking water, cascade aeration may be used to remove chlorine and to eliminate or reduce offensive taste and odour (e.g. Corsi et al., 1992).

In summary, stepped cascades are very efficient means of aeration because of the strong turbulent mixing, the large residence time and the substantial air bubble entrainment (e.g. Fig. 6.1).

6.2 BASIC MASS TRANSFER EQUATIONS

Fick's Law states that the mass transfer rate of a chemical across an interface normal to the s-direction and in a quiescent fluid varies directly as the coefficient of molecular diffusion D_{gas} and the negative gradient of gas concentration:

$$\frac{\partial}{\partial t} M_{gas} \propto -D_{gas} * \left(\frac{\partial}{\partial s} C_{gas} \right) \tag{6.1}$$

where C_{gas} is the concentration of the dissolved chemical in liquid and t is the time. The analysis of the fluid layers surrounding a gas bubble is very complex because of the bubble shape, the presence of laminar or turbulent flow, a mobile interface in the case of large air bubbles and the interactions between concentration boundary layers from adjacent bubbles. Generally the gas transfer of a dissolved chemical (e.g. oxygen, nitrogen) across an air-water interface is written as:

$$\frac{\partial}{\partial t} M_{gas} = K_M * A * \left(\frac{P_{gas}}{H_{gas}} - C_{gas} \right) \tag{6.2}$$

where K_M is the mass transfer coefficient, A is the gas-liquid interface area, P_{gas} is the partial pressure of the chemical in air and H_{gas} is the Henry's Law constant.

When the particular chemical is volatile (e.g. oxygen, chlorine), the transfer is controlled by the liquid phase and the coefficient of transfer is almost equal to the liquid film coefficient K_L (i.e. $K_M \sim K_L$). The concentration of dissolved gas in water at equilibrium C_s[1] is defined as:

[1]Also called saturation concentration (see App. 1).

$$C_s = \frac{P_{gas}}{H_{gas}} \qquad (6.3)$$

where H_{gas}, called Henry's Law Constant, is a function of the salinity, temperature, surfactants and, to a lesser extent, the pressure.

The mass transfer equation for volatile gas in a liquid becomes:

$$\frac{\partial}{\partial t} C_{gas} = K_L * a * (C_s - C_{gas}) \qquad (6.4)$$

where a is the specific surface area defined as the air-water interface area per unit volume of air and water. When C_s is greater than C_{gas}, the gas will go into solution (i.e. dissolution): e.g. re-oxygenation. When C_{gas} is greater than C_s (i.e. supersaturation), the gas will desorb: e.g. denitrification, VOC stripping.

On stepped chutes, the air-water mass transfer of chemicals is enhanced by the aeration of the flow, the strong turbulent mixing and the greater residence time. The latter results from the smaller flow velocity than on smooth chutes for identical flow rate. The large amount of entrained air bubbles increases the air-water interface area due to the cumulative bubble surface area. For example, if the mean bubble diameter is 1mm and the air content is 10%, the specific interface area is 600 m^2 per cubic metre of air and water.

For a hydraulic structure, the overall gas transfer may be measured by the deficit ratio r defined as:

$$r = \frac{C_s - C_{US}}{C_s - C_{DS}} \qquad (6.5)$$

where C_{US} is the upstream dissolved gas concentration and C_{DS} is the dissolved gas concentration at the downstream end of the channel. Another measure of aeration is the aeration efficiency E:

$$E = \frac{C_{DS} - C_{US}}{C_s - C_{US}} = 1 - \frac{1}{r} \qquad (6.6)$$

The temperature effect on the air-water gas transfer was examined by several researchers (Table 6.2). Apha et al. (1989) suggested to use an exponential relation to describe the temperature dependence on oxygen transfer:

$$\frac{Ln(r)}{Ln(r_{T_o})} = 1.0241^{(T-T_o)} \qquad (6.7)$$

where r is the deficit ratio at temperature T, T_o is a reference temperature and the constant 1.0241 was obtained by Elmore & West (1961). A more general formulation was proposed by Gulliver et al. (1990).

Coefficient of mass transfer K_L

Mass transfer at bubble interface is a complex process. Kawase & Moo-Young (1992) reviewed several correlations for the mass transfer coefficient calculations in turbulent gas-liquid flows. They showed that the mass transfer coefficient K_L is almost constant regardless of bubble sizes and flow situations. The transfer coefficient may be estimated by:

$$K_L = 0.28 * D_{gas}^{2/3} * \left(\frac{\mu_\omega}{\varrho_\omega} \right)^{-1/3} * \sqrt[3]{g} \qquad (d_{ab} < 0.25 \text{ mm}) \quad (6.8a)$$

$$K_L = 0.47 * \sqrt{D_{gas}} * \left(\frac{\mu_\omega}{\varrho_\omega} \right)^{-1/6} * \sqrt[3]{g} \qquad (d_{ab} > 0.25 \text{ mm}) \quad (6.8b)$$

for gas bubbles affected by surface active impurities, where μ_w and ρ_w are the dynamic viscosity and density of the liquid, D_{gas} is the coefficient of molecular diffusion, d_{ab} is the gas bubble diameter, g is the gravity acceleration, all variables being expressed in SI units (App. 1). Equation 6.8 was compared successfully with more than a dozen experimental studies.

Gas transfer calculations in air-water flows

In air-water turbulent flows, Equation 6.4 may be numerically integrated between two successive cross-sections as:

$$\frac{\Delta C_{gas}}{C_s - C_{gas}} = \frac{K_L * a}{U_w} * \Delta s \qquad (6.9a)$$

where a and U_w are respectively the mean specific interface area and flow velocity over the distance Δs between cross-sections, and the transfer coefficient K_L is a function of the physical and chemical fluid properties only (Eq. 6.8). It yields:

$$\Delta C_{gas} = \left(C_s - C_{gas} \right)_1 * \left(1 - \exp\left(-\frac{K_L * a}{U_w} * \Delta s \right) \right) \qquad (6.9b)$$

where the subscript 1 refers to the upstream cross-section.

Table 6.2. Temperature dependence of air-water gas transfer.

Reference (1)	Temperature dependence (2)	Comments (3)
Gameson et al. (1958)	$\frac{Ln(r_T)}{Ln(r_{15})} = 1 + 0.027*(T-288.15)$	273.15 < T < 313.15 K Oxygen transfer.
Elmore & West (1961)	$\frac{Ln(r_T)}{Ln(r_{20})} = 1.0241^{(T-293.15)}$	278.15 < T < 303.15 K Oxygen transfer. Adopted by APHA et al. (1989).
Holler (1971)	$\frac{Ln(r_T)}{Ln(r_{T_o})} = \Theta^{(T-T_o)}$	in the range 1.014 to 1.047. Oxygen transfer.
Essery et al. (1978)	$(r_T - 1) = (r_{20} - 1)*[1 + 0.0335*(T-293.15)]$	283.15 < T < 303.15 K Oxygen transfer.
Daniil & Gulliver (1988)	$\frac{Ln(r_T)}{Ln(r_{20})} = \sqrt{\frac{T}{293}} * \frac{\nu_{20}}{\nu} * \sqrt{\frac{\varrho_{20}}{\varrho}}$	273.15 < T < 313.15 K Gas transfer controlled by liquid phase.
Gulliver et al. (1990)	$\frac{Ln(r_T)}{Ln(r_{T_o})} = \sqrt{\frac{T}{T_o}} * \left(\frac{\mu_{T_o}}{\mu_T}\right)^{3/4} * \left(\frac{\varrho_{T_o}}{\varrho_T}\right)^{17/20} * \left(\frac{\sigma_{T_o}}{\sigma_T}\right)^{3/5}$	273.15 < T < 313.15 K Liquid phase controlled gas transfer.
Daniil et al. (1991)	$\frac{Ln(r_T)}{Ln(r_{20})} = 1 + 0.02103*(T-273.15) + 8.261E-5*(T-273.15)^2$	273.15 < T < 313.15 K Oxygen transfer.

Notes: T_o: reference temperature; r_{15}: deficit ratio at 15 Celsius; r_{20}, ν_{20}, ρ_{20}: deficit ratio, kinematic viscosity and density at 20 Celsius.

6.3 NAPPE FLOW REGIME

In a nappe flow regime, the mass transfer on each step results from the flow aeration and mixing in the free-falling nappe, at the plunge point and possibly at the downstream hydraulic jump (Fig. 6.2). For a series of N_{step} steps, the aeration efficiency of the complete cascade can be expressed in terms of the aeration efficiency of each step:

$$E = 1 - \prod_{i=1}^{N_{step}} (1 - E_i) \tag{6.10}$$

where E_i is the aeration efficiency of the i-th step. For a stepped chute with N_{step} steps of equal flow characteristics, the aeration efficiency becomes:

$$E = 1 - (1 - E_1)^{N_{step}} \tag{6.11}$$

6.3.1 *Nappe flow with hydraulic jump (NA1)*

In a nappe flow with hydraulic jump (NA1), most aeration takes place at the plunge point and in the hydraulic jump. The aeration efficiency at one individual step may be modelled as:

$$E_1 = E^{jet} + E^{HJ} * (1 - E^{jet}) \qquad \text{Nappe flow NA1 (6.12a)}$$

where E^{jet} is the aeration efficiency of the plunging jet flow and E^{HJ} is the aeration efficiency of the hydraulic jump. For deep pooled steps, most of the gas transfer results from the plunging jet process:

$$E_1 = E^{jet} \qquad \text{Deep pooled steps (6.12b)}$$

Tables 6.3 to 6.6 regroup a number of empirical correlations to estimate E^{jet} and E^{HJ}.

For design purposes, a crude estimate of the oxygen transfer may be obtained using the correlations of Avery & Novak (1978). The results must be double-checked with overall correlations (e.g. Department of Environment, 1973). The approximations might provide an order of magnitude of the downstream (DO) content if the flow conditions are within the range of validity of the empirical correlations. Combining the correlations of Avery & Novak (1978) with the nappe flow properties (Chapter 3), the aeration efficiency of a channel with N_{step} identical steps becomes:

$$E_{15} = 1 - \left(1 + 0.9525 * k_{AN} * \left(g * \frac{h^3}{\nu_w^2}\right)^{0.265} * \left(\frac{d_c}{h}\right)^{-0.534}\right)^{-N_{step}} * \left(1 + 6.965 * k' * \left(g * \frac{h^3}{\nu_w^2}\right)^{0.375} * \left(\frac{d_c}{h}\right)^{0.259}\right)^{-N_{step}} \qquad (6.13)$$

where k_{AN} and k' are given in Tables 6.4 and 6.6 respectively. Equation 6.13 is compared with laboratory data (Essery et al., 1978) and prototype data at a Calumet waterway station (Robison, 1994; Table 6.1) in Figure 6.3. The results suggest that Equation 6.13 overpredicts the oxygen transfer in nappe flow situations because the gas transfer by plunging jet is limited by the thin water depth on the receiving steps, while the correlation of Avery & Novak (1978) was developed for overfalls with deep receiving pools.

6.3.2 *Nappe flow without hydraulic jump (NA3)*

In a nappe flow without hydraulic jump, most flow aeration takes place in the free-falling nappe, at nappe impact and in the downstream supercritical flow. Detailed measurements of the air-water flow properties were conducted at the University of Queensland (Table 6.7). Careful measurements of specific interface area distributions were conducted for each experiment (e.g. Fig. 6.4).

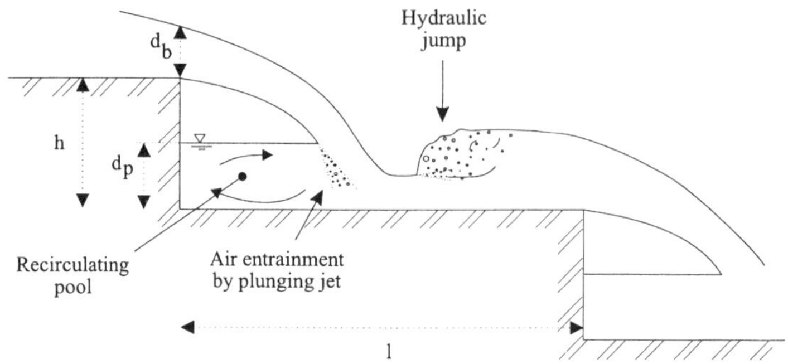

Figure 6.2. Flow aeration in nappe flow regime (sub-regime NA1).

Table 6.3. Free-overfall experiments of gas transfer.

Ref. (1)	q_w M²/s (2)	Fall height m (ᵃ) (3)	Data (4)	Formulation (5)	Remarks (6)
Mastropietro (1968)		1.22 to 4.6	Oxygen transfer	$r = \frac{1}{1-0.141*\Delta H}$	Prototype data. ?H = head loss (in m). 20 < T < 25 Celsius
Holler (1971)		3.7, 8.4, 10.7	Oxygen transfer	$r_{20} = 1 + 0.211*\Delta H$	Prototype data. Tailwater depths < 4.9 m.
Foree (1976)	Qw = 0.057 to 0.4 m³/s	0.79 to 7.3	Oxygen transfer	$r_{25} = \exp(0.525*\Delta H)$	Prototype data.
Avery & Novak (1978)	0.003 to 0.05	0.25 to 2.1	Oxygen transfer	$r_{15} = 1 + k_{AN}*\left(\frac{\rho_\omega}{\mu_\omega}*(D_H)_i*\sqrt{\frac{g*\Delta H}{8}}\right)^{0.53} * \left(\frac{2*\Delta H}{(D_H)_i}\right)^{0.89}$	Rectangular jets. Model data. W = 0.1, 0.22, 0.3 m
Nakasone (1987)	0.011 to 0.741	0.24 to 1.98	Oxygen transfer	$Ln(r_{20})$ $k_{N1}*\Delta H\ k_{N2}*\theta_\omega\ k_{N3}*\delta_t^{0.310}$	Weirs. Model data. ?H and d_t in m, q_W in m²/s. W = 0.2, 0.3 m.
	0.027 to 7.8	1.52 to 5.8	Oxygen transfer	$Ln(r_{20}) = 5.92*\frac{\Delta H^{0.816}}{q_w^{0.363}}*d_t^{0.310}$	Weirs. Prototype data. ?H > 1.2 m and q_W > 0.065 m²/s.

Notes: (ᵃ): difference between water levels; $(D_H)_i$: jet hydraulic diameter at impact; d_t: tailwater depth; $CDCI_3$ = deuterated chloroform; EDB = ethylenedibromide.

Figure 6.4 presents distributions of air concentration, velocity and bubble frequency on the centreline and next to the sidewall at the first drop. The data were integrated at each longitudinal cross-section to obtain the cross-sectional average specific interface area (Fig. 6.5). Figure 6.5 presents longitudinal distributions of dimensionless average interface area along the step downstream of the first drop. The location of the nappe impact is highlighted. The data show depth-averaged specific interface area ranging from 20 to over 120 m^{-1}

Table 6.4. Coefficients k_{AN} (Avery & Novak, 1978).

Water (1)	k_{AN} (2)
Tap water	0.627E-4
Tap water + 0.3% of $NANO_3$	0.869E-4
Tap water + 0.6% of $NANO_3$	1.243E-4

Table 6.5. Oxygen transfer correlations at hydraulic jumps.

Reference (1)	Formula (2)	Remarks (4)
Holler (1971)	$r_{20}-1=0.0463*\Delta V^2$	Model experiments: $0.61 < \Delta V < 2.44$ m/s $277.15 < T < 299.15$ K
Avery & Novak (1978)	$r_{15}-1=k'*Fr_1^{2.1}*Re^{0.75}$ k' is function of the salinity (Table 6.6).	Model experiments: $1.45E+4 < Re < 7.1E+4$ $\nu_W = 1.143E-6$ m²/s $W = 0.10$ m
Wilhelms et al. (1981)	$r_{15}-1=4.924E-8*Fr_1^{2.106}*Re^{1.034}$	Model experiments (W = 0.381 m) based upon Krypton-85 transfer: $1.89 < Fr_1 < 9.5$ $2.4E+4 < Re < 4.3E+4$
Johnson (1984)	Empirical fit based of field data from 24 hydraulic structures.	Prototype data. Working well for deep plunge pools.
Chanson (1995a)	$Ln(r) \approx 13.39*\frac{K_L}{g}*V_1^{4.93}*(V_1-0.414)*\frac{\left(\sqrt{1+8*Fr_1^2}-1\right)*(Fr_1-1.5)}{Fr_1^2}$	Hydraulic jump with partially-developed inflow. Semi-analytical calculation validated with the data of Wilhelms et al. (1981).

Notes: ΔV: difference between the upstream and downstream velocities (m/s); d_1: upstream flow depth (m); Fr_1: upstream Froude Number defined as: $Fr_1 = q_w/\sqrt{g*d_1^3}$; q_w: discharge per unit width (m²/s); Re: Reynolds Number defined as: $Re = q_w/\nu_w$; r_{15}, r_{20}: oxygen deficit ratio at 15 Celsius and 20 Celsius; W: channel width (m); ΔH: head loss in the hydraulic jump (m of water).

Table 6.6. Coefficient k' (Avery & Novak, 1978).

Water (1)	k' (2)
Tap water	1.0043E-6
Tap water + 0.3% of $NANO_3$	1.2445E-6
Tap water + 0.6% of $NANO_3$	1.5502E-6

typically along each step, associated with maximum bulk specific interface area of up to 160 m^{-1} at the largest flow rate. At each step, the maximum mean interface area was observed at the impact of the free-falling nappe and in the following spray region.

The mass transfer equation (Eq. 6.4) was integrated to predict the aeration

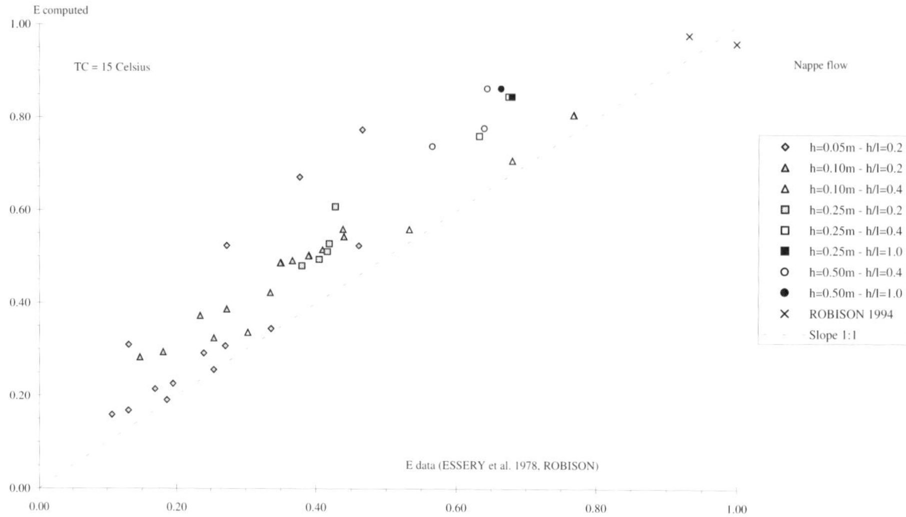

Figure 6.3. Oxygen transfer in nappe flows with fully developed hydraulic jump at 15 Celsius. Comparison between laboratory data (Essery et al., 1978), prototype data (Robison, 1994) and computations (Eq. 6.13).

efficiency of the cascade in terms of dissolved oxygen. In Equation 6.4, the liquid film coefficient was calculated using Kawase & Moo-Young's (1992) correlations developed for gas bubbles affected by surface active impurities. The calculations yield aeration efficiencies for a single step ranging from 3 to 4%[2] and between 18 and 27% for the 24 m long cascade (Table 6.7). Dissolved oxygen concentrations were measured and the data give a measured aeration efficiency of 42% (Table 6.7).

Discussion

The results highlight the aeration potential of stepped cascades at low to medium flows (e.g. Fig. 6.1). For the study of Toombes & Chanson (2000) and Chanson & Toombes (2000), previous investigations, conducted in the same 24 m long chute equipped with a smooth-invert, showed an aeration efficiency of about 3% for $q_w = 0.15$ m²/s (Table 6.7). In terms of metre length of channel, the stepped chute design is at least 10 times more efficient than the smooth chute for the same flow rate. The experimental results (Table 6.7) demonstrate further that the numerical integration of the mass transfer equation compares well with the observed DOC transfer.

[2]in terms of dissolved oxygen at standard pressure and 20 Celsius.

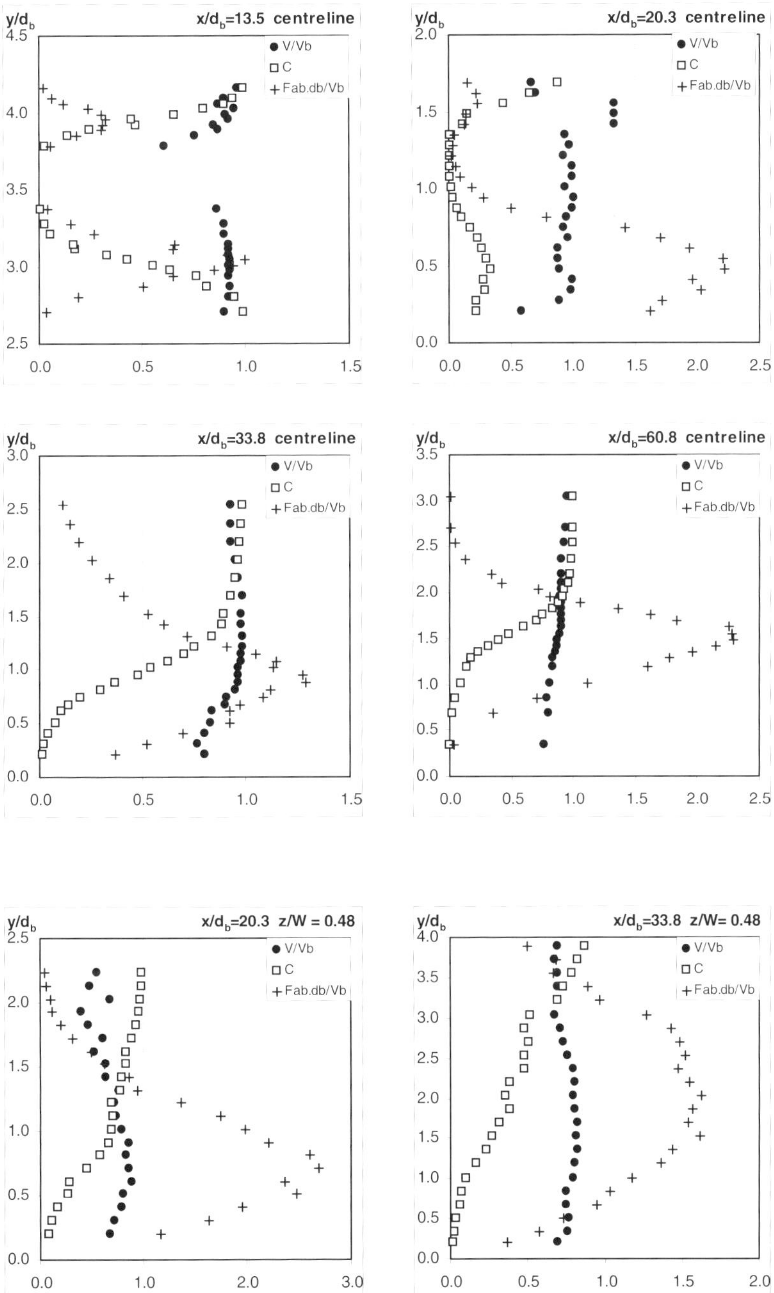

A.

B.

Figure 6.4. Dimensionless distributions of air content V, air-water velocity V/V_b, bubble count rate $F_{ab}{*}d_b/V_b$ (after Toombes & Chanson, 2000). $q_w = 0.1052$ m^2/s, $d_b = 0.0296$ m, 1st drop (Step No. 2). A. Centreline data (z/W = 0). B. Sidewall data (z/W = 0.48, 5mm from glass wall).

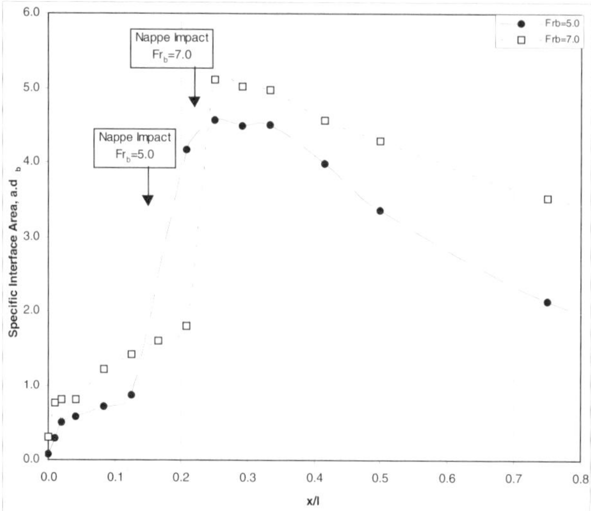

Figure 6.5. Longitudinal distributions of dimensionless mean specific interface area $a_{mean}*d_b$ along a step (after Toombes & Chanson, 2000). h = 0.143 m, W = 0.25 m, 1st drop (Step No. 2). (a) Fr_b = 5.0, q_w = 0.0838 m²/s, d_b = 0.0306 m. (b) Fr_b = 7.0, q_w = 0.1052 m²/s, d_b = 0.0296 m.

Table 6.7. Aeration efficiency at a stepped cascade with nappe flow without hydraulic jump (NA3), in terms of DO at standard pressure ($P = P_{std}$) and T = 20 Celsius.

Ref.	d_c/h	d_b/h	E^a Integration	E^b Measured	Comments
(1)		(2)	(3)	(4)	(5)
Stepped cascade					
University of Queensland (Toombes & Chanson, 2000)					Nappe flow NA3. α = 3.2°, h = 0.143 m, W = 0.25 m.
	0.607	0.206	0.0393	–	1st drop (Step No. 2).
	0.607	0.213	0.0373	–	1st drop (Step No. 2).
	0.666	0.167	0.0409	–	1st drop (Step No. 2).
	0.727	0.277	0.0344	–	1st drop (Step No. 2).
	0.864	0.199	0.0395	–	1st drop (Step No. 2).
					Nappe flow NA3. α = 3.4°, h = 0.143 m, l = 2.4 m, W = 0.5 m, 10 steps.
	0.607	–	–	0.42	For the 24m long cascade.
University of Queensland (Chanson & Toombes, 2000)					Nappe flow NA3. α = 3.4°, h = 0.143 m, l = 2.4 m, W = 0.5 m, 10 steps.
	0.607	0.235[1]	0.18	–	For the 24m long cascade.
	0.750	0.257[1]	0.25	–	For the 24m long cascade.
	0.923	0.285[1]	0.27	–	For the 24m long cascade.
Smooth chute					
University of Queensland (Chanson, 1997b,c)					Smooth-invert chute. α = 4.0°, W = 0.5 m, L = 24 m.
	q_w = 0.15 m²/s	N/A	0.03	–	For the 24m long chute.

Notes: d_b: supercritical inflow depth (or brink depth); [1]: measured at first brink; [a]: integration of the mass transfer equation based upon detailed measurements of the specific interface area; [b]: based upon measured dissolved oxygen concentrations.

6.4 SKIMMING FLOW REGIME

6.4.1 *Introduction*

In skimming flows, the upstream flow region is not aerated and the mass transfer is nearly zero upstream of the inception point of air entrainment. For a given chute geometry (length L, slope α, step height h), free-surface aeration is observed for flow rates greater than a characteristic discharge $(q_w)_c$ for which the growing boundary layer reaches the free surface at the spillway end. The flow rate $(q_w)_c$ may be deduced from boundary layer growth results (Section 5.2.2):

$$(q_w)_c = 0.129 * L^{1.1403} * (\sin\alpha)^{0.388} * (h * \cos\alpha)^{0.0975} \tag{6.14}$$

where L is the spillway length.

6.4.2 *Fully-developed bubbly flow properties*

Downstream of the inception point of air entrainment, the air-water interface area may be estimated from the air-water flow properties (Chapter 5). For any bubble shape, bubble size distribution and chord length distribution, the mean chord length size[3] equals $C*V/F_{ab}$ where V is local air-water velocity and F_{ab} is the bubble count rate[4]. The specific air-water interface area derives from the mass conservation for air:

$$a = \frac{4 * F_{ab}}{V} \tag{6.15}$$

Equation 6.15 is valid in bubbly flows (i.e. $C < 0.3$ to 0.5). In high air content regions ($C > 0.3$ to 0.5), the flow structure is more complex and the result is not exactly the true specific interface area. It becomes simply proportional to the number of air-water interfaces per unit length of air-water mixture (i.e. $a \propto 2*F_{ab}/V$). The velocity distribution satisfies:

$$\frac{V}{V_{90}} = \left(\frac{y}{Y_{90}}\right)^{1/N} \tag{5.12}$$

where y is the distance normal to the invert, Y_{90} is the distance where $C = 90\%$, V_{90} is the characteristic velocity at $y = Y_{90}$ and $N = 3.7$ (Section 5.3.2). The bubble frequency is related to the air content by a parabolic law:

[3]That is, the number mean size.
[4]That is, the number of bubbles impacting the probe per second.

$$\frac{F_{ab}}{(F_{ab})_{max}} = 4*C*(1-C) \tag{3.23}$$

where the maximum bubble frequency $(F_{ab})_{max}$ is always observed for C = 50% (Section 5.3.1; Fig. 5.13). The parabolic shape was observed first in smooth-chute flows, in high-velocity water jets and in hydraulic jumps (Chanson, 1997b; Brattberg et al., 1998; Chanson & Brattberg, 2000) and it holds also skimming flows[5].

The distribution of void fraction may be described by a diffusion model:

$$C = 1 - \tanh^2\left(K' - \frac{y}{2*D'*Y_{90}}\right) \tag{1.7}$$

where tanh is the hyperbolic tangent function, D' is a dimensionless turbulent diffusivity and K' is an integration constant (App. 6; Chanson, 1997a). D' and K' are functions of the mean air content C_{mean} only.

A typical distribution of air-water specific area (Eq. 6.15) is presented in Figure 6.6 where it is compared with Equations 1.7, 3.23 and 5.12. All the curves are plotted in dimensionless form. Defining the depth-averaged mean specific interface area as:

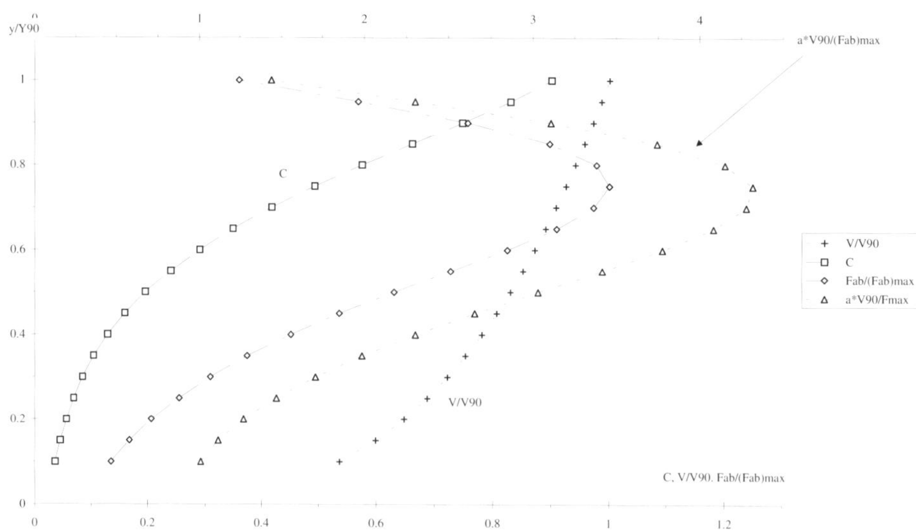

Figure 6.6. Dimensionless air-water specific area distribution in skimming flow.

[5]Experimental observations: University of Queensland (α = 22°).

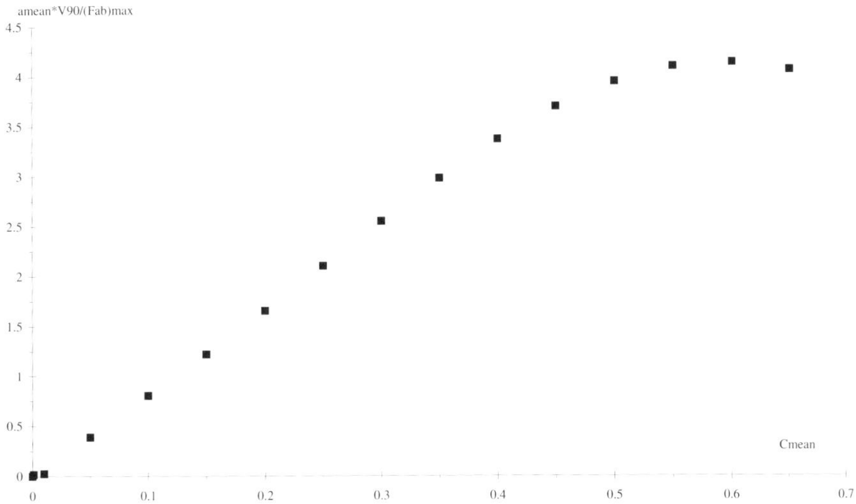

Figure 6.7. Dimensionless mean specific area in skimming flow.

$$a_{mean} = \frac{1}{Y_{90}} * \int_0^{Y_{90}} a * dy \tag{6.16}$$

the results imply that the dimensionless average specific interface area $a_{mean}*V_{90}/(F_{ab})_{max}$ is a function of the mean air concentration only (Fig. 6.7). The results are best correlated by:

$$\frac{a_{mean}*V_{90}}{(F_{ab})_{max}} = 8.54*C_{mean} \qquad C_{mean} < 0.45 \tag{6.17a}$$

$$\frac{a_{mean}*V_{90}}{(F_{ab})_{max}} = 7.2*C_{mean} + 9.43*C_{mean}^{\ 2} - 16.24*C_{mean}^{\ 3} \qquad C_{mean} < 0.7 \tag{6.17b}$$

6.4.3 *Aeration efficiency in skimming flow*

In the fully-developed flow region, the mass transfer equation (Eq. 6.9) may be numerically integrated by calculating the air-water flow properties including the air-water interface area at any location down the chute. Complete calculations were performed for slopes ranging from 22 to 53°, d_c/h between 1.5 and 10, q_w between 1 and 10 m^2/s, and assuming $u_r = 0.4$ m/s. Typical results are presented in Figure 6.8 in terms of oxygen aeration efficiency as a function of the dimensionless dam height H_{dam}/d_c. Further details are given in Appendix 8. In Figure 6.8, the calculations are compared with DO measurements

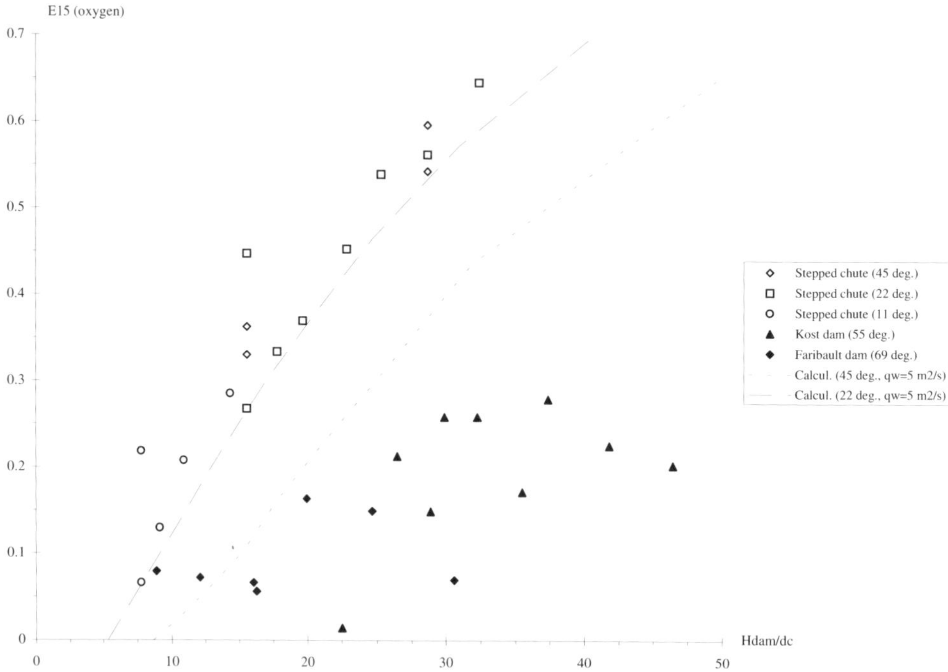

Figure 6.8. Aeration efficiency in terms of dissolved oxygen at standard pressure and 15 Celsius. Comparison between calculations (Eq. 6.9), stepped chute data (Essery et al., 1978) and prototype smooth chute data (Rindels & Gulliver, 1989). Black symbols: smooth-invert chute (prototype). White symbols: stepped chute (laboratory).

in skimming flow (Essery et al., 1978) and DO data on prototype smooth chutes (Kost and Faribault dams; Rindels & Gulliver, 1989).

Figure 6.8 highlights greater aeration efficiencies on stepped chutes with skimming flow (white symbols) than those on smooth-invert chutes (black symbols). With a stepped cascade, a greater aeration per metre drop in elevation is achieved with flat stepped chutes ($\alpha < 30°$). For a given geometry, the largest aeration is obtained at low flow rates.

Discussion

Specific interface area calculations are based upon an estimate of the maximum bubble count rate $(F_{ab})_{max}$. Experimental observations suggest:

$$(F_{ab})_{max} = 20.65 * U_w^{1.74} \qquad 2 \leq u_w \leq 8 \quad (6.18)$$

where F_{ab} and U_w are in SI units (App. 8).

Free-surface aeration can contribute to a large part of the air-water transfer of atmospheric gases along a smooth chute (e.g. Chanson, 1995c). On a stepped chute, the air-water gas transfer is further enhanced by the early apparition of "white waters" (i.e. free-surface aeration), the slower mean velocity and the associated increase of residence time. Overall, Figure 6.8 provides a simple method to estimate an order of magnitude for the air-water gas transfer (at 15 Celsius).

6.5 DISCUSSION

6.5.1 *Comparison between flow regimes*

Although the characteristics of nappe and skimming flow regimes differ significantly, a comparative analysis of the overall efficiency is relevant. Figure 6.9 compares skimming flow data (Essery et al., 1978) and nappe flow data (Essery et al., 1978; Toombes & Chanson, 2000). The results indicate that the aeration efficiency of nappe flows is primarily a function of the discharge and total drop height (Fig. 6.9). The oxygen transfer increases with increasing drop in elevation for a given discharge. Also, the oxygen transfer increases with decreasing discharges for a given drop in elevation and channel slope.

In skimming flows, the aeration efficiency is nearly zero as long as no free-surface aeration occurs. When free-surface aeration take place, the oxygen

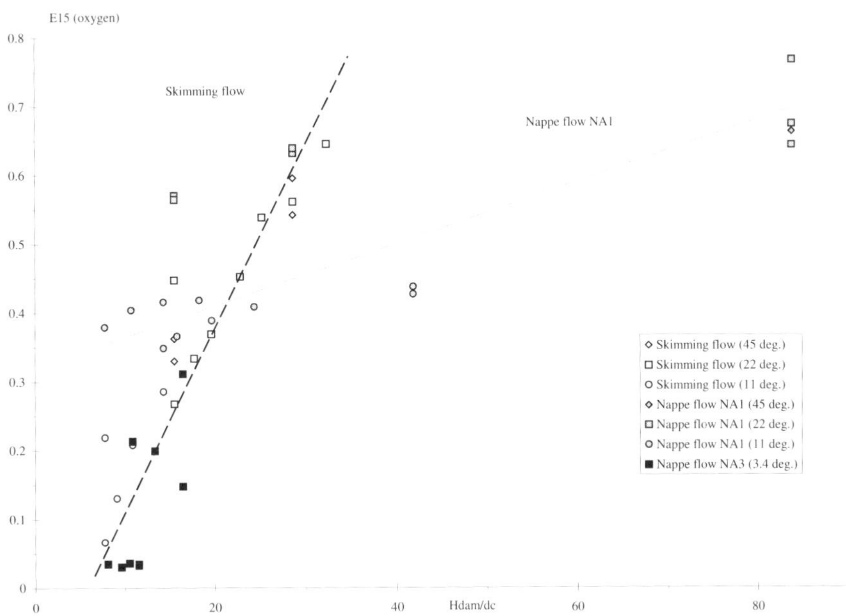

Figure 6.9. Aeration efficiency in terms of dissolved oxygen at standard pressure and 15 Celsius. Comparison between nappe flow NA1 and NA3, and skimming flow data. Black symbols: nappe flow NA3 (Toombes & Chanson, 2000). Gray symbols: nappe flow NA1 (Essery et al., 1978). White symbols: skimming flow (Essery et al., 1978).

transfer increases sharply with the dimensionless drop in elevation H_{dam}/d_c and the trend is approximated by a thick dotted line in Figure 6.9. The largest aeration efficiencies are obtained with the smallest discharges. The analysis shows also that the channel slope and dimensionless step height h/d_c have little effects on the aeration efficiency of skimming flows.

In summary, the air-water gas transfer is maximum for small discharges with skimming flows; and it is maximum for relatively large discharges in nappe flow situations. This apparent contrast may be explained by the different mechanisms of air entrainment and air-water gas transfer between the two flow regimes.

6.5.2 *Mass transfer in the downstream energy dissipator*

At the downstream end of a stepped spillway, additional mass transfer takes place in the energy dissipator. For deep plunge pools, Johnson (1984) developed an empirical fit of prototype data that works well although Rindels & Gulliver (1986) showed that Johnson's method does not apply to the re-aeration of waters with low DO contents and with shallow plunge pools. With a hydraulic jump stilling basin, the prototype data of Rindels & Gulliver (1989) re-analysed by Chanson (1995c) are best fitted by the correlation of Wilhelms et al. (1981). For flip bucket energy dissipators, mass transfer takes place in the free-jet and downstream of the nappe impact into the receiving pool.

CHAPTER 7

Design of stepped channels, chutes and spillways

7.1 INTRODUCTION

7.1.1 *Presentation*

The storage of water is essential for providing drinking water and irrigation water reserves. An artificial water storage system (e.g. reservoir behind a dam) must be developed when the hydrology of a stream does not provide enough supply all year round. Hydrological studies provide information on the storage capability and as well as on the maximum (peak) flow in the system. The basic design approach must be based upon a *system approach*. A hydraulic structure must be analysed as part of the surrounding catchment. Impacts on the upstream and downstream valleys must be considered: e.g. sediment trap, fish migration, downstream water quality (e.g. dissolved oxygen content), modifications of the water table and associated impacts (e.g. salinity). Structural, geotechnical and hydraulic considerations must be taken into account.

With small dams and weirs, a consequent component of the overall cost is the spillway designed to pass safely the maximum peak flood. An overflow spillway includes typically three sections. The crest is designed to maximise the discharge capacity of the spillway. The chute is designed to pass the flood waters above the dam, and the energy dissipator is designed to dissipate the residual kinetic energy of the flow at the downstream end of the chute.

In any design exercise, the first questions to ask are: 1. What is the purpose and the duty of the structure? 2. What are the constraints? Once the decision to select a stepped channel is done, all the parameters must be considered to decide the preferential type of flow regime (nappe or skimming flow) for the design discharge and for unusual (emergency) flow conditions (in drought or/and flood periods). Usually a transition flow regime is not a satisfactory design flow condition, although it might be suitable in artistic fountains and water treatment structures. The flow conditions in the transition flow regime are characterised by transitory fluctuations that might induce improper or dangerous flow behaviours and unnecessary vibrations of the structure (Chapters 4 and 8). In this chapter, new design procedures for various applications of stepped chutes are described. Specific comments are detailed in each case.

This chapter does not intend to review all possible cases but the main applications.

7.1.2 *Selection of the flow regime: a discussion*

The previous chapters have been developed for prismatic channels with constant discharges and flat step geometry. The flow regime has been considered either nappe, transition or skimming flow depending upon the discharge per unit width (q_w) and step geometry (h, l) (Section 1.3). Flow conditions in the transition flow regime exhibit some instabilities leading to deflecting nappes, fluctuating hydrodynamic loads on the steps. The transitory flow may be characterised by rapid changes of pressure distribution from zero pressure gradient (free-falling nappe) to quasi-hydrostatic, by the disappearance of the air cavity behind the vertical step face and by the establishment of a wake region downstream of the step edges. Designers must avoid such conditions for the design flow rate and consider additional hydraulic and structural tests if they cannot avoid a transition flow regime at design conditions.

In gradually-varied flow, the transition between nappe flow and skimming flow regime is a function of the dimensionless critical depth d_c/h and the channel slope (Chapter 1). A modification of the channel width, an incoming or outcoming discharge (e.g. side channel) or a change of channel slope will entrain a modification of the discharge per unit width q_w and critical depth d_c, and possibly a change in flow regime. The designer might consider variable step height h, step length l, channel width W or a combination of these methods along the stepped channel to prevent a flow regime transition if the discharge is constant. With variable flow rates, a transitory regime might not be avoidable. A particular example is the Bank of China Cascade in Hong Kong (Fig. 7.3): incoming and outcoming discharges in a network of channels with identical width and step geometry provide a display of nappe, transition and skimming flows.

Application: side stepped channel at the chute toe

Considering a side stepped channel at the bottom of a spillway chute (Fig. 7.1), the stepped channel collects the flow discharging over the spillway (or weir). The discharge per unit width changes with the incoming discharge and the modification of the channel width. At the upstream end of the side channel, the flow is a nappe flow regime and it might become a skimming flow at the downstream end. Prototype examples include the Muela Dam, Lesotho (H_{dam} = 55 m, 1996), and the Shimagawa Dam, Japan (H_{dam} = 89.5 m, 1998) (Figure 7.1B). Often the training wall is equipped with a deflector at the top to minimise the wall height (e.g. Hakoishi & Sumi, 2000).

For the example sketched in Figure 7.1A, the total discharge and the channel width vary both spatially. The discharge per unit width q_w can be deduced from the continuity equation:

$$\frac{\partial q_w}{\partial s} = q_w' * \frac{\cos\alpha}{W} - \frac{q_w}{W} * \frac{\partial W}{\partial s} \tag{7.1}$$

where s is the direction along the channel, q_w' is the discharge per unit width of the spillway (i.e. incoming discharge), α is the stepped channel slope and W is the stepped channel width.

For a spillway chute of constant slope α' and constant discharge per unit width q_w', and for a stepped channel of constant slope α, the critical depth at each position s equals:

$$\frac{d_c}{h} = \left(\frac{Q_w'}{W_1 * \sqrt{g * h^3}} * \left(\frac{\frac{s * \cos\alpha}{W'}}{1 + \frac{W'}{W_1} * \frac{\tan\alpha}{\tan\alpha'} * \left(1 - \frac{s * \cos\alpha}{W'}\right)} \right) \right)^{2/3} \tag{7.2}$$

where Q_w' is the total spillway discharge, W' is the horizontal spillway width, W_1 is the stepped channel width at the downstream end and h is the step height (Fig. 7.1). A side spillway flow is greatly affected by the inflow from the main spillway. The supercritical lateral inflow is subjected to a rapid change in flow direction and the flow is extremely turbulent.

7.1.3 *Spillway design*

For the design of a spillway system, the significant design parameters are the probable maximum flood (PMF) discharge, the maximum discharge capacity and the upstream reservoir free-surface elevation at design flow conditions. A recent investigation of the causes of dam failures worldwide since 1950 showed that more than 40% of the failures were caused by overtopping resulting from inappropriate spillway discharge capacity (Lemperiere, 1993). This figure highlights the importance of an appropriate estimation of the PMF discharge and a suitable selection of the maximum discharge capacity. In Europe, with temperate climates, the maximum spillway discharge capacity is much lower than the PMF discharge. In tropical countries (e.g. Northern Australia, Taiwan), the maximum discharge capacity is slightly smaller that the PMF. The standards depend also upon the risks of life danger.

Stepped spillways must be designed to sustain maximum design discharges without major damage to the spillway and to the dam itself. For embankment dams, the major flood risk relates to possible erosion of the dam, while the flood risk for concrete and masonry gravity dams concerns the stability rather than erosion. A gravity dam is designed for a given maximum headwater level and might overturn for an extra rise of headwater. A discussion of the risks of failures is further developed in Chapter 8.

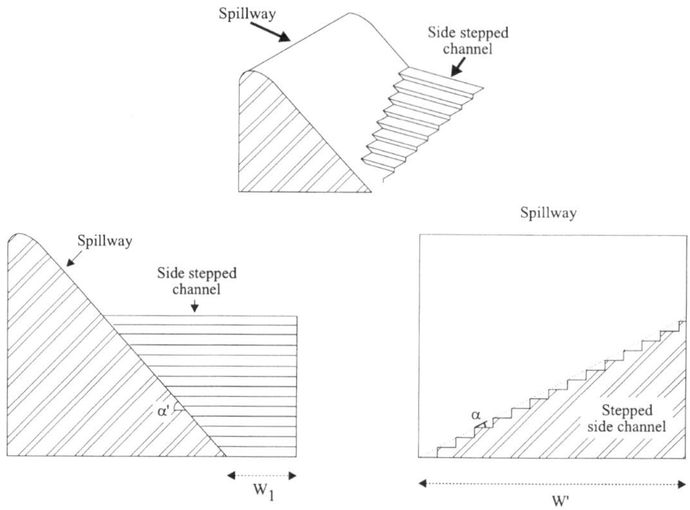

A.

B.

Figure 7.1. Side stepped channel at a spillway bottom. A. Definition sketch. B. Shimagawa Dam, Japan (1998, H_{dam} = 89.5 m). Main spillway: $\alpha = 53.1°$, W = 79 m, Design flow: 1,000 m^3/s. View from the dam crest in November, 1998 with the two side stepped spillways; the stilling basin in the middle and on-going works in the right side stepped spillway.

Crest design

The crest of an overflow spillway is designed to maximise the discharge capacity of the structure: i.e. to pass safely the design discharge at the lowest cost. The common shapes include the broad-crest, the circular crest and the ogee crest. For a given specific energy, maximum flow rate is achieved for critical flow conditions[1] (Bélanger, 1828). For a broad-crested weir (rectangu-

[1]Flow conditions for which the specific energy is minimum are called critical flow conditions.

lar cross-section), the maximum discharge per unit width may be deduced from the continuity and Bernoulli equations:

$$q_w = \sqrt{g} * \left(\frac{2}{3} * (H_{max} - H_{dam}) \right)^{3/2} \quad \text{Broad-crest (7.3)}$$

where g is the gravity acceleration, H_{max} is the upstream total head and H_{dam} is the spillway crest elevation. At circular and ogee crests, the observed discharge differs from Equation 7.3 because the pressure distribution on the crest is not hydrostatic. The flow rate is expressed as:

$$q_w = C_D * \sqrt{g} * \left(\frac{2}{3} * (H_{max} - H_{dam}) \right)^{3/2} \quad (7.4)$$

where C_D is a discharge coefficient.

The design of circular-crested weirs was introduced at the end of the 19th century. These were found to have a larger discharge capacity (for identical upstream head) than broad-crested weirs. Experimental observations show that the discharge coefficient C_D increases with increasing dimensionless head on crest, C_D being larger than unity except for very low overflows (e.g. Chanson & Montes, 1998).

The basic shape of an ogee crest is that of the lower nappe trajectory of a sharp-crested weir flow for the design flow conditions. The definition implies that there is only one optimum ogee design per design flow conditions. The pressures on the crest invert should be atmospheric at design head. Ideally the crest profile should start tangentially to the upstream apron, with a smooth and continuous variation of the radius of curvature (Montes, 1992b). At design flow conditions, the discharge is larger than that for a broad-crested weir with identical upstream head above crest. For large dam heights, the ogee crest design can discharge 28% more than a broad-crest. For discharges less than the design flow rate, the discharge coefficient is less than the design value and tends to unity at low flow rates. For discharges greater than the design flow rate, the pressure on the crest invert becomes sub-atmospheric[2].

Optimum crest design for stepped chutes

The ogee shape was developed for a smooth-invert crest. It is unsuitable when steps are introduced into the ogee development because the presence of the steps modify the flow field next to the invert and the pressures differ from the atmospheric optimum. Experience suggests that a broad-crest design is most suitable for stepped chutes with slopes less than 45º. The first drop may be

[2]E.g. Chow (1959: 363-365), Henderson (1966: 180-182), Chanson (1999: 324-329).

identical to the following steps and the risks of jet deflection in skimming flow are small (Section 5.2.3).

If a circular crest or ogee crest is selected, it should be designed with a smooth-invert development: i.e. the steps must be installed downstream of the circular or ogee profile. The risk of jet deflections at the first steps must be assessed (Section 5.2.3). The kinetic energy of the accelerated flow is important, and the residual energy at the downstream end of the chute might still be considerable. The increase in discharge capacity associated with the circular or ogee shape might not be a real advantage if greater energy dissipation is to take place at the downstream end.

Possibly, a wide crest (i.e. broad-crest) might be preferred for construction ease. With RCC dams (Section 7.3), a wide crest facilitates the truck traffic for the supply of concrete and for the compacting process by rollers. At the Upper Stillwater Dam (Table 1.2), a 9.1 m wide crest was preferred to the original 4.6 m wide crest to ease the construction. In Queensland (Australia), some stepped diversion weirs are designed with provision for an inflatable rubber dam. The crest is designed with a relatively long flat crest inclined upwards to avoid negative pressures (e.g. Bucca Weir; Figs 7.5 & 7.6).

Discussion

A labyrinth crest may be added to increase the spillway discharge capacity with minimum water level rise. Completed in 1948, the Milton Seaman Dam was equipped with a new stepped spillway and a labyrinth crest in 1996. Although the chute is only the 34 m wide ($\alpha = 33.7°$, h = 0.61 m), the labyrinth crest length is 159 m long and designed for $Q_w = 1{,}130$ m^3/s (Toms & Gauthiere, 1997).

7.1.4 *Comparison between flat and pooled steps*

This monograph has analysed primarily flat horizontal steps. The presence of a pool height at the step brink modifies substantially the flow pattern (i.e. streamlines and flow net) next to the edge of the step. This is true in any flow regime (nappe, transition, skimming). It is believed that a pool height, even a very small (but finite) height, perturbs considerably the flow characteristics: the results obtained with flat steps cannot be applied to a pooled step geometry.

Application to nappe flow regime

In a nappe flow regime, the presence of a pool height at the step edge modifies the nappe contraction, the lower nappe free-surface and the free-fall trajectory. Figure 7.2 presents a comparison of the nappe trajectory downstream of a 1. flat step, 2. a finite pool height d_t, and 3. a sharp-crested weir. Further photographic evidence may be found in Rouse (1938, Fig. 163), showing substantial disturbance of the streamlines caused by a tiny pool height. This example

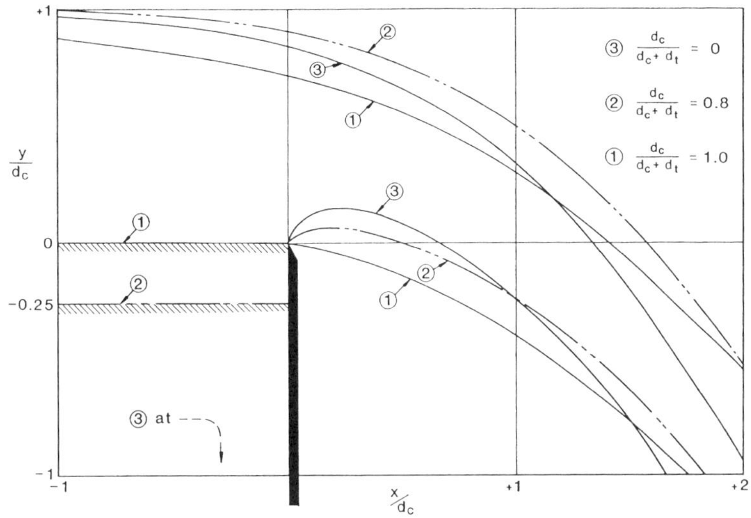

Figure 7.2. Comparison of nappe trajectories between 1. a flat step; 2. a finite pool height; 3. a sharp-crested weir. Ideal fluid flow solution (after Rouse, 1938). Note: d_t = pool height.

highlights that the introduction of pool height is a major hydraulic decision. Pooled-step channels cannot be designed using guidelines developed for flat-step chutes. With a pooled-step geometry, designers are advised to conduct proper hydraulic investigations to provide them with some guidance (Section 8.2).

7.2 DESIGN OF STEPPED FOUNTAINS AND WATER STAIRCASES

7.2.1 *Introduction*

In decorative architecture, waterfalls and water staircases are used to please the eye and the ear, providing a focal point of attention (Fig. 7.3). Cascades enhance the perception of a site by providing an alteration of environmental factors to increase human comfort. In crowded cities, running waters can mask aggressive noise and help to provide a peaceful atmosphere. Water is a significant element in interior landscapes where water provides simultaneously tranquillity and vitality. For landscape designers, stepped fountains and water staircases combine simple architectural forms with a large amount of splashing. A "white water" effect is produced by the large amount of entrained air bubbles and diffraction of the beams of natural light which gives a white appearance to the flow. The stepped geometry is well suited to a steep topography.

Aesthetical applications of stepped cascades include water staircases in parks and gardens, stepped fountains in cities and indoor fountains (Fig. 7.3; Table 7.1). Rushing waters along stepped cascades may reflect some parallel with

Table 7.1. Characteristics of stepped cascades, water staircases and public fountains.

Name	Slope (deg.)	Cascade height (m)	Max. disch. (m^3/s)	Step height (m)	Nb of steps	Type of steps	Remarks
(1)	(2)	(3)	(4)	(5)	(6)	(7)	(8)
Modern public fountains							
Bank of China, Hong Kong	17	–	0.002 to 0.022	0.04	–	Flat steps inclined downwards.	Nappe and skimming flows. Incoming & outcoming discharges. W = 0.25 m. l = 0.13 m.
Central Plaza, Hong Kong	–	–	–	–	–	Flat steps, drops and pooled steps.	Nappe flow regime. Circular fountain.
Chater Garden, Hong Kong	–	–	–	0.5 to 1	3	Pooled steps.	Square fountain. Thin free-falling nappes.
City Riverside Fountain, Brisbane, Australia	23.2	3.45	0.011	0.15	6	Horizontal steps followed by a pool.	Nappe flow regime. W = 2 up to 3 m.
				0.15	17	Horizontal steps.	W = 5.5 up to 7.5(?) m.
The Forum, Central, Hong Kong	–	–	–	–	9	Pooled steps.	
Hong Kong Park, Admiralty, Hong Kong	26.6	1.75	–	0.35	5	Flat steps.	Nappe flow at first steps becoming skimming flow.
Keio Plaza Hotel, Tokyo, Japan	–	–	–	–	6	Pooled steps.	Nappe flow regime.
Peak Tramway fountain, Hong Kong	~ 50	–	–	–	23	Horizontal steps.	Skimming then nappe flow regime. W = 0.8 to 5 m.
Taipei World Trade Centre Fountain, Taiwan	–	2.4	–	0.4(?)	6	Pooled steps.	Succession of overfalls. No hydraulic jump.
Multi-purpose cascades							
Calumet Waterway re-aeration cascades, Chicago, USA	–	3.6 to 4.6	1.6 to 16.4	0.91 or 1.52	3 to 4	Pooled steps.	5 re-oxygenation cascades. Nappe flow regime.
Niigta Prefecture park, Japan	–	–	–	–		Flat steps built from stones.	Stepped river training and vertical drops in an attraction park.
Modern great cascade							
Sydney 2000 Olympic Games opening ceremony	33	36	0.380	0.4	77	Flat horizontal steps followed by a 4.8m high drop.	Cascade surrounding the Olympic Cauldron. Nappe flow regime. Water depth ~ 50 to 60 mm. Made in timber plywood. W = 16 down to 10 m.

mountain rivers or natural cataracts: e.g. the cataracts of the River Nile, the Zambesi rapids in Africa, or the Rhine waterfalls. Stepped cascades can be also designed as multipurpose systems: e.g. for river training, water quality systems (Fig. 1.2). The historical development of water staircases and stepped fountains is discussed in Section 2.5.

7.2.2 *Design procedure*

Most of the attraction is provided by the water splashing and the "white waters". Since Antiquity, architects tried to estimate the most aesthetical fountain by trial-and-error, empirical correlations or personal experience. Recent developments in stepped chutes and cascades (Chapters 3 to 5) provide new information on their hydraulic characteristics, in particular on the apparition of "white waters". This is a result of the entrainment of multiple air bubbles by turbulent self-aeration at the free surface and by hydraulic jump.

In designing a new cascade, the first step is selecting the flow regime: nappe, transition or skimming flow. For cascading waterfalls (nappe flow) the water depth over the upper step should be sized to provide no more than one-third of

A.

B.

Figure 7.3 Examples of stepped cascades. A. Chater Garden, Admiralty, Hong Kong in 1993. Thin free-falling nappes and intermediary pools. B. Brisbane riverside, Australia in 1993. Horizontal steps.

C.

D.

Figure 7.3 Examples of stepped cascades. C. Examples of stepped cascades. Indoor cascade (with pooled steps). Keio Plaza Hotel, Tokyo, Japan in 1993. D. Bank of China, Hong Kong in 1994. Inclined downward steps.

E.

F.

G.

Figure 7.3 Examples of stepped cascades. E. Detail of the flow in the Bank of China fountain in 1994. Note the nappe flow in the left channel and skimming flow in the right channel. F. Central Plaza, Hong Kong in 1994. Combination of steps and drops. G. Taipei World Trade Centre, Taiwan in 1994. Horizontal steps.

H.

I.

J.

Figure 7.3 Examples of stepped cascades. H. Taipei World Trade Centre, Taiwan in 1994. Note the skimming flow at upstream end and nappe flow regime at downstream end a ~ 27°, l ~ 0.7 m. I. Hong Kong Park, Pacific Plaza in 1994. J. Franklin Roosevelt Memorial, Washington DC in June, 1998. Pooled step design.

the step height (i.e. Table 3.4). For a turbulent flow (skimming flow), the water depth over the crest must be larger than the step height (Section 1.3). Special effects might be obtained also by combining two flow regimes or using the transition flow regime: e.g. the Bank of China and the Taipei World Trade Centre fountains.

In a *nappe flow* regime, most splashing occurs near the impact of the falling nappe on the step and possibly at the downstream hydraulic jump: e.g. the Sydney 2000 Olympic Games opening ceremony cascade (Table 7.1). The main design parameters are: the location of the free-falling jet impact L_d and the length of the hydraulic jump L_r (Section 3.3). These parameters are deduced as a function of the critical flow depth (i.e. $d_c = \sqrt[3]{q_w^2/g}$) and the step height h (Chapter 3).

In a *skimming flow* regime, the upstream flow is non-aerated and translucent. When the turbulence generated by the bottom boundary layer reaches the free-surface, air is continuously entrained through the air-water interface and the free-surface becomes "white". The distance L_I between the start of the stepped cascade and the apparition of the "white waters" is deduced from Equation 5.7.

$$\frac{L_I}{h} = 9.719 * (\sin\alpha)^{-0.277} * (\cos\alpha)^{-0.0695} * \left(\frac{d_c}{h}\right)^{1.0695} \tag{5.7}$$

where α is the slope of the cascade ($\tan\alpha = h/l$).

7.2.3 *Practical considerations*

Crocker (1987) indicated that waterfalls and cascades required a minimum discharge of about 0.006 m^2/s (i.e. $d_c > 15$ mm) to give a reasonable effect. In case of high wind speeds, he suggested to use a minimum discharge of 0.019 m^2/s (i.e. $d_c > 33$ mm). Erbe (1974) advised to use waterfall discharges less than 0.17 m^2/s to reduce the depth of the pools (i.e. $d_c < 143$ mm).

For stepped cascades, the first step must be carefully designed to avoid water jumping off the step. Designers may consider any combination of drops and steps (e.g. Central Plaza, Hong Kong; Figure 7.3). They may choose variable channel widths and/or water discharges to display both nappe and skimming flows: e.g. variation of discharge at the Bank of China Cascade or a change of channel width at the Taipei World Trade Centre Fountain (Fig. 7.3).

With a thin overfalling nappe, nappe oscillations might develop and create air movements if the free-falling nappes are not adequately ventilated. In such cases, the sheet of water oscillates back and forth at a frequency of several Hertz. The danger to the structure is small but the associated noise might be deafening and is always unwelcome (Casperson, 1993). If the nappe cannot be aerated, splitters can be used to control the oscillations of the nappe. A maximum spacing between splitters of about two-thirds of the fall height (i.e. 2/3*h) is advised.

7.3 CONCRETE STEPPED SPILLWAYS AND WEIRS

7.3.1 *Introduction*

Stepped spillways are used to assist in energy dissipation and to help in reducing the size and the cost of the downstream stilling basin. Typical examples are presented in Table 1.2, and Figures 1.2 and 7.5 to 7.10. The comparison of the rate of energy dissipation between flow regimes (Section 5.4) has shown that:

a) for short stepped chute, a nappe flow with fully-developed hydraulic jump (sub-regime NA1) enables more energy dissipation;
b) with both nappe and skimming flows, the rate of energy dissipation decreases with increasing flow rate for a constant dam height;
c) nappe flow situations tend to provide a greater energy dissipation but require larger step heights and shallower slopes.

Hence the maximum discharge capacity, the length of the channel (and the dam height) and chute slope will affect the preferential flow regime for optimum operation.

7.3.2 *Design procedure*

A typical design procedure must include the following steps:

1. Assessment of the design flood.

The construction of a dam across a stream will modify the upstream and downstream flow conditions. The dam crest elevation must be selected carefully to provide the required storage of water or upstream water level rise. The spillway and stilling basin must operate safely for a wide range of flow rates and tailwater flow conditions.

2. Assessment of the chute geometry.

The main characteristics are the channel width and slope, and the chute height. The spillway slope might be a design variable in some cases. A steep slope can minimise the volume of an overflow structure but greater energy dissipation is obtained with nappe flows on flat slopes.

3. Selection of the optimum step height.

The step height is chosen as a function of the discharge, channel slope and possibly construction technique to enable the selected flow regime (nappe or skimming flow usually), the appropriate level of energy dissipation (Chapters 3 to 5), and an acceptable operation at maximum discharge capacity.

4. Selection of the step length.

If the channel slope is not fixed by the topography or by the method of construction, the step length and the channel slope must be selected to provide the optimum flow conditions and/or energy dissipation characteristics.

5. Calculation of the hydraulic characteristics.

For the selected step geometry, the flow depth, velocity, the amount of air entrainment and the energy dissipation are calculated using the information provided in Chapter 3 (nappe flow), Chapter 4 (transition flow) or Chapter 5 (skimming flow). In particular, the effects of free-surface aeration must be taken into account for the design of the chute sidewalls and for the rate of energy dissipation.

6. Calculation of the downstream dissolved gas contents.

For the climatic conditions of the site, and the chemical properties of the water, the aeration efficiency of the structure may be estimated (Chapter 6). The most usual parameters are the dissolved oxygen and nitrogen gas contents (DOC and DNC). The prediction of the downstream dissolved gas contents must be performed for a wide range of discharges and atmospheric conditions.

Maximum discharge capacity

Modern stepped spillways have been designed with maximum discharge capacity less than 20 to 30 m^2/s (Table 1.2) although there are few exceptions. For example, Bucca (55 m^2/s, h = 0.6 m), La Grande 2 (132 m^2/s, h = 9 to 12 m). Some ancient designs were designed and have operated with large discharges. Classical examples include Mestalla Weir (A.D. 960, $q_w = 85\ m^2/s$, h = 0.35 to 1 m), Bilberry Dam (1867, $q_w = 16\ m^2/s$), Quinson Dam (1870, $q_w = 30\ m^2/s$, h = 0.75 m), Pas-du-Riot Dam (1873, $q_w = 22\ m^2/s$, h = 2 to 3 m), Cantref Dam (1892, $q_w = 17\ m^2/s$, h = 0.305 m) (Tables 2.1 & 2.6, Fig. 2.12). More recently, prototype experience with large overflows have included the Dneiper Hydroplant tests in 1976 ($q_w \sim 60\ m^2/s$, V = 23 m/s, h = 0.405 m, precast block system) (Fig. 8.9), the Cunningham Weir overflow in 1956 ($q_w = 28.6\ m^2/s$, h = 0.9 m, timber crib structure), the Kerrville Dam spill in 1986 ($q_w = 20\ m^2/s$, h = 0.3 m, unprotected RCC construction) with flood flow carrying logs, trees and debris (Table 8.4). An interesting case is the Shuidong Dam (RCC) in China. For three days (1-3 May, 1994), a major overflow took place with q_w up to 90 m^2/s without damage.

Design guidelines

In a *nappe flow*, a large flow rate is only achievable on high steps. The maximum flow capacity must be associated with a maximum acceptable step height h. The experience with unlined rock spillway (Section 7.7) suggests that a maximum acceptable step height might be less than 12 to 15 m corresponding to an impact velocity in excess of 17 m/s.

In *skimming flow*, a maximum flow rate is related to a corresponding step height. A detailed review of large roughness flows demonstrated that the flow resistance decreases with decreasing relative roughness height (Chanson, 1995a: 92-96). If the relative roughness $h^*\cos\alpha/D_H$ is small, the flow resistance is significantly reduced and, at the limit, the steps become a negligible roughness. A well-known example is the Moody Diagram expressing the rough-turbulent friction factor in pipes. It shows a drastic reduction in friction factor

Table 7.2. Maximum flow capacity: design guidelines.

Flow conditions (1)	Guideline (2)	Remarks (3)
Nappe flow	h < 15 m	Based upon inclined rock cascades.
Skimming flow	$\frac{d_c}{h} < 0.1193 * \cos\alpha * (\sin\alpha)^{0.259} * \left(\frac{L}{h * \cos\alpha}\right)^{0.935}$	For fully-developed flow conditions at the downstream end.
	AND	
	$\frac{d_c}{h} < 15 * \cos\alpha$	Based upon uniform equilibrium flow conditions.

Note: L: chute length.

for $k_s/D_H \leq 0.03$ to 0.05 that is consistent with the original experiments of Nikuradse (1933)[3]. In uniform equilibrium flow, such a limit yields $d_c/(h*\cos\alpha) \leq 15$ to 30 for skimming flows. A further limit exists for skimming flows: i.e. fully-developed flow conditions must be achieved before the downstream end of the chute. It may be expressed as a function of the chute length and channel slope:

$$\frac{d_c}{h} < 0.1193 * \cos\alpha' \ (\sin\alpha)^{0.259} * \left(\frac{L}{h * \cos\alpha}\right)^{0.935} \tag{7.5}$$

Design guidelines are summarised in Table 7.2. In a nappe flow regime, the guideline is based upon a maximum acceptable step height. For skimming flows, the flow must be fully developed before the downstream end of the chute, and the step roughness must be large enough.

Nappe aeration

In a nappe flow regime, the ventilation of long free-falling nappes is essential to avoid unsteady fluctuations resulting from nappe oscillations (Section 3.2.3 and Chapter 8). This precaution is particularly important for wide rectangular channels with large step heights at low flow rates.

In general, the most convenient method to ventilate the nappe is the use of sidewalls deflectors which provide space between the sidewalls and the deflected nappe. The design of these deflectors is somewhat similar to the design of aeration devices (Chanson, 1989a; Wood, 1991). It is possible also to aerate the nappe by small ducts feeding air from beyond the sidewalls to near the top of each drop. If these techniques are not feasible, splitters may be used to break the sheet of free-falling water, allowing space for aeration and preventing nappe fluctuations (Section 3.2.3).

[3]Also Schlichting (1979: 656), Haugen & Dhanak (1966), Sams (1952).

Remarks
In some cases, the selection of the step height and channel slope might be determined by external factors rather than hydraulic considerations. Step heights larger than 2 m might be selected to prevent people climbing up and down the stepped chute. Steep slopes (i.e. $\alpha > 45°$) or large step heights (i.e. $h > 2$ m) might be considered also to prevent motorcycles driving up and down the steps, as observed at the Brushes Clough Reservoir (Baker & Gardiner, 1994) (Fig. 7.9). In one particular instance, one spillway step of a diversion weir ($h \sim 2$ m, $l \sim 6$ m) was designed as a roadway for local farmers. It is used regularly during the dry periods. The roadway step must be wide (l-3 m) and adequately reinforced.

7.3.3 *Roller compacted concrete (RCC) dams*

Roller compacted concrete (RCC) is defined as a no-slump consistency concrete that is placed in horizontal lifts and compacted by vibratory rollers. Since the late 1970s, RCC has become a popular material for the construction of gravity dams (Tables 1.2 & 7.3). The primary advantages of RCC gravity dams are cost effectiveness, reduced foundation surface area compared to earthfill or rockfill structures of same height, and the short duration of construction works.

The low cost of RCC gravity dams results from the small material volumes, using vertical upstream and steep downstream faces, compared with earthfill or rockfill structures, the lower cost per unit volume than conventional concrete[4], the construction technique (little formwork, placement with standard earthwork techniques), and the reduced cost of auxiliary structures (intakes attached to the vertical face, stepped spillway on the downstream face) in comparison with rockfill and earthfill embankments. Another application of RCC is the protection of the downstream slope of embankment dams (Section 7.4).

Discussion
Figures 7.5 and 7.6 show construction details of a RCC diversion dam (Bucca Weir, Australia). The structure was built between June 1986 and May 1987. It was made of 0.3 m thick RCC overlays with conventional concrete blocks protecting the downstream face. Figure 7.6C shows the dam after completion. Examples of construction techniques are summarised in Table 7.3 and further photographs of RCC structures are shown in Figures 1.2 and 7.6.

During the construction of RCC dam with overflow stepped spillway, the steps can be made of un-formed exposed RCC, formed exposed RCC, formed conventional cast-in-place concrete, pre-cast concrete facing elements or grout enriched RCC (Ditchey & Campbell, 2000). Examples are shown in Figure

[4]Ditchey & Campbell (2000) estimated that RCC is about 25 to 30% cheaper than conventional concrete.

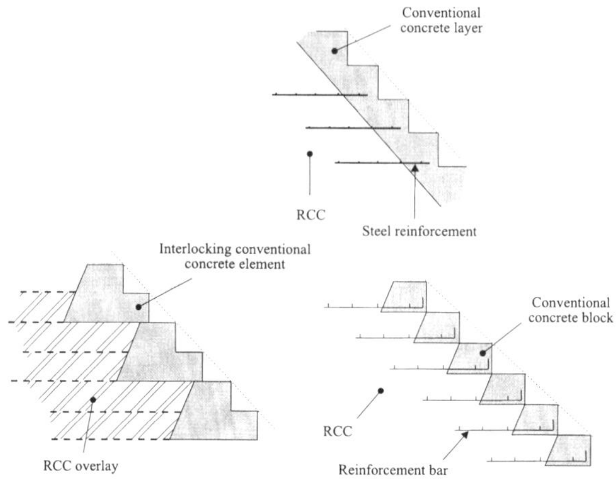

Figure 7.4. Details of construction of steps on RCC dams.

7.4. Pre-cast concrete facing elements tend to be the preferred option in terms of costs and durability (Table 7.3). The construction technique is well-suited to cold climate and frequent overflows. In practice, the joints between RCC and a conventional concrete protection layer must be made with great care during construction. Model tests (Frizell & Mefford, 1991; Lejeune & Lejeune, 1994) showed that the vertical face of the steps can be subjected to sub-atmospheric pressures, and it is essential to incorporate into the design a drainage system below the erosion-resistant layer. Drains may be located behind the vertical step faces. Exposed RCC has a lower durability and it is more suitable to secondary spillway designs with infrequent operation.

RCC strength results were obtained at the New Victoria Dam (Table 7.3, column 6). At 3 years, the test results showed a 100% increase of the RCC strength obtained at 180 days (Wark et al., 1998). Note that most masonry and concrete stepped spillways built prior to 1920 were reinforced with cut stones or granite blocks protecting the horizontal face of the steps (e.g. New Croton Dam, Pedlar River Dam) (Chapter 2). At that time, the strength of concrete was much lower than today[5]. Cut stones and granite blocks were placed to protect the steps in a similar purpose as a conventional concrete protection layer on modern RCC dams.

A review of lessons learnt from RCC dam construction highlighted two problems: uncontrolled seepage and cracks. The former may be minimised by the installation of a geotextile membrane. The latter problem can be minimised by using special cements and ice (Bass, 1998).

[5]Prior to 1900 most concretes had a maximum allowable stress of about 0.5 to 1 MPa comparable to low cement content RCC.

Table 7.3. Examples of stepped spillway construction for RCC dams.

Dam (1)	Ref. (2)	Slope deg. (3)	Step height m (4)	Revetment layer (5)	Comments (6)
RCC dam with protective layer					
De Mist Kraal, South Africa, 1986	(HD)	59	1	Conventional concrete.	RCC dam (117 kg cement/m^3). RCC overlay: 0.25 m thick.
Bucca Weir, Australia, 1987	(QW)	63.4	0.6	Conventional concrete placed before RCC and drilled by drainage holes.	RCC weir (90 kg cement & flyash/m^3). RCC overlay: 0.3 m thick. σ_c (90 days) = 20 MPa. Steps inclined upwards (δ=+2.9°).
Les Olivettes, France, 1987	(B)	53.1	0.6	Concrete (350 kg cement/m^3) with steel anchorage.	RCC dam (130 kg cement/m^3). RCC overlay: 0.3 m.
Riou, France, 1990	(HC, B)	59	0.43	Precast concrete blocks.	RCC dam (120 kg cement/m^3). σ_c (90 days) = 6 MPa.
New Victoria Dam, Australia, 1991	(Wa1)	72 & 51.3	0.6	Slip-formed interlocking facing elements made of conventional concrete.	RCC dam (225 kg cement & flyash/m^3). RCC overlay: 0.3 m thick. σ_c = 12 MPa (28 days), 25 MPa (180 days), 48 MPa (3 years), 53 MPa (5 years).
Burton Gorge Dam, Australia, 1992	(Fo)	51.3	1.2	Precast concrete blocks.	RCC dam (85 kg cement/m^3, no flyash). H_{dam} = 24 m. σ_c = 8 MPa (90 days).
North Dandalup Dam, Australia, 1994	(Wa2)	56	0.6	Conventional concrete.	RCC spillway crest (155 kg flyash/m^3). RCC overlay: 1.2 m thick. σ_c = 26 MPa
Shuidong Dam, China, 1994	(HZ)	57.0	0.9	Precast interlocking concrete blocks.	RCC dam (52 kg cement /m^3, 88 kg flyash/m^3). σ_c (90 days) = 12.4 MPa.
Loyalty Road Dam, Australia, 1996	(Fo)	51.3	0.9	Precast concrete panels and conventional concrete.	RCC dam (90 kg cement/m^3). H_{dam} = 30 m. Design flow: 1,040 m^3/s.
Cadiagullong, Australia, 1997		53.1	0.6	Conventional concrete.	RCC dam H_{dam} = 44 m. Steps inclined upwards (δ=+1.9°). Design flow: 4.46 m^2/s.
Masonry dam reinforcement					
Santa Cruz Arch Dam				Unprotected RCC overtopping protection and concrete buttresses over existing concrete arch dam (completed in 1929).	H_{dam} = 46 m, W = 22.9 m. Design flow: 56 m^3/s.
Camp Dyer Diversion Dam, USA, 1992	(DF)			Unprotected RCC overtopping protection and RCC buttresses over existing masonry gravity dam (completed in 1926).	H_{dam} = 24 m, W = 265.8 m. Air-entrained RCC (134 kg cement & pozzolan/m^3, 2.5% air content). Overlays: 0.3 m thick. σ_c (90 days) = 13.8 MPa. Design flow: 0.77 m^2/s.

Notes: H_{dam}: dam or weir height; s_c: compressive strength.
References: (B) Bouygue et al. (1988); (DF) Dolen & von Fay (1993); (Fo) Forbes (1995); (HC) Present study; (HD) Hollingworth & Druyts (1986); (HZ) He & Zeng (1995); (QW) Queensland Water Resources (1988); (Wa1) Wark et al. (1991,1992); (Wa2) Wark et al. (1998).

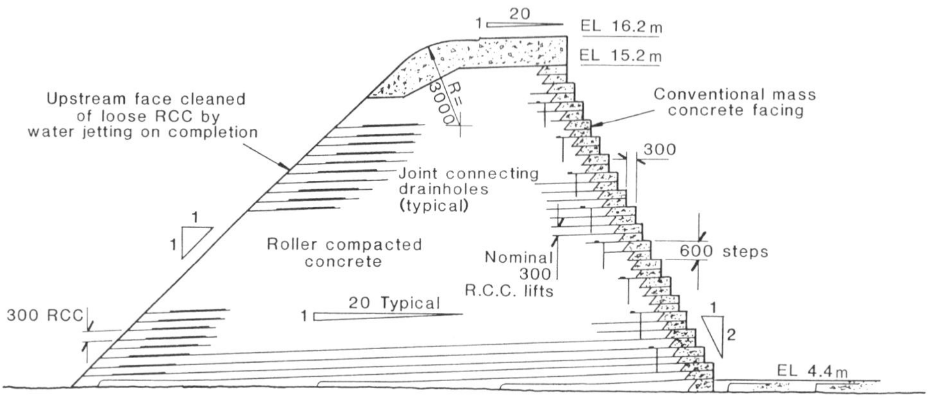

Figure 7.5. Cross-section of Bucca Weir spillway (Australia). Note the downstream stepped face (h = 0.6 m, h/l = 2) and the concrete protection layer.

Figure 7.6. RCC dams with stepped spillways. A. Construction of the Bucca Weir on the Kolan River (Australia) (courtesy of Mr. Beitz, Queensland Department of Natural Resources). α = 63.4°, h = 0.6 m, W = 130.8 m, H_{dam} = 11.8 m. View from downstream (3 January 1987).

B.

C.

Figure 7.6. RCC dams with stepped spillways. B. Bucca Weir on 13 March 1987. View from downstream: note the completed portion of the crest on the left. C. Bucca Weir after completion on 27 May 1987. Note the concrete apron downstream of the spillway toe.

D.

E.

Figure 7.6. RCC dams with stepped spillways. D. Camp Dyer Diversion Dam (courtesy of the US Bureau of Reclamation). Note few smaller steps downstream of the broad-crest. E. Loyalty Road Flood Retarding Dam (courtesy of Patrick James). Note the ogee crest followed by few smaller steps.

7.4 OVERFLOW EMBANKMENTS: (1) CONCRETE PROTECTION

7.4.1 *Presentation*

In recent years, the design floods of a number of dams were re-evaluated and the revised flows were often larger than those used for the original designs. In many cases, occurrence of the revised design floods would result in dam overtopping because of the insufficient storage and spillway capacity of the existing reservoir. A number of overtopping protection systems were developed (e.g. Asce, 1994). Concrete overtopping protection systems are presented in this section (Section 7.4.2).

7.4.2 *Concrete overtopping protection*

Concrete overtopping protection allows an increase of spillway capacity (Fig. 7.7). In North America, a number of dam overtopping rehabilitations were conducted primarily on embankment structures with dam heights ranging from 4.6 to 33.5 m. It is believed that the first ones were the Ocoee No. 2 Timber Crib Dam in 1980 and the Brownwood Country Club Earth Dam in 1984. Various construction techniques were used (Table 7.4). Current trends favour the use of roller compacted concrete (RCC)[6].

Roller compacted concrete is placed in a succession of overlays of 0.2 to 0.4 m thickness and with a width greater than 2.5 m for proper hauling, spreading and compacting. Exposed RCC is frequently used for secondary spillways with infrequent spills of less than 5 m^2/s. Experience of overflows over unprotected RCC faces is described in Table 8.4. In colder climates, unprotected RCC surfaces can experience fracturing or chipping during freeze-and-thaw cycles. Some binders enable to entrain air in roller compacted concretes and 1.5 to 2% air content is sufficient to protect concrete from frost deterioration (Delagrave et al., 1994). Alternatively, a conventional concrete protection overlay may be applied after the RCC or at the completion of construction works to protect the RCC.

With both RCC and conventional concrete protection, a drainage system beneath the concrete layers is essential to prevent uplift. Its purpose is to relieve pore pressure at the interface between the embankment and RCC overlays. In some cases, the drainage installation may be replaced or supplemented with drain holes formed through the RCC during placement. At the downstream end a cutoff wall must be built to prevent undermining of the concrete layer during overtopping.

A case study: Melton Dam

The Melton Dam[7] is an embankment dam located 40 km west of the city of Melbourne (Australia). In 1994, the reservoir was equipped with an overtopping protection system acting as a secondary spillway. It was designed to assist in passing the PMF flood with overtopping of the entire dam (crest length:

[6]RCC dam rehabilitation accounts for nearly two-thirds of the RCC dam construction in USA in the, 1990s.

[7]Completed in 1916, the dam was heightened twice because of heavy siltation during the, 1930s and, 1940s caused by cattle and sheep grazing, gold and coal mining and rabbit colonies in the catchment (Chanson, 1998b).

Table 7.4. Examples of concrete protection on the downstream slope of embankment dams.

Dam (1)	Ref. (2)	Slope deg. (3)	Step height m (4)	Revetment layer (5)	Comments (6)
RCC without protective layer					
Brownwood Country Club Dam, USA, 1984	(MH)	26.6	0.23	Unprotected RCC over existing embankment.	Embankment dam (H_{dam} = 6 m). 0.23m thick RCC overlay (184 kg cement & flyash/m^3).
White Cloud Dam, USA, 1990	(MH)	21.8	0.15	Unprotected RCC over existing embankment.	Embankment dam (H_{dam} = 4.6 m). 0.15m thick RCC overlay (148 kg cement/m^3)
Horsethief Dam, USA, 1992	(MH)	26.6	0.30	Unprotected RCC over existing embankment.	Embankment dam (H_{dam} =, 19.8 m). 0.305m thick RCC overlay (193 kg cement/m^3)
Tongue River Dam, USA, 1940-1999	(MY)	17.8	0.66	Unprotected RCC. Secondary spillway over existing embankment.	1940 embankment dam (H_{dam} = 22 m). W =, 198 m. Design discharge: 8.6 m^2/s.
Conventional concrete protection					
Melton Dam, Australia, 1994		26.5	0.6	Secondary spillway made of conventional concrete built over existing embankment.	1916 embankment dam heightened in 1937 and, 1967 (H_{dam} = 35 m). α = 26.6° (but α=32° for upper 4m), W = 181 m. Design discharge: 2,800 m^3/s. σ_c = 40 MPa.

Notes: H_{dam}: dam or weir height; σ_c: compressive strength.
References: (MH) McLean & Hansen (1993); (MY) McDonald & Yadon (1998).

181 m). The downstream face of the dam was protected with mass concrete laid in a stepped fashion (h = 0.6 m)[8] (Fig. 7.7).

Two unusual features of the design were the selection of conventional concrete and the large overflow capacity. Although the design specifications called for a RCC construction, the cheapest tender proposed conventional concrete. It was suggested that the supplier preferred a stronger concrete resistance to allow access and passage for heavy equipment. The proximity of Melbourne CBD might have also influenced the low plain-concrete cost. The overflow chute (or secondary spillway) is designed to pass 2,800 m^3/s at PMF with the reservoir level being 4.4 m above the dam crest. At PMF the primary spillway on the left bank will pass 1,700 m^3/s.

7.4.3 *Alternatives for embankment stepped overflow*

Alternative overtopping protection systems include timber cribs, sheet-piles, riprap and gabions, and reinforced earth (Fig. 7.8). Timber crib overflows were

[8]The concrete step faces are irregularly shaped to look like cut-stones. Note that the staircase shape was adopted for ease of construction primarily.

A.

B.

Figure 7.7. Melton Dam (Australia, 1916) on 30 January 2000. Secondary spillway over existing embankment: design flow: 2,800 m^3/s, W = 181 m (at crest), h = 0.6 m, $\alpha = 26.5°$ but upper 4m. A. View from the right bank. B. Concrete step details. The concrete step faces are irregularly shaped to look like cut-stones.

used in Russia since the 18th century (Chapter 2; Fig. 2.2). Some recent structures are still in use (Section 7.8). A number of weirs were designed with steel sheet-piles and concrete slabs in Russia and Australia (Fig. 1.2). An experimental structure was built with a reinforced-earth stepped overflow (Fig. 7.8A). A related design consists of pre-cast concrete panels set on very coarse concrete acting as drainage layer (Fig. 7.8B). Section 7.5 describes the pre-cast concrete block overflow system developed in Russia, followed by a discussion of overflow systems made with rockfill and gabions in Section 7.6.

A.

B.

Figure 7.8. Alternatives for embankment dam stepped overflows. A. Jordan II, Gatton QLD (Australia, 1992) on 22 February, 1998. H_{dam} = 5.3 m, h = 1.4 m, α = 17.7°, earthfill embankment with a rockfill upstream protection and reinforced earth downstream slope construction. View from downstream. B. Stepped weir made of pre-cast concrete units over an embankment. Storm water system around Robina Shopping Centre (Australia, 1996) (courtesy of Ross Tomkins). α = 27°, h = 0.5 m. Small overflow on 30 July, 1996 (Q_w = 70 L/s).

7.5 OVERFLOW EMBANKMENTS: (2) EARTH DAM SPILLWAY WITH PRECAST CONCRETE BLOCKS

7.5.1 *Introduction*

Soviet engineers were among the first to propose a stepped concrete chute design on the downstream face of embankment dams under the leadership of P.I. Gordienko (e.g. Gordienko, 1978) (Table 1.2). The choice of a stepped structure allows the use of individual blocks interlocked with the next elements, and the design assists in the energy dissipation (Pravdivets & Bramley, 1989) (Fig. 7.9). For new dams, a stepped spillway made of concrete blocks may be considered as the primary flood release structure of the embankment (Miller et al., 1987). The design concept was more recently tested in USA and UK, although it did not prove cost-effective. An interesting feature of the concrete block system is the flexibility of the stepped channel bed allowing differential settlements of the embankment. Individual blocks do not need to be connected to adjacent blocks. A further advantage is the short construction time on site.

7.5.2 *Protective layer*

For an earth dam with overflow spillway, the uppermost important criterion is the stability of the dam material. Seepage may occur in saturated embankment and the resulting uplift pressures might damage or destroy the stepped channel and the dam (Section 8.2). An adequate drainage is therefore essential. In a typical design, the blocks lay on a filter and erosion protection layer (Fig. 7.10). The layer has the functions to filter the seepage flow out of the subsoil and to protect the subsoil layer from erosion by flow in the drainage layer. Furthermore, the protection layer reduces or eliminates the uplift pressures acting on the concrete blocks. A geotextile membrane is normally laid on the embankment before the placing of the layer, and another covers the protection layer before the installation of the blocks.

Suction of the fluid from underneath the concrete steps can be produced by the pressure differential created by the high velocity flow over the vertical face of the step. Drains placed in areas of sub-atmospheric pressure will function to relieve uplift pressures (Fig. 7.13B). In any case, the location of the embankment drains must be appropriately selected to avoid reverse flow in the drains and dynamic pressures associated with hydraulic jumps (in case of nappe flow regime). Grinchuk et al. (1977) recommended that the total area of the drainage holes should be 10-15% of the exposed step area. More recently, Baker et al. (1994) suggested that an open area of 2-5% could be optimum. Frizell (1992) studied flows past horizontal steps and sloped-downward steps with $\delta = -10^{\circ}$ and -15°. Her model tests suggested that the aspiration (on the vertical face of the steps) increased with the downward slope of the steps. The seepage flow in the embankment dam must be predicted accurately to make the appropriate provision for drainage and evacuation of seepage flow through the blocks. The seepage flow may be influenced by the infiltration into the downstream slope caused by the spillway flow, in addition to the flow through the embankment.

A.

B.

Figure 7.9. Earth dams. Stepped spillways with precast concrete blocks. A. Wedge concrete block spillway of the Brushes Clough Dam, UK (1859/1991) (courtesy of Mr. Gardiner, NWW). α = 18.4°, h = 0.19 m, inclined steps (δ = -5.6°), trapezoidal cross-section (2.2m bottom width, 1V:2H sideslope). Low flow spill in July, 1993 (nappe flow regime). B. Details of the drainage holes (127×29mm) at Brushes Clough Dam (courtesy of Mr. Gardiner, NWW). Each concrete block weights about 120 kg.

C.

D.

Figure 7.9. Earth dams. Stepped spillways with precast concrete blocks. C. Sosnovsky Farm Dam (Russia, 1980) (courtesy of Prof. Y. Pravdivets). H_{dam} = 11 m, design flow: 3.3 m^2/s, S_o = 0.167, W = 12 m, overlapping precast concrete blocks (1.5 m × 3 m × 0.16 m). D. Volymia Dam spillway (courtesy of Prof. Y. Pravdivets). Experimental earth dam in the Magadan region, Siberia. H_{dam} = 20 m, α = 26.6°, chute ending with a ski jump.

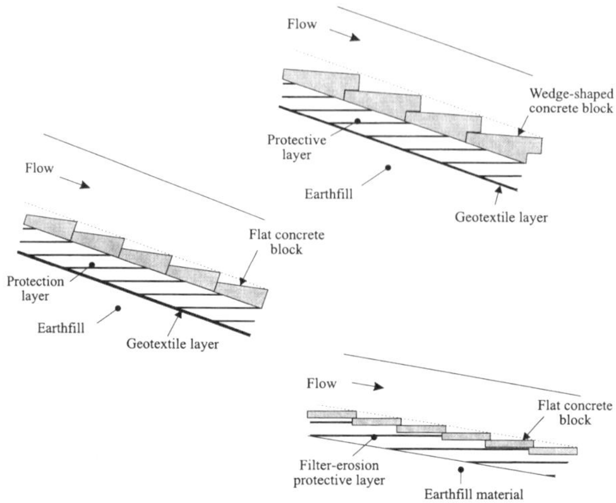

Figure 7.10. Details of concrete blocks on earth dam spillway. (Top) flat concrete blocks (rectangular cross-section). (Middle) flat concrete blocks of trapezoidal cross-section (non-overlapping block system). (Bottom) wedge-shaped concrete blocks (overlapping block system).

7.5.3 *Block geometry and configuration*

Soviet engineers developed a strong expertise in the design of concrete wedge blocks. Pravdivets & Bramley (1989) described several configurations (Table 7.5). Blocks are normally laid parallel to the slope on the top of the protective layer (Fig. 7.10). Individual blocks do not need to be connected to adjacent blocks. The channel bed is very flexible and allows differential settlements. For large discharges, the blocks may be tied to adjacent blocks and they would be made normally of reinforced concrete. Baker et al. (1994) recommended a step height-to-length ratio in the range 1:4 to 1:6 to ensure maximum stability of the blocks.

A related design option is the use of reinforced grass waterways (Collett, 1975; Hewlett et al., 1987). Hewlett et al. (1987) tested a variety of reinforcements: textile, nylon and concrete blocks. Their results indicated that grass waterways, reinforced with concrete blocks, could sustain velocities up to 4 to 8 m/s without damage for a 3 hour test (Table 7.5). Several grass species can be used: e.g. bent, couch, fescue, kikuyu, meadow grass, perennial ryegrass. In any case, Collett (1975) advised continual operation for no longer than 7 days, otherwise the grass could be adversely affected[9].

[9]Vegetation cover may provide limited erosion resistance on embankment dam slopes. Prototype overtoppings ($q_w < 1$ m²/s) of short durations (< 1 to 2 hours) took place without major damage (Asce, 1994: 96-97).

Table 7.5. Design and stability of concrete blocks for embankment dam spillway

References (1)	Max. disch. (m^2/s) (2)	Slope degrees (5)	Block shape (4)	Block thickness m (5)	Comments (6)
Grinchuk et al. (1977)	up to 60 (V ? 23 m/s)	8.7	Wedge-shape reinforced concrete block.	0.7	Dneiper hydro plant. Horizontal steps.
Pravdivets & Bramley (1989)	2 to 27	7 to 15	Wedge-shaped block construction.	0.15 to 0.45	Inclined steps downward.
Baker (1990)		21.8	Wedge-shaped blocks overlapping and non-overlapping (? =. 8.3°).		Model tests. No failure reported for: V ≤ 6.5 m/s.
<u>Reinforced grass water way</u>					
Collett (1975)	V < 2.2 to 2.7 m/s		Grass reinforced with wire		Reinforced grass water way.
Hewlett et al. (1987)	V < 4 to 7 m/s	21.8	Interlocking non-tied concrete blocks.		Reinforced grass water way. 3-hour tests.
	V < 7 to 8 m/s	21.8	Interlocking tied concrete blocks.		

7.5.4 *Design considerations*

There are two fundamental design rules for precast concrete block systems: a skimming flow in a straight prismatic chute. The step block system was developed for a skimming flow regime: maximum block stability can only be achieved in skimming flows (e.g. Baker, 2000). All but one Russian applications were designed for relatively small discharge capacity: $q_w \leq 3$ m²/s. Pravdivets & Bramley (1989) urged that "the alignment of the spillway should be straight from the crest to the toe. Any curvature of the spillway in plan, or change in cross-section, will cause an uneven distribution of flows within the spillway which, in general, should be avoided".

Usually, the channel sidewalls are flat inclined slopes (i.e. trapezoidal spillway cross-section). The slopes of the sidewalls can be designed as inclined stepped surfaces (in the flow direction) and may use the same concrete block system as the main channel. Pravdivets & Bramley (1989) suggested typical sidewall slopes of about 1V:3H (i.e. 18°). A known construction weakness is the joint between the chute invert and the sidewalls. At Brushes Clough, two longitudinal concrete guides were built to facilitate the installation of the blocks and the connection with the stone-pitched sidewalls (Fig. 7.9). At the downstream end, the residual energy of the flow must be dissipated with a small flip bucket arrangement and a conventional concrete pool. Laboratory tests showed high risks of block uplift and failure under a hydraulic jump (Baker, 2000).

The Russian experience with overflow earth dam spillways (α = 9.4 to 26.6°) showed the potential of the concept and highlighted that the quality of the drainage layer is uppermost in importance. Pravdivets (1992) and Baker et al. (1994) related failure cases resulting from improper drainage revetment (Section 8.2). If the drainage requirements are fulfilled, the stepped spillway can sustain large floods[10] and discharges even ice debris. In Siberia, the Magadan Experimental Dam (Table 1.2) has resisted for over 15 years without accident (Fig. 7.9D). Additional information on design criteria can be found in Baker et al. (1994) and Baker (2000).

Remarks
In one case (Brushes Clough Dam spillway) numerous acts of vandalism were reported, including destruction of concrete blocks. In practice it is recommended to use concrete blocks heavy enough to avoid their displacement by individuals and to enhance their strength against acts of vandalism. At the very least, another construction technique (e.g. RCC overlays) should be selected if man-made destruction cannot be prevented. In Russia, the cheapest construction method uses precast concrete slabs which are made for the building industry. The slabs are rectangular (3 m long, 1.5 m wide and 0.160 m thick) and they are installed in an overlapping arrangement with mild steel spacers (Fig. 7.9C). A long-term effect of this construction method would be the corrosion of the spacers.

7.6 GABION STEPPED WEIRS AND CHUTES

7.6.1 *Introduction*

Gabions are extensively used for earth retaining structures and in hydraulic structures (e.g. weirs, channel linings). Their advantages are: 1. their stability, 2. low cost, 3. flexibility and 4. porosity. The porosity of gabions is an important characteristic preventing the building-up of large uplift pressures. Typical design of gabion weirs are shown on Figure 7.11. Some existing structures are presented in Figures 1.2, 3.6 and 7.12.

Originally a gabion[11] was a basket filled with earth or stone for use in fortification and engineering. The technique has been employed for thousands of years. The original gabions used by Egyptians were made of rushes and papyrus, whereas in China, hydraulic engineers have used extensively gabions of bamboo basketwork[12].

[10]Dr. R. Baker suggested that the most critical test of a concrete block chute is the first operation at low flow because the stability of the blocks is much lower than for high flow rates.

[11]The word gabion originates from the Italian "*gabbia*" cage, meaning large cage.

[12]Bamboo gabions (or "*chu-lung*") are made of woven bamboo slats packed with stones. Writings and paintings indicate that they were used before A.D. 1000 as illustrated by Needham (1971). The technique is still used today, for example; in China, Taiwan and Japan.

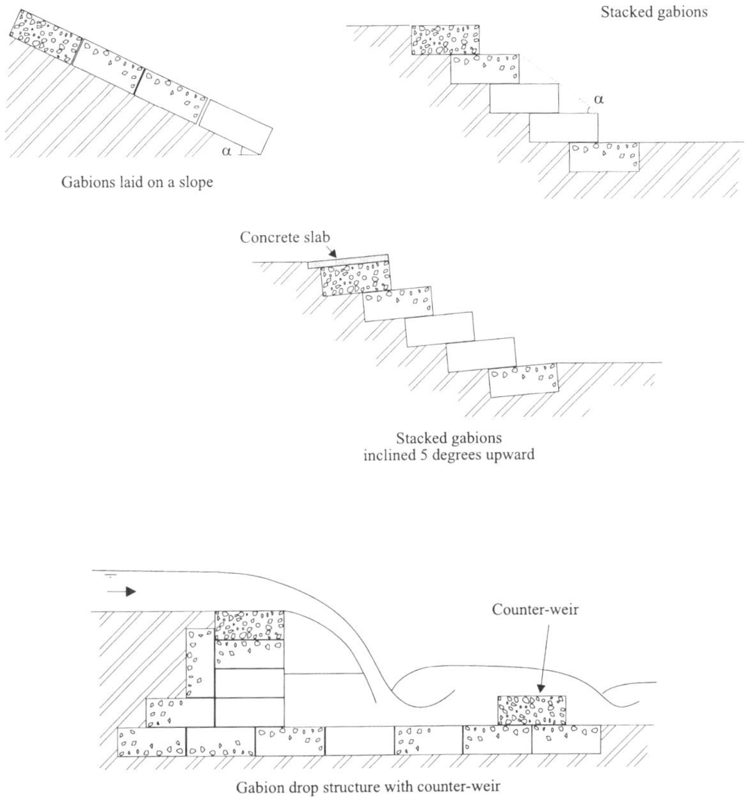

Figure 7.11. Typical gabion weirs.

7.6.2 *Description of the material*

Gabions used as construction material consist of rockfill material enlaced by a basket or a mesh. *Box gabions* are rectangular cages with rockfill material. Typical gabion dimensions are heights of 0.5 to 1 m, a width equal to the height and length-to-height ratio between 1.5 to 4. *Long gabions* are usually subdivided into cells by inserting diaphragms made of mesh panels to strengthen the gabion.

Wire and mesh

The wire is normally a soft steel with a zinc coating. In practice, the durability of gabion structures relies strongly upon the quality of the mesh and wires. The strength of the gabion container might be reduced by corrosion, gabion flexing and debris impacts. High quality and strengthened mesh wires can be used: e.g. galvanised wire, plastic (or PVC) coated wire. The Chinese experience showed that bamboo baskets are durable and can last five to ten years at least.

Table 7.6 Density of different types of rocks (after Agostini et al., 1987).

Type of rock (1)	Density kg/m^3 (2)
Basalt	2,900
Granite	2,600
Hard limestone	2,600
Trachytes	2,500
Sandstone	2,300
Soft limestone	2,200
Tuff	1,700

Filling material

The gabion filling consists of loose or compacted rocks. The stone size of the rockfill must equal at least 1 to 1.5 times the mesh size but should not be larger than 2/3 of the minimum dimension of the gabion. The use of small-sized stone (i.e. 1.5 times the mesh size) permits more economical filling of the cage and it allows a better adaptability of the gabions to deformation. The structural stability of gabions structures depends upon the weight of the gabions and the density of the filling material. The apparent unit mass of gabion is

$$[(1-n_{gabion})*\rho_s] \qquad \text{(dry gabion)}$$

where n_{gabion} is the gabion porosity and ρ_s is the density of the filling material (Table 7.6). The porosity of the gabion n_{gabion} is generally between 0.30 to 0.40. The density of gabion saturated with water equals

$$[(1-n_{gabion})*\rho_s + n_{gabion}*\rho_w] \qquad \text{(gabion saturated with water)}$$

7.6.3 *Risks of damage and destruction*

Hydraulic erosion and destruction

The hydraulic performances of gabion stepped chutes are limited by the gabion resistance to abrasion and destruction, and their stability. Various researchers proposed criteria to prevent the destruction of gabion spillways (Table 7.7). For gabions laid parallel to the flow, the stability of the gabion revetment depends upon the gabion dimensions but also upon the gap between gabions. Despite conservative guidelines (Table 7.7), Huanxiong & Caiyan (1991) reported large overflow discharges (i.e. up to 34 m^2/s) on prototype structures (gabions laid parallel to flow) without major damage. Stephenson (1979b, 1980) discussed the structural stability of stacked and aligned gabions.

For stacked (stepped) gabions, Peyras et al. (1992) recommended to use discharges less than 1.5 to 3 m^2/s, depending upon the mesh resistance. Solid material carried by the stream and large debris may eventually abrade and fracture the gabion mesh. With large-size debris (e.g. trees), it is usual to pro-

A.

B.

Figure 7.12. Examples of gabion stepped chutes. A. Gabion stepped weir in Yemen (1983) (courtesy of Officine Maccaferri). Note the concrete slabs on the steps to protect the gabions. B. Gabion stepped weir in Ontario, Canada (1980) (courtesy of Officine Maccaferri). Canalised water course near Highway No. 401.

C.

D.

Figure 7.12. Examples of gabion stepped chutes. C. Gabion stepped spillway of the Guariraba Weir at Campo Grande, Brazil (1984) (courtesy of Officine Maccaferri). Note the seepage flow and the absence of overflow. D. Gabion river training, Loy Yang Mine, Victoria, Australia (courtesy of Mr. Williams, Geolab Group). Note the partial concrete protection of the horizontal steps.

Figure 7.12. Examples of gabion stepped chutes. E. Gabion torrent training in Saudi Arabia (1982) (courtesy of Officine Maccaferri). Canalisation of a torrent to protect a bridge on Route 54. Note the pooled steps.

tect the step surfaces with timber, steel sheets, concrete facing (0.05 to 0.1 m thick) or even reinforced concrete slab (0.2 to 0.3 m thick) (Agostini et al., 1987; Peyras et al., 1992). Figures 1.2 and 7.12 show some examples of concrete facing. With bamboo gabions, bamboo matting may be used to protect the step surfaces.

Gabion dams and gabion structures might be subjected to another risk of destruction. In some countries (e.g. parts of Africa or south America), the inhabitants might use the gabion wires for other purposes (e.g. fences), or the gabion fillings as construction material. In such places, gabions should not be selected as construction material.

Discussion: comparison with rockfill chutes

It is interesting to note that the discharge stability criteria for gabion spillways are of the same order of magnitude as the recommendations for unprotected rockfill channels (Table 7.8). Protected rockfilled slopes can achieve larger overflow discharges (Table 7.8) but their cost is five times more expensive than unprotected rockfill (Lemperiere, 1991). Thomas (1976), Allen (1984) and Lawson (1987) described several overtopping cases of rockfill dams. For short flood events (few days), discharges as large as 15 m^2/s were achieved over reinforced rockfill slopes without major damage.

Table 7.7. Stability criterion for gabion chutes

Reference (1)	Criterion of stability (2)	Comments (3)
Stacked gabions in steps		
Stephenson (1988)	$V < 4$ m/s	
Peyras et al. (1991)	$q_w < 3$ m^2/s	Gabions with strengthened mesh and lacing.
Peyras et al. (1992)	$q_w < 1.5$ m^2/s	Normal gabions.
Gabion laid parallel to the channel		
Simons et al. (1984)	$V < 2.1$ m/s (0.05 m thick mattress) $V < 3.2$ m/s (0.15 m thick mattress)	Reno mattresses (0.05 to 0.15 m thick). Onset of rock movement in mattress.
Peyras et al. (1991)	$q_w < 1$ m^2/s	Risk of overturning for larger discharges.
Asce (1994)	$q_w < 3.7$ m^2/s	USBR tests. $\alpha = 9.5$ and 14°.

Table 7.8. Stability criterion for rockfill channels

Reference (1)	Criterion of stability (2)	Comments (3)
Unprotected rockfill		
Olivier (1967)	$q_w < 0.2335 * k_s^{1.5} * \left(\frac{\varrho_s}{\varrho_w} - 1\right)^{1.667} * (\tan\alpha)^{-1.167}$	Model data (W = 0.56 m): $4.6 < \alpha < 11.3°$, $0.016 < k_s < 0.06$ m. Gravel, pebbles, crushed granite (SI units).
Knauss (1979)	$q_w < \sqrt{g} * k_s^{1.5} * (2.7 - 3 * \sin\alpha)$	Simplified formula based on model data (SI units).
Abt & Johnson (1991)	$q_w < 3.39 * k_s^{1.79} * (\tan\alpha)^{-0.768}$	Model data (W = 6.1 m): $0.6 < \alpha < 11.3°$, $0.025 < k_s < 0.15$ m. $\rho_s = 2650$ kg/m^3 (SI units)
Lemperiere (1991)		Theories and models.
Rockfill protection		
Lawson (1987)		Mesh-protected rockfill dams.
Lemperiere (1991)		Pervious downstream protection (reinforced rockfill).

Notes: k_s: median boulder size; ρ_s: density of rockfill; W: channel width.

7.6.4 *Design of gabion stepped chutes*

The dimensions of the gabion and the design discharge determine normally the type of flow regime: nappe, transition or skimming flow. Typically the step height h equals the height of the gabion. In some cases, h might equal twice or three times the gabion height. Usually the step length equals 1/2 to 1/4 of the gabion length. With these dimensions, the type of flow regime for the design discharge is determined from Figure 1.7. The dimensions of the gabion and/or the width of the chute may be modified to achieve a suitable flow regime either by changing the step height h or altering the discharge per unit width q_w.

For a structure made solely of gabions, the choice of a steep slope with a skimming flow regime may reduce the number of gabions and the cost of the structure. For an earthfill structure protected by gabions, a flat slope associated with a nappe flow regime may be more appropriate for the stability requirements of the embankment. The design considerations for the stability of gabion weirs are generally the same as for any gravity structure. The calculations of structural stability involve checking the stability of the weir against overturning, sliding and uplift.

Inclined (upward) gabion-stepped spillways may also be used. Larger energy dissipation is achieved but their construction requires greater care. A recommended gabion tilt is: $\delta = +5°$.

Discussion

With vertical drop structures, a counterweir is often used downstream of the primary structure to reduce the effect of erosion (Figs 3.6 & 7.11). The counterweir creates a natural stilling pool between the two structures which reduces the scouring force below the falling nappe impact. In comparison with concrete stepped chutes, the flow above a stepped gabion weir is characterised by: 1. some interaction between the seepage flow and the free-surface flow, and 2. the rougher surface of the gabion steps. The hydraulics might be further complicated by the presence of timber or concrete lining (e.g. Kells, 1995). With gabion steps, the rougher surface of rocks and mesh (in comparison with concrete) enhances: a) the energy dissipation at the jet impact, below the hydraulic jump and downstream of the jump in the nappe flow regime, and b) the energy dissipated by friction between the recirculating eddies (observed below the pseudo-bottom formed by the step edges) and the step faces for skimming flows.

The seepage flow is influenced by the infiltration caused by the spillway flow as well as by the flow through the gabion structure. The seepage will modify spatially the total discharge past the steps. Curtis & Lawson (1967) and Kells (1993) discussed some associated problems. Typically, new gabion structures are porous. But after few years, small debris may obstruct the voids and the structures become less and less porous with time. Figure 7.12C shows the seepage flow through a recent gabion weir in absence of overflow. Note the seepage flow increasing from the weir crest down to the downstream end.

7.7 UNLINED ROCK CASCADES

Unlined rock cascades are used when the abutment geology is suitable (Table 7.9). This design has been common with small weirs as well as high dams. Examples of large dam cascades include La Grande 2 and Dartmouth (Figs 7.13 & 8.2). With rockfill dams, flood waters may be discharged over the quarry face if the quarry may be located downstream of the dam: e.g. Dartmouth Dam, Bellfield Dam. Prototype experience suggests that a maximum acceptable step height must be less than about 12 to 15 m (Table 7.9). The Dartmouth Dam Cascade was damaged during low overflows (Section 8.2) and it was suggested that the step height was too large.

Table 7.9. Comparative characteristics of unlined rock stepped cascades

Name (1)	Year compl. (2)	(3)	W m (4)	h m (5)	Comments (6)
Ternay, France	1868	–	–	0.3 to 0.8	Still in use. Dam height: 41 m.
La Tâche, France	1891	–	2.2	2.4 to 5.4	Still in use. Dam height: 51 m.
Junction Reefs, NSW	1897	37.9	20	2 to 4	Still in use. Dam height: 19 m. Reservoir fully-silted.
Bellfield VIC, Australia	1966	20.9	23	12.2	Dam height: 55 m.
Dartmouth VIC, Australia	1977	2,755	300 to 350	15	Dam height: 180 m.
La Grande 2, Canada	1982	16,140	122	9.1 to 17.8	Dam height: 134 m. Average step height: 10 to 12 m.
North Dandalup WA, Australia	1994	1,300	52	14	Dam height: 62 m.

Note: (–): data not available.

Discussion

Unlined rock cascades might have to be reinforced if the rock is of poor quality or if the cascade becomes damaged. A concrete cascade, built over the rock foundation, might be a suitable remedial solution. An example is the Gold Creek Dam (Brisbane, Australia). Completed in 1885, the 26 m height embankment dam was equipped with an unlined rock chute. In 1887 and 1890, the spillway was damaged during large overflows and the world's first concrete stepped cascade was therefore laid in place in 1890[13] (Chapter 2).

Note that a cascade spillway may not be acceptable for environmental reasons. In general, quarries are located upstream of the dam to cause the least disfigurement of the country side.

7.8 TIMBER AND TIMBER CRIB DAMS

In countries and regions where transportation is difficult and timber plentiful, timber dams are serious competitors of concrete dams. In Russia, Gordienko (1944) reported a 12 m high rockfill dam with a timber crib overflow spillway built around A.D. 1700 (Fig. 2.2B). In America, the north-east benefited from the experience of northern European settlers and timber dams were reported as early as A.D. 1600. Prior to A.D. 1800, timber dams were usually 3 to 4.5 m high. At a later date, some much bigger ones were built successfully to a height of 30 m but most timber dams were less than 6 m high.

Today timber crib dams are still built. They are unexpansive structures, often favoured in developing countries (e.g. Africa, Asia). They are used also in

[13]The concrete aggregate was obtained from the original spillway and hand labourers were recruited to break the rocks (Chanson & Whitmore, 1998).

A.

B.

Figure 7.13. Unlined rock cascades. A. La Grande 2, Baie-James, Québec (Canada, 1982) (courtesy of Michel Lefebvre, Canada). Design discharge: 16,140 m^3/s, W = 120 m, h = 9 to 12 m. Cascade operation in 1983. B. Dartmouth Dam spillway (Australia, 1977) (courtesy of Mr. Jeffery, Goulburn-Murray Water). Design flow: 2,700 m^3/s, h = 15 m. Cascading flow during the, 1996 overflow ($Q_w \sim 225$ m^3/s). View from right spillway bank.

developed countries with abundant timber resources (e.g. Fig. 7.14). Timber crib is often regarded as an environmentally friendly construction material. Typical examples are detailed in Table 1.2 and Figures 7.14 to 7.16. Timber crib weirs are porous structures. The dam porosity can be used as a means to regulate the downstream flow. It can be reduced by use of geotextiles and of an upstream clayfill embankment.

The life of a well-built timber weir can be estimated at 20 to 30 years,

A. 

B. 

Figure 7.14. Timber debris dams in Austria (courtesy of Dr. G. Fiebiger, Austria). A. During winter. B. Low overflow.

although dams reputed to be 80 to 100 years old are also cited (see Etcheverry, 1916; Creager et al., 1945). The author inspected a number of timber crib weirs built between 1935 and 1959 that are still in use today (e.g. Figs 1.2 & 7.16). One particular case, the Kobila Dam (Table 2.1), operated for 352 years! The construction cost of a timber dam is lower than a concrete structure, but the maintenance charges are substantial, particularly at sites where large floods and ice and debris runs are frequent. Leakage is also frequent and might be very large. A serious construction consideration is the tightening of the foundation.

Design experience

Etcheverry (1916) recommended to design timber dams with rockfill cribs and a stepped downstream face. He advised using step length larger than 3 m. For $d_c/h<4$, he suggested a downstream slope of h/l = 1/3, and h/l = 1.6 for 4 $d_c/h<6$. To facilitate the passage of ice and debris, he advised a curved crest with an inclined upstream face (slope of 1V:4H to 1V:2H). Such a typical crib weir is shown in Figure 7.15. The Feather River Weir was 5 m high and 85 m wide (Figure 7.15). The downstream face consisted of 4 steps (h = 2.4 m, l = 4.9 m). The structure was made of 0.09 m^2 square timber and the steps were covered by one layer of 0.15 m lumber. The design maximum discharge was about 73 m^2/s. One year after the construction, the weir discharged a flood of 54 m^2/s without damage.

Timber crib weirs were commonly built in Australia up to the 1960s. An unusual series of timber crib weirs were built in Queensland, Australia, with relatively standardised design (Table 7.10; Fig. 7.16). Most weirs were timber crib structures held by timber piles with a downstream stepped face. A clayfill embankment was placed upstream of the weir for water tightness. Recent weir refurbishments included the replacement of the clayfill by a vertical upstream concrete wall. A scour outlet was usually installed next to a bank for easy access. The basic design (Type I) was the simple staircase shape (e.g. Greenup, Cunningham). Usually the weir was lean and $h/l < 0.4$ where h is the step height and l is the step length. The second design (Type II) was characterised by diagonal brass aligned in the flow direction and connecting the pile heads of different rows. The design was used mainly for steep weirs (e.g. Silverleaf, Whetstone). Altogether the timber stepped weirs have been a successful design. Some weirs have been in operation for over 60 years, sustained successfully very large overflows and are still in good condition (e.g. Murgon, Cunningham). The presence of the stepped geometry was adopted for stability and simplicity of construction. Nonetheless, the steps contributed to the dissipation of the flow energy and no stilling basin was required downstream of the weirs. At low flows, most of the energy was dissipated on the steps.

7.9 DEBRIS DAMS

In mountain areas, where debris torrents might have catastrophic (and dramatic) impacts, "check dams" (or debris dams) can be used to reduce the impact of debris torrents. Debris dams are common features in Europe, North America and Far East Asia. Illustrations are shown in Figures 7.14 and 7.17. The term "debris dam" is used to describe both consolidation dams and sediment retention structures. The former is generally a wall-type structure. It is designed to elevate the torrent bed to fix and to stabilise the bottom profile. The latter type of structure is commonly an open structure (i.e. grid dam, beam dam and split dam) designed for the trapping of medium- to large-size debris (rocks, boulders, logs). Armanini et al. (1991) described several examples of open structure check dams. In the following section, we shall consider only the consolidation dams as this type of structure is particularly equipped with stepped overflows. Furthermore, the dam construction decision process is not discussed. It is assumed that the choice of debris dam(s) building and location are sound and optimal enough to prevent debris flow catastrophes[14].

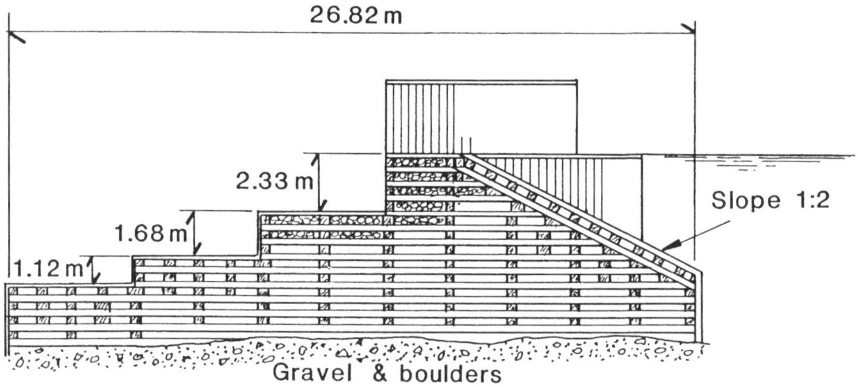

Figure 7.15. Temporary timber dam on Feather River (USA, 1912) (after Etcheverry, 1916).

A.

Figure 7.16. Examples of timber crib stepped weirs in north-eastern Australia. A. Silverleaf Weir, Australia, during construction in 1952 (courtesy of the Queensland Department of Natural Resources). View from the right bank looking upstream.

[14]In Taiwan, a series of debris dams were dug into debris six months after construction. Discussions with experts suggest that the debris control system was ill-designed.

B.

C.

Figure 7.16. Examples of timber crib stepped weirs in north-eastern Australia. B. Silverleaf Weir, Australia (1953). Overflow shortly after completion (courtesy of Mr. J. Mitchell). C. Silverleaf Weir, Australia (1953). Recent photograph showing the crest heightening in 1995 with a 1 m high concrete wall (photograph taken in 1997).

D.

E.

Figure 7.16. Examples of timber crib stepped weirs in north-eastern Australia. D. Whetstone Weir, Australia (1951) (photograph taken in 1998). E. Emerald Weir, Australia (1947) (courtesy of the Queensland Department of Natural Resources). Small overflow in 1951.

Figure 7.16. Examples of timber crib stepped weirs in north-eastern Australia. F. Greenup Weir during an overflow on 10 March, 1958, shortly after completion (courtesy of the Queensland Department of Natural Resources).

Design considerations
A debris dam can be made of woven fences, timber, earth, stone, riprap or concrete (Table 7.11). *Timber debris dams* are cheap to build (e.g. Fig. 7.14). The dam consists of a timber truss filled with stone. The crest can be paved or faced with logs. Typical life period of timber check dams range from 5 to 20 years. *Gabion check dams* provide reasonably-priced structures. The dam can be a simple vertical drop or a stepped weir. Culverts can be installed in the middle of the structure (e.g. Fig. 7.17). Usually the dam crest is protected with timber, steel or concrete facing. *Earth debris dams* are more appropriate for wide valley. The dam possesses a spillway section made of concrete or stones with sidewalls at least 3 m long. The upstream slope of the earth structure must be protected with riprap or pavement. *Stone dams* are more aesthetic. If the material is available in the vicinity of the dam location, the dam can be made of large boulder on cement. In cold regions, the quality of construction of the joints between boulders is very important: the action of the frost might damage (extensively) the masonry structure. *Concrete debris dams* are the most usual in developed countries. They are built as arch structures and are cheaper than stone structures. Mamak (1964) recommended the following geometry: vertical upstream face, inclined downstream face (4V:1H), crest width of about 2 m.

Simple check dams have drop heights from 2 to 7 m (e.g. Fig. 1.2). For higher dam height, it is advised to perform a stability analysis. Additional baffle blocks can be used to enhance energy dissipation. Trashracks can be

Table 7.10. Timber crib weirs built in Queensland, Australia.

Name	Year H_{dam}	Weir height (m)	Weir length W (m)	Step height h (m)	Nb of steps	Construction	Remarks
(1)	(2)	(3)	(4)	(5)	(6)	(7)	(8)
Theodore Weir	1925	5.8 (*)	65	–	–	Timber cribs held by timber piles. Diagonal brassing.	On the Dawson River. Intermediate raisings in 1929 and, 1931. 65 m long. Replaced by a concrete structure in 1984.
Sunnygirl Weir	1930?	–	–	–	1	Timber log drop.	On Macintyre Brook.
Orange Creek Weir	1932	5.8	60.4	–	–	Timber cribs held by timber piles.	On Dawson River. Still in use.
Bingera Weir	1933	1.8	195	–	–	Timber cribs held by timber piles.	On Burnett River. Removed in 1960.
Murgon Weir	1935?	–	–	–	3	Timber cribs held by timber piles.	On Barambah Creek. Still in use.
Sandy Creek	1944	2.7	13.4	–	–	Timber cribs held by timber piles.	On Sandy Creek. Replaced by a sheet-pile cascade in 1978 or 1983.
Moura Weir	1946	5.9 (*)	165	–	3	Timber cribs held by timber piles.	On Dawson River. Combined bridge and weir. Refurbished in 1985. Still in use.
Emerald Creek Weir	1947	5.2	79.5	–	3	Timber cribs held by timber piles.	On Emerald Creek. Disused since, 1977 (blown up or submerged).
Hilliard Creek Weir	1948	3.4	66	–	2	Timber cribs held by timber piles.	On Hilliard Creek. Repaired in 1963. Very poor condition because of lack of maintenance.
Whetstone Weir	1951	5	–	1.5	3	Timber cribs held by timber piles. Diagonal brassing.	On the Macintyre Brook. l = 1.9 m, h/l= 0.8. Refurbished in 1990s. Still in use.
Mulgeldie Weir	1952	3.1	40	–	2	Timber cribs held by timber piles. Diagonal brassing.	On Three Moon Creek. Small structure. Still in use.
Binda Weir	1953	5.2	73.1	–	5	Timber cribs held by timber piles. Diagonal brassing.	On Dawson River. Destroyed in 1986 because unsafe.
Bonshaw Weir	1953	3.7	144.5	0.9	4	Timber cribs held by timber piles.	On Dumaresq River. Design flow: 1,560 m^3/s. Failure in 1956. New sheet-pile weir built in 1958.
Silverleaf Weir	1953	5.1	87.8	–	4	Timber cribs held by timber piles. Diagonal brassing.	On Barambah Creek. Refurbished in 1995. Still in use.
Cunningham Weir	1954	4	58.5	0.9	4	Timber cribs held by timber piles.	On Dumaresq River. Design flow: 1,560 m^3/s,. l = 2.4 m, h/l = 0.375. Still in use.
Greenup Weir	1958	5	–	1.3	4	Timber cribs held by timber piles.	On Macintyre Brook. l = 2.74 m, h/l = 0.47. Still in use.

Note: (*) after refurbishment.

installed upstream of the drops to stop coarse debris. Mamak (1964) suggested that the overflow discharge q_w should be no larger than 1.1 m^2/s for timber structure and less than 8.9 m^2/s for reinforced concrete.

A more sophisticated structure at Harvey Creek, Canada was described by Hungr et al. (1987). This debris dam is an earth embankment with a double

Table 7.11. Examples of debris dams.

Name	Ref.	Dam height m	Slope degrees	Construction	Comments
(1)	(2)	(3)	(4)	(5)	(6)
Cubatao weirs, Brazil, 1986	(AG)			Three successive gabion stepped weirs.	PVC-coated gabion wires.
		9	63		Step height: h = 1 m.
		1	63	Equipped with 5 culverts (Ø = 0.6 m).	Step height: h = 1 m. Figure 7.19.
		8	63	Equipped with 4 bottom culverts and 9 mid-level culverts.	Step height: h = 1 m.
Harvey Creek Dam, Canada, 1987	(HU)			Earth embankment equipped with a double drainage culver.	Concrete stepped overflow for debris flows. 16 steps.
Lammbach Brienz, Switzerland		9 to 21		7 drop structures made of cemented dry masonry,	Slope of torrent bed: 18.3 deg.
		5 to 6		6 concrete dams	
Shiraiwa Sabo Dam, Japan, 1931-1939	(NA)	63		Concrete structure.	12 steps. Can trap up to 1 Mm^3 of sediment.

Notes: (AG): Agostini et al. (1987); (HU): Hungr et al. (1987); (NA): Nakao (1993).

central drainage culvert and a overflow stepped debris spillway. The culvert is designed to carry the 200-year flood discharge and the upstream channel slope is covered with removable reinforced concrete beams. The beams can be lifted to inspect and clean the structure. It is worth noting that a stepped bottom geometry is recommended also for avalanche corridors (*couloir d'avalanche*) (Jamme, 1974).

Sabo works in Japan

Japan is characterised by steep unstable topographical features with frequent volcanic activities and earthquakes. Debris flows are frequent and numerous disasters have occurred. The original purpose of the Sabo[15] structures was to reduce the excess sediment discharge to prevent river degradation further downstream (e.g. Fig. 7.17) and to enable ship navigation in the downstream streams. More recently the emphasis of Sabo works shifted to the control of debris flows without much sediment trapping.

A major debris structure is the 63 m-high Shiraiwa Sabo Dam[16] (Table 7.12). The main dam is equipped with a 12-step overflow spillway (Nakao, 1993) and it is designed to trap up to 1 Mm^3 of sediment. Another large Sabo structure is the Mount Tokachidake Sabo Works in the Furano River catchment. After completion, the Sabo system includes 72 check dams, 11 slit dams and 71 consolidation dams. The total cost was 0.8 billion Yen (in 1993). An interesting example of Sabo works system is the Kakurajima Volcano Sabo Works,

[15]The Japanese word *Sabo* means mountain protection system.
[16]Also called Siraiwa dam.

A.

B.

C.

Figure 7.17. Examples of debris dams. A. Debris dam near La Motte-du-Caire, France, in a catchment of "marnes noires" (blacklands) (June, 1998). B. Cascade of debris dams in Bavaria (courtesy of Dr. M.R. Gourlay). C. Debris (Sabo) dams near the town of Omachi, Japan (November 1998). Sabo works on the right slope of Kagokawa River. Note the heavy gravel bed load.

D. 

E. 

Figure 7.17. Examples of debris dams. D. Sabo dam (H=12.5m, L=117m) and stepped storm waterway (built in 1992) in Toyohashi, Japan (Mar. 1999). The Sabo works protect Sekiganji temple and a kindergarten. The stepped channel includes 18 steps; the footpaths being designed to act as floodplains during extreme events. E. Ancient Sabo works near Matsumoto, Nagano Prefecture, Japan (November 1998). The artificial stepped channel (construction: 1916-18) was modelled on some Durance catchment works in France during the 19th century.

F.

G.

Figure 7.17. Examples of debris dams. F. Soil retention check dam with a gabion stepped spillway (h = 1 m), Congohas, Brazil (1987) (courtesy of Officine Maccaferri). The colour of the flow indicates large soil-debris contents. G. Check dam with gabion stepped overflow near Cubatao, Sao Paulo, Brazil (1986) (courtesy of Officine Maccaferri). The dam is equipped with five culverts (0.6 m diameter pipes) to allow the passage of low flows, and the gabion wires are PVC-coated because of the extensive air and water pollution of the area. H_{dam} = 11 m, h = 1 m.

Table 7.12. History of construction of the Shiraiwa Sabo works.

Date (1)	Description (2)
1906	1st dam
1916	2nd dam
1922	Destruction of the dams by debris flood.
1922-1929	Railway construction from Kanbadaira to Mizutani.
1931	Railway extension to Shiraiwa
1931-1939	Construction of a new dam (H = 63 m, L = 75 m, trap capacity ~ 1 E+6 m^3)
	Construction of subdams: Shiraiwa subdams No. 1 to 5
1950?	Construction of Shiraiwa subdams No. 6 and 7
1960-1966	Rail replacement

located on Kakurajima Island. The island is 12 km wide by 19 km long and the volcano is still active. Large-scale volcano eruptions took place in 1471-1476, 1779-1785, 1914 and 1946. All 18 rivers are equipped with Sabo works and debris flow occurred each year. During debris runoff, the velocity of debris flow were observed to reach 40 to 70 km/h (Nakao, 1993).

CHAPTER 8

Accidents and failures: Learning from experience

8.1 PRESENTATION

Most hydraulic structures, including stepped channels and chutes, are designed for an optimum use at the most economical cost. A safety analysis must be conducted, and design and non-design flow conditions must be considered in the analysis of the safety of a dam and its spillway. Major accidents are analysed below to assess possible consequences and to devise means to prevent or mitigate these consequences. The optimum design must minimise the *real* total cost: i.e. the cost of construction + maintenance cost + cost of safety. The cost of a failure is not always measurable: damage to properties is assessable, damage to the environment is real and has often a political value, but loss of life is a matter of suffering. In the past, a significant number of hydraulic structures failed including stepped chutes. Professional engineers must be aware of the consequences of an improper design. General reviews were prepared by ANCOLD, ICOLD, USCOLD, and lessons can be learned from this experience. Several accidents involving hydraulic structures equipped with stepped channels and related failures will be discussed within Section 8.2. In Section 8.3, considerations for a safe design of stepped channels will be developed. Finally, a discussion will highlight some important design aspects which must be considered by designers and operators of stepped chutes.

8.2 FAILURES AND ACCIDENTS

Over the past two hundred years, several accidents and failures have been experienced at hydraulic structures (e.g. dams, weirs) equipped with stepped channels. A summary is given in Table 8.1. Each accident is investigated and discussed case by case. In Section 8.2.2, severe accidents are discussed (Table 8.2).

Table 8.1. Summary of accidents and failures of stepped chute (sorted by year of construction).

Dam/Reservoir (1)	Ref. (2)	Year(s) of construction (3)	Year(s) of accident (4)	Accident/Failure (5)	Lives lost (6)
Stepped spillways					
Kobila Dam, Slovenia	(BRE)	1586	1948	Dam destruction by a flood (caused by lack of maintenance).	–
Puentes Dam, Spain	(JA, SC, SM)	1785-1791	1802	Dam break (caused by a foundation failure).	608
Sreeramadavara Weir, India	(SA)	–	1842, 1844, 1859, 1860, 1863	Failures.	–
Muddoor Weir, India	(SA)	–	1850s	Failure.	–
Narora Weir, India	(BU)	1877	1898	Failure of stilling basin during very large overflow.	–
Arizona Canal Dam	(EN)	1887	1891 and, 1905	Partial destruction of the dam during a flood (caused by foundation problems and timber deterioration).	–
Goulburn Weir, Australia	(CU)	1891	1978	1. Gate failure (caused by corrosion). 2. Foundation stability.	None
Minneapolis Mill Dam, USA	(WE)	1893-1894	1899	Dam break during a small spill (caused by cracks resulting from ice pressure on the dam).	–
New Croton Dam, USA	(FO, GO, MA, W7)	1892-1905	1955	Spillway damage during flood releases.	None
Lahontan Dam, USA	(DG, JA, RH)	1915	1930-40	Damaged spillway concrete.	None
Warren Dam, Australia	(JO)	1916	1917	Dam overtopped (no damage).	None
St. Francis Dam, USA	(JA)	1926	1928	Dam break (caused by foundation failures).	450
Hilliard Creek Weir, Australia	(CH)	1948	1990s	Ruins (lack of maintenance).	None
Binda Weir, Australia	(CH)	1950-1953	1986	Weir deliberately destroyed because unsafe (lack of maintenance).	None
Silverleaf Weir, Australia	(CH)	1951-1953	1995	Weir overtopping during refurbishment works (no damage).	None
Bonshaw Weir, Australia	(CH)	1952-1953	1956	Weir failure resulting from left abutment bypass during major flood.	None
Dartmouth Dam, Australia	(CH)	1977	1996	Unlined rock steps damaged by flow concentration during low flow.	None
Moscovite earth dams, Russia (former USSR)	(BA, PR)	1978	1978-80	Failure of two overflow earth dams caused by incorrect drainage layer construction.	–
Bucca Weir, Australia	(CH)	1986-1987	1987	Nappe deflection at first step associated with loud noise and vibrations at low overflows.	None
Gyrandra Weir, Australia	–	1985-87	1989	Complete failure of 3rd and 4th steps during medium overflows.	None
Debris dams, Taiwan	(CH)	1994	1994	Ill-designed debris control system.	None

Notes: (–): unknown information

Ref.: (BA) Baker et al. (1994); (BRE) Breznik (1984); (BU) Buckley (1905); (CH) Present study; (CU) Cummins et al. (1985); (DG) Douma & Goodpasture (1940); (EN) Engineering News (1905a); (FO) Ford (1957); (GO) Goubet (1992); (JA) Jansen (1983); (JO) Johnson & Templar (1974); (MA) Marsh (1957); (PR) Pravdivets (1992); (RH) Rhone (1990); (SA) Sankey (1867); (SC) Schuyler (1909); (SM) Smith (1971); (W7) Wegmann (1907); (WE) Wegmann (1911).

8.2.1 *Failures and accidents of stepped channel structures*

Arizona Canal Dam (USA)

The Arizona Canal Dam was located 45 km north of the city of Phoenix (USA). It was designed to divert water from the Salt River for irrigation purposes. Completed in 1887, the dam was a 280 m long structure made of timber cribs filled with loose rocks and gravels. The dam height was 10 m and the base of the dam was about 11 to 14.6 m wide. At each end, the dam was fastened to bed rock. Beyond the rock, excavation was made in the river in which mud sills were laid upon which the cribs were built. The timber cribs were fastened together by wire cables.

The western half of the dam was an overflow stepped spillway. It included 3 horizontal steps. At the time of construction, the maximum flood discharge observation was about 8,100 m^3/s (i.e. $q_w \sim 33$ m^2/s). It is believed that the overflow section was designed for that discharge.

In 1891 the central section of the dam settled considerably and portions of the dam were washed away. Prompt repairs were made. In 1899 signs of timber deterioration were observed in some upper portions of the dam. The dam was repaired again and strengthened. On 13 April, 1905, a major flood caused a dam break. At the time of failure, the discharge was about $q_w = 11.3$ m^2/s. Portions of the centre of the dam were torn away and the dam breach was about 91 m wide.

It is believed that the failure was caused by unstable foundation problems and some timber deterioration. At the time of break, the dam was 18 years old. During the 1905 flood overflow, the ratio d_c/h was about 0.7: i.e. the flow regime was a transition flow (Section 1.3). The unstable flow situation might have induced additional fatigue to the structure.

Binda Weir (Australia).

The Binda Weir was a timber crib piled weir on the Dawson River QLD, Australia. Built between March 1950 and May 1953, the weir was 5.2 m high and 73.1 m long, and the overflow section had 5 steps. The weir was used for more than 30 years until it was destroyed (blown up) in 1986 because it had become unsafe. The destruction of this weir was a form of failure, caused by a lack of maintenance.

Bonshaw Weir (Australia)

The (first) Bonshaw Weir was completed in May, 1953 (Fig. 8.1). It was designed for irrigation water, the primary industry at the time being tobacco. The weir was located on the Dumaresq River, which marks the boundary between the Queensland and New South Wales states and is a tributary of the Macintyre and Darling rivers. The 3.7 m high weir was a timber crib piled structure (144.5 m long), designed with 4 steps ($h = 0.9$ m). In elevation, the weir formed a broken line: one straight overflow section of 92 m long on the left, followed by a second straight overflow section of 44 m long up to the right abutment.

A major flood occurred in January and early February 1956. Intense rainfall was recorded in the catchment (e.g. 185 mm on 21 January and 225 mm on 22 January, 1956). The Dumaresq River was in flood for more than 2 months. The maximum recorded stream heights were 11.13 m in January 1956 and 8.76 m in February 1956 at the Cunningham Weir, located 60 km downstream. No discharge was recorded. The nearby town of Texas was flooded and inhabitants could not cross the river for nearly two months. The Bonshaw Weir failure took place during this hydrological event. Both abutments failed. The left abutment was bypassed and it was heavily scoured (Fig. 8.1). The right abutment and a section of the weir were washed away.

In 1958, a new weir was built in steel sheet-piles and concrete. Interestingly, the new weir has a straight profile (in elevation) and it is only 2.4 m high. It is worth comparing the operation of the Bonshaw Weir with that of the Cun-

Figure 8.1. Bonshaw Weir (courtesy of the Texas Historical Society and Queensland Department of Natural Resources). A. View of the stepped chute shortly after completion in 1953.

B.

C.

Figure 8.1. Bonshaw Weir (courtesy of the Texas Historical Society and Queensland Department of Natural Resources). B. Small overflow on 19 April, 1955; view from right bank. C. View of the left abutment after the flood. Note the extent of the crack.

D.

E.

Figure 8.1. Bonshaw Weir (courtesy of the Texas Historical Society and Queensland Department of Natural Resources). D. View of the damaged weir from the left bank. Note the missing right abutment in the background. E. View from the left bank: construction of the second weir. Note the ruins of the first weir downstream.

ningham Weir. Completed in 1954, the Cunningham Weir was not so extensively damaged during the 1956 flood and it is still in good condition today (Fig. 6.1). Suggestions were made that the construction quality of the Bonshaw Weir was not as good as that of the Cunningham Weir.

Bucca weir (Australia)

Completed in 1987, the Bucca Weir is a 12 m high RCC weir (Figs 7.5 & 7.6). The weir is equipped with an overflow stepped spillway (h = 0.6 m, l = 0.3 m) and the maximum design discharge is 7,250 m^3/s (Table 1.2). The crest design is an inclined upward broad crest followed by a 1.6 m drop and seventeen 0.6 m high steps (2V:1H chute slope). Shortly after completion, loud noises were noted during low overflows. At the end of the crest, the water was deflected away from the stepped chute and the nappe was not ventilated. The resulting nappe instabilities generated loud noise (which was not acceptable to nearby residents), and low-frequency vibrations that were felt throughout all the structure (Section 3.2.3). The problem was remedied by the installation of splitters to ventilate the nappe at low-flows. The splitter devices are tiltable and drop at large overflows.

Dartmouth Dam (Australia)

Completed in 1977, the Dartmouth Dam is a 180 m high earth and rockfill structure on the Mitta Mitta River (Figs 1.2, 7.13 & 8.2). The spillway is located on the left bank. It consists of an approach channel, a concrete-lined crest and chute (W = 91.4 m), followed by a 10-step unlined rock cascade (W = 300 to 350 m). The rock is granitic gneiss and the cascade excavation was used for the dam rockfill. The steps are 15 m high. The design discharge capacity is 2,755 m^3/s.

During the 1996 flood ($Q_w \sim 225$ m^3/s), the unlined rock stepped cascade was partially damaged. Although the discharge was much lower than the design flow rate, significant erosion was observed at several steps. Erosion resulted from flow concentration in the right side of the cascade. A programme of repairs was undertaken in 1996 and 1997: provision of shotcrete protection, construction of anchored concrete retaining walls in the three upstream steps. Further walls were built to prevent concentration of low-level flows in some areas.

Debris dams in Taiwan

Three debris dams were completed in 1994. The structures are straight concrete walls and they were all located in the same catchment on the eastern coast of Taiwan. During a typhoon in 1994, the dams failed to control the debris flows and to protect the downstream catchment. All three dams were digged into (i.e. covered by) debris material (mud, gravel, rocks) after torrential floods and debris avalanches in the catchment less than six months after completion. Photographs taken after the flood showed only the crest of the

A.

B.

Figure 8.2. Dartmouth Dam spillway during the 1996 overflow (Q_w ~ 225 m^3/s) (courtesy of Mr. Jeffery, Goulburnmurray Water). A. Spillway view from downstream. B. View of the concrete intake and first unlined rock steps.

Figure 8.2. Dartmouth Dam spillway during the 1996 overflow (Q_w ~ 225 m^3/s) (courtesy of Mr. Jeffery, Goulburnmurray Water). C. Details of the cascading waters and the surrounding fog. View from right spillway bank.

dams emerging from the newly-formed channel bed. Discussions with experts suggest that the debris control system was ill-designed. A completely new system had to be built again.

Goulburn Weir (Australia)

The Goulburn Weir is located in north-central Victoria (south-east of Australia). It is an important structure in the irrigation system of northern Victoria, being the primary diversion of the Goulburn River to nearly 50,000 hectares of irrigated land. Built in 1891, the weir is a concrete gravity dam (15 m high, 141 m wide). Flood waters discharge over the structure. The downstream face consists of stepped horizontal granite blocks (h = 0.5 m) (Figs 5.6 & 8.3). The maximum discharge capacity is about 1,980 m^3/s. The structure was initially equipped with 21 cast and wrought iron overshoot gates (3 m high, 6 m wide), supported by cast iron piers.

In 1978 one gate failed to operate. The accident had no direct consequences, but it initiated extensive inspections of the 90 year-old structure. Inspections revealed serious deterioration to a number of gates and other damage in the recesses. Furthermore, the weir is founded on erodible weathered mudstone foundation. The inspections highlighted the need to stabilise the foundation and to prevent the risk of sliding.

Repairs were undertaken between 1980 and 1986. All the gates and piers were replaced by a system of 9 radial gates (3.65 m high, 12.87 m wide) supported by new concrete piers. The stability of the structure was secured by six pre-stressed anchor bars per pier, and ground anchors at the spillway toe.

Figure 8.3. Goulburn Weir (Australia, 1891) (courtesy of the Victoria Rural Water Corporation). View from the right bank . Photograph taken after the replacement of the gates in 1986.

Following the replacement of the original overshoot gates by radial gates (i.e. undershoot gates), flows at small-gate opening are deflected past the steps and impact onto the new downstream concrete apron (Fig. 5.6). Hence, the steps do not contribute to the dissipation of the kinetic energy at low flows.

Gyrandra Weir
Located in Central Queensland, the Gyrandra Weir is an overflow stepped weir designed for in-stream water storage (crest length: 147 m). Built between 1985 and 1987, the 12 m high earthfill structure is equipped with a stepped overflow cascade (h = 2 m, l = 6 m) designed to pass 5,120 m^3/s. It is a steel sheet-pile and concrete slab construction (slab thickness between 0.15 to 0.23 m).

On 1 April 1989, the 3rd and 4th steps failed completely during a medium overflow (Q_w = 1,250 m^3/s). Consequently, it was found the failure was caused by uplift pressures on concrete slab and fill erosion. The stepped design was tested in laboratory with flat horizontal steps. At a later stage, pile cappings (0.2 m high) were added for structural and construction purposes, but not tested in physical model. Subsequent model studies showed that the addition of the capping generated a major change onto the nappe trajectory and downstream hydraulic jump flow (Section 7.1.4). Yet during the overflow event, the flow regime was a transition flow (d_c/h = 0.97). It is possible that fluctuating loads on the concrete slabs and sheet piles might have accelerated the failure. Overall, the failure is an example of lack of communication between structural and hydraulic engineers.

Figure 8.4. Hilliard Creek Weir, Australia (1948). Photograph in 1997 from the left bank looking at the weir. Note the piles in the foreground and the overgrown vegetation on the weir.

Hilliard Creek Weir

The Hilliard Creek Weir was completed in 1948 on Hilliard Creek, less than 1 km from the river mouth. It was built for water supply and to prevent salt intrusion in the creek. The 3.4 m high weir is a timber crib piled weir. Repaired once in 1963, the weir is now in ruins because of the lack of maintenance. The timber crib skeleton and some rockfill are visible, showing the construction technique but the weir is not holding water: i.e. it is porous (Fig. 8.4). Warning signs are posted around the weir indicating that it is unsafe to approach or to walk on the structure. Surprisingly, the weir was still listed as a storage weir in the 1997 Annual Report of the Queensland Department of Natural Resources.

Kobila Dam (Slovenia)

The Kobila Dam was a timber crib structure filled with rocks. Built in 1586 across the Idrijca River, the dam was over 10 m high and the crest length was about 20 m. The structure was equipped with an overflow stepped spillway. The spillway crest was protected with wooden planks from floating logs. The dam operated continuously up to 1948 when it was destroyed by a flood after 362 years of existence. During World War Two and the subsequent years, the dam was not maintained. It is believed that the break resulted from a lack of maintenance and care over a period of nearly ten years.

Figure 8.5. Lahontan Dam left spillway, Newlands Project, Nevada (courtesy of the US Bureau of Reclamation). A. Aerial view in 1972 (photograph by J.C. Dahilig). B. Spill on 31 May, 1922: $Q_w = 54\ m^3/s$ (photograph by G.G. Engle). C. Spill on 31 May, 1922: $Q_w = 54\ m^3/s$ (photograph by G.G. Engle). Stilling basin at the downstream ends of the stepped chutes. Note the right stepped chute in foreground and the left stepped spillway in the background.

Lahontan Dam (USA)

Built in 1915, the Lahontan Dam is an earth structure, 49 m high across the Carson River (Nevada). The dam is owned by the US Bureau of Reclamation. It is equipped with two stepped spillways on the left and right abutments (Fig. 8.5). The maximum discharge capacity is 424.8 m^3/s per spillway. The right spillway consists of two drops followed by a long curved flat chute with converging walls leading to a series of 7 steps ($h = 3.35$ m, $\alpha = 22.9°$, $W = 45.72$ m). The left spillway consists of a series of 6 steps ($h = 3.05$ m, $\alpha = 26.6°$, $W = 76.3$ m), a converging flat chute section and a curved stepped channel (3 steps, $h = 3.05$ m, $l = 6.096$ m, $W = 45.72$ m) with a curvature radius ranging from 39 to 50 m. The right and left spillways meet in a circular stilling basin at the dam toe.

In the late 1930s, the concrete of the spillways was in poor condition, particularly in the right spillway, although only two floods were recorded since construction (Douma & Goodpasture, 1940). It was suggested that the concrete construction was damaged by freeze and thawing (Jansen, 1983). Hydraulic tests showed also that a hydraulic jump took place at the end of the flat chutes (Douma & Goodpasture, 1940). A re-analysis shows further that the right spillway would experience a transition flow regime at maximum flow rate ($d_c/h = 0.62$). It is thought that the spillway damage was initiated by scour associated with the presence of hydraulic jumps and further enhanced by flow instabilities during the transition flow operation in the right spillway. The spillway system was repaired in 1940, and the dam and cascade are still in use.

Minneapolis Mill Dam (USA)

The Minneapolis Mill Dam was built between 1893 and 1894. The structure was a masonry dam (5.5 m high, 163 m long). The crest was 1.6 m wide and the base of the dam was 3.66 m wide. The spillway was a 51.8 m long section of the dam, with 9 steps ($h = 0.61$ m, $\alpha \sim 75°$). In Minneapolis, the winter climate is severe and the ice of the reservoir adhered to the masonry structures during the winters. In February 1899 a slight motion of the dam was noticed. It was believed that the movement was caused by ice pressure. No cracks or leaks were discovered and steel anchors were placed to strengthened the structure. On 30 April 1899 the spillway was discharging 0.036 m^2/s when the anchors snapped and the weir slid out. The failure was supposed to be initiated by cracks under the dam. These cracks were opened when the dam was slightly moved in February 1899 and uplift pressures caused the final collapse of the dam.

Moscovite earth dams (Russia)

In 1978, Soviet engineers built five overflow earth dams in the Moscow region (Pravdivets, 1992). The dams were about 7 to 15 m high. The spillway systems comprised an overflow with precast concrete blocks (Section 7.5). They were designed to pass between 30 to 60 m^3/s. Two structures failed during the first year of operation. Their protective layer (Section 7.5.2) did not perform adequately and the earthfill embankments were eroded. In one case (i.e. Zhilevski Dam), the smallest stone size of the drainage layer was smaller than the drainage holes in the concrete blocks. In absence of a geotextile fabric[1], drainage material and earthfill were sucked through the drainage holes and washed away during a storm leading to the failure.

Narora Weir (India)
Completed in 1877, the Narora Weir is part of the Lower Ganges Canal. It was a 3.05 m high 1,160 m wide drop structure in brick works, designed to pass 3,400 m^3/s (Buckley, 1905). The weir failed in March 1898. Concrete and stonework of the stilling basin were damaged over a 107 m wide section. The weir was discharging 13,800 m^3/s at the time of failure: i.e. four times the design flow rate! The weir was strengthened in 1898 and 1899.

New Croton Dam (USA)
The New Croton Dam is a masonry dam built between 1892 and 1905. The dam was constructed across the Croton River, 65 km north of the city of New York. The structure is 90.5 m high and the base thickness is 63 m. The spillway is an overflow stepped channel at the right-bank extremity of the dam (Figs 2.10 & 8.6). The spillway face consists of granite blocks with mortared joints. The spillway was designed for a maximum discharge of 1,550 m^3/s (Wegmann, 1907). The spillway crest is 305 m wide and the slope is about 53°. The step height is 2.13 m. During the construction, the spillway crest was raised by 1.2 m: 229 m of the spillway crest were completed in masonry but 76 m of crest close to the dam were obstructed by 1.2 m high flashboards and concrete slabs. Around 1907-1908, 0.6 m of additional flashboards were added over the entire crest width to increase the reservoir capacity. The flashboards (and concrete slabs) were left permanently in place for 50 years.

Following a storm event from 14 to 16 October 1955, the spillway was damaged during a water release of 651 m^3/s, with an overflow depth of about 1.2 m. After the flood, inspections of the spillway revealed that the masonry close to the dam was badly cracked with several longitudinal cracks. The structure was no longer watertight and waters were escaping in significant amounts through the cracks (up to 65 to 90 l/s). Extensive repairs were carried out between, 1955 and 1957. Marsh (1957) suggested that the masonry at the end of the spillway (close to the dam) was subjected to severe vibrations resulting from improper flow conditions below the added flashboards and concrete slabs. It is certain that, at the time of the flood, the spillway crest was incorrectly profiled. But the severity of the damage, for a discharge smaller than the maximum design discharge, suggests some concerns about the quality of the construction and the behaviour of the stepped spillway (Goubet, 1992).

[1]In Russia, the embankment overtopping protection systems have been built without geotextile fabric.

Figure 8.6. New Croton Dam (USA, 1906) (courtesy of the American Society of Civil Engineers). View of the spillway from downstream, about 1910. Note the masonry dam on the right, and the bridge connecting the dam crest to the right bank over the spillway weir.

During the spill, the ratio of critical depth to step height (d_c/h) was 0.36. For a similar slope (α=51.3°), Beitz & Lawless (1992) observed the transition between nappe and skimming flow for $(d_c)_{onset}/h$ = 0.4 (Table 1.3). The comparison suggests that the spillway of the New Croton Dam operated with a transition flow regime. The resulting flow instabilities might have added extra loads to the spillway structure particularly below the flashboards. The fluctuating hydrodynamic loads must have accelerated the crack growth.

Puentes Dam (Spain)

The Puentes Dam was built between 1785 and 1791 on the Guadalantin River (Spain). The gravity dam was a rubblemasonry structure and the faces were made of cut stones. It was 50 m high, 282 m long, 11 m thick at the crest and 22.2 m thick at the base. The upstream face of the dam was vertical. The downstream face consisted of four steps (h = 4.2 m, l = 3.3 m) followed by an uniform slope down to the base (Fig. 8.7). Flood waters were to be discharged over the dam crest (uncontrolled overflow). The dam was equipped with a large scouring gallery (6.7 m wide, 7.5 m high) at the base and two outlets (1.6 m wide, 1.95 m high each), one near the base and the second 30.5 m below the crest. The masonry structure was founded on wooden piles driven into a bed of alluvial soil and sand. Hundred of wood piles (6.1 m long, 0.5 m diameter each) were driven into the alluvium and braced with horizontal timbers at their tops. From 1791 to 1802, the reservoir was never filled and the depth of water behind the dam did not exceed 25 m. At the beginning of 1802,

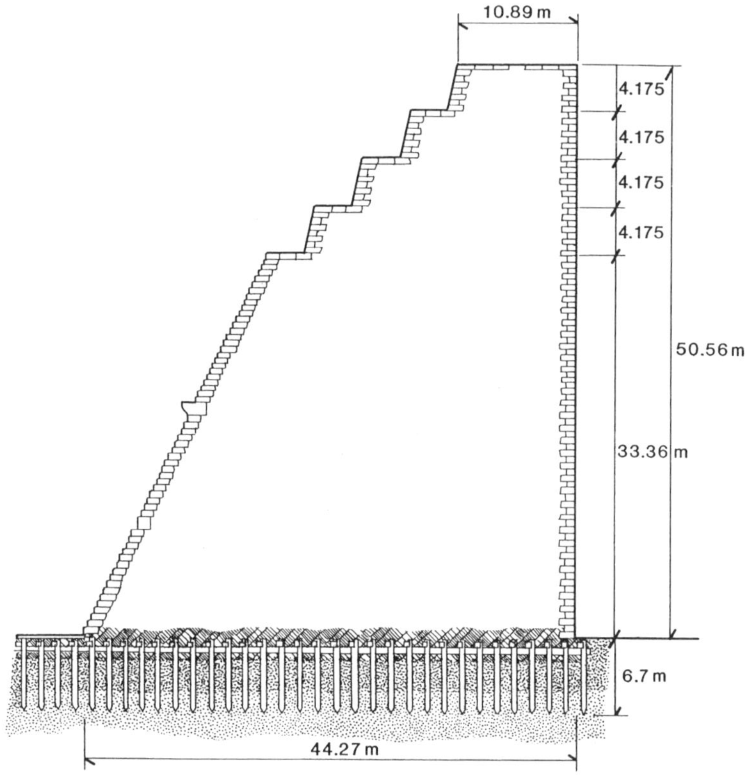

Figure 8.7. Cross-section of the Puentes Dam (after Schuyler, 1909). Note the wooden pile foundation (6.7 m deep) below the masonry structure.

the water level began to rise and reached 46.8 m above the base of the dam in April. The spillway was not used before the dam collapse.

The failure occurred on 30 April 1802. At the time of the accident, the water depth was 33.4 m and the mud deposited in the reservoir was 13.4m deep. The water level was located below the dam crest. Although the dam in itself was sound enough, the upstream hydrostatic pressure blew a hole all the way underneath the dam. Immediately after, the central lower part of the masonry collapsed and the stored water (~ 8 Mm^3) was quickly discharged. The reservoir was drained in less than one hour. The flood hit the downstream city of Lorca and 608 people were drowned.

After the rupture, the dam wall and abutments were still standing on the hillsides for many decades (see photograph in Schnitter, 1994). The dam breach was 17 m wide and 33 m high. The collapse was caused by the failure of the dam foundation.

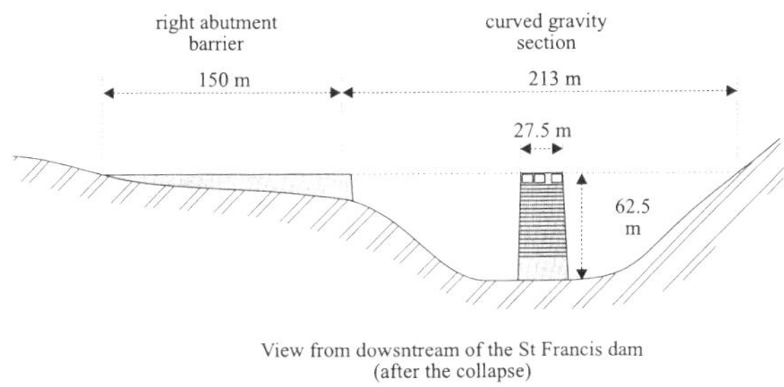

A.

B.

Figure 8.8. Catastrophe of the St. Francis Dam, California (USA, 1928). A. View from downstream of the dam after the collapse. B. View from downstream of the remnant part after the dam collapse.

St. Francis Dam (USA)

The St. Francis Dam, completed in 1926, was located in the San Francisquito Canyon, 72 km north of Los Angeles. The structure was a 62.5 m high curved concrete gravity dam. The dam thickness was 4.9 m at the crest and 53.4 m at the lowest base elevation. The dam was equipped with 11 spillway openings (2.8 m^2 each) and 5 outlet pipes (∅ 0.7 m). The spillways discharged onto the downstream stepped face of the dam (h = 0.4 m). The overflow spillways were never used before the failure. The reservoir began filling in 1926 and was nearly full on 5 March, 1928. The volume of the reservoir was 46.9 Mm^3 at that date. The dam site was investigated on the morning of 12 March, 1928; nothing in the performance of the structure was judged "hazardous" (Outland, 1963).

On the evening of 12 March, 1928, the collapse came all at once between 11:57 and 11:58 pm. The dam broke completely, releasing the 46.9 M^3 of water over 70 minutes. The peak discharge just below the dam reached 14,200 m^3/s. The resulting surge devastated bridges, highways, railroads, a construction camp and the town of Santa Paula (70 km downstream) before reaching the sea. The death toll was about 450 deaths. During the collapse, a single erect monolith survived essentially unmoved from its original position (Fig. 8.8). After investigations, the remnant part was blasted by dynamite on 23 May, 1929.

The results of the investigation showed that the dam failure was caused by a combination of a massive landslide on the left abutment and uplift pressure effects. Afterwards, geological and geotechnical investigations revealed the very poor quality of the geological setting. Jansen (1983) stated that "in view of the many deficiencies of the site, the survival of the structure for 2 years is remarkable indeed!". Discussions (Noetzli, 1931; Wiley, 1931) suggested undoubtedly the presence of large cracks in the interior of the gravity dam. Although the spillway was never used, the writer wonders what could have been the effects of water release on the downstream stepped concrete face! Stepped spillway flows induce large hydrodynamic forces on the chute bottom and cavity recirculation creates fluctuating pressures on the step faces. The interactions between the large fluctuating efforts and the internal cracks could have been disastrous.

Silverleaf Weir (Australia)

The Silverleaf Weir is located on the Barambah Creek, a tributary of the Burnett River about 14 km north-west of Murgon, QLD, Australia (Fig. 7.16). The dam was built between 1951 and 1953 for stock water. The 5.1 m high weir was designed as a basic timber crib piled weir with concrete protection of the steps (4 steps). The upstream wall was sealed with a (yellow) clay embankment held by boulders set in mortar. Note that concrete was added later to consolidate the clay protection. The dam was equipped with one bottom outlet. Silverleaf Weir is located today downstream of a major dam (Bjelke-Petersen Dam, 1988). It used to regulate irrigation along Barambah Creek, as a buffer volume between the Bjelke-Petersen Dam and downstream users. The weir was refurbished in 1995 with installation of a new outlet, construction of an upstream vertical concrete wall to replace the clay fill and to heighten the weir crest by an additional concrete step, and construction of a counterweir downstream.

During the 1995 refurbishment, an earthfill cofferdam was built 20 m upstream to de-water the weir. At one point in the repair work, a major flood event took place. On the evening of 7 February, 1995, the cofferdam was overtopped. The resulting flood wave submerged the weir. Altogether, the structure was overtopped for 24 hours and seepage through the timber cribs lasted a further 12 hours before repairs could start. After the event, inspections showed that the weir sustained the surge wave without major damage. The overtopping was filmed and the document highlights the good stability of the structure, the significant energy dissipation during large overflow and seepage through the rockfill material during the receding flood surge. The Silverleaf Weir refurbishment was completed as planned and the weir has been in use since. Overall, the experience gained from the accident suggests that the weir design was sound and the structure very stable despite its age.

Sreeramadavara and Muddoor weirs, Mysore, India

Sankey (1867) described the accidents of stepped weirs in Mysore, India. The Sreeramadavara Weir was originally a mass of rubble and large stones. The downstream face consisted of stones "well shaped and laid in regular courses, each course projecting about 2½ feet beyond the upper one". The step height and length were respectively 0.348 m and 0.762 m (i.e. $\alpha = 21.8°$). The weir was breached five times and repaired in 1842, 1844, 1859, 1860 and 1863-64. Interestingly, the stepped design was abandoned and the latest weir had a steep smooth downstream face. Another structure, the Muddoor Weir, failed and was reconstructed with a masonry stepped overflow. The newer design had a 2.74 m long broad-crest followed by 11 steps ($H_{dam} \sim 4.4$ m, $h \sim 0.4$, $l \sim 0.3$ m).

Warren Dam (Australia)

The Warren Dam is a concrete gravity dam built in 1916. The dam was raised in 1926 and again modified in 1974. The original structure was 17.4 m high and the spillway was a staircase channel (4 steps, $h = 0.37$ m). The maximum discharge capacity was approximately 100 m^3/s, which corresponds to a 1-in-40 year flood. In 1917, the structure was subjected to a record flood of 128 m^3/s during which the dam was overtopped. No failure or damage was observed. But the inadequacy of the spillway capacity was clearly highlighted. In 1926, the structure was raised by 1.5 m and the spillway crest was modified to become an ogee profile followed by a smooth chute. The maximum discharge capacity remained 100 m^3/s. More recently, investigations of the safety of the dam suggested that the structure should pass a design flood of 177 m^3/s (1-in-100 year flood). In 1974, the spillway was modified again. The spillway crest was lowered by 1.07 m to allow the 177m^3/s discharge without overtopping the dam.

Table 8.2. Summary of severe accidents (sorted by year of construction).

Dam/Reservoir	Ref.	Year(s) of construction	Date of accident	Design/Structure	Accident/Failure	Lives lost
(1)	(2)	(3)	(4)	(5)	(6)	(7)
Smooth spillways						
South Fork (Johnstown) Dam, USA	(SM)	1839	1889, May	Earthfill dam (H_{dam}=22 m).	Dam break following overtopping because of inadequate spillway capacity.	over 2,000
Habra Dam, Algeria	(SM)	1873	1881, December	Masonry gravity dam (H_{dam}=38 m). Reservoir capacity: 30 Mm^3.	Dam wall breach caused by inadequate spillway capacity.	209
Pinet Dam, France	(LE, SO)	1927-1929	1982, November	Concrete gravity dam (H_{dam}=40 m). Gated spillway (3,500 m^3/s).	Spillways blocked by large debris (e.g. trees, camping cars).	None
Palagnedra Dam, Switzerland	(BRU, JA)	1952	1978, 7 Aug.	Curved gravity dam (H_{dam}=72 m). Spillway: 450 m^3/s.	Dam overtopping caused by a combination of large flood (Q_w =1,000 m^3/s) and large amount of wooden debris (>25,000 m^3).	24
Belci Dam, Romania	(DI)	1958-1962	1991, 28-29 July	Earthfill dam (H_{dam}=18 m). Concrete gated spillway (1,400 m^3/s).	Dam overtopping and breach (caused by a failure of gate mechanism).	97
Glen Canyon Dam, USA		1966	1983	Concrete arch dam (H_{dam}=216 m). Tunnel spillways (Ø=12.5 m, 2,900 m^3/s each).	Scour initiated by cavity damage.	None
Karun Dam, Iran		1975	1975-1993	Concrete arch dam (H_{dam}=200 m). Gate overflow spillways (3,000 m^3/s).	Large scour of flip bucket initiated by cavitation damage.	None
Tous Dam, Spain	(UT)	1977	1982, 20 October	Embankment dam (H_{dam}=36.5 m). Gated concrete spillway (7,000 m^3/s).	Dam break following overtopping. Collapse caused by electrical failure to open gates and human error.	None

Notes: (–): unknown information
Ref.: (BRU) Bruschin et al. (1982); (DI) Diacon et al. (1992); (JA) Jansen (1983); (LE) Lefranc (1992); (SM) Smith (1971); (SO) Soyer (1992); (UT) Utrillas et al. (1992).

8.2.2 Related severe accidents

Stepped chutes may be also be subjected to severe accidents usually experienced with any type of spillway design (Table 8.2). This may include dam overtopping, piping through embankment dams, foundation failure (e.g. Malpasset Dam in 1959). Dam overtopping may be caused by inadequate spillway capacity (e.g. Warren, Habra, Tous, Johnstown), by a reduction of spillway capacity induced by debris obstruction (e.g. Palagnedra, Pinet) and it may even result from human error (e.g. Belci). High-head spillways may be subjected to severe cavitation damage (e.g. Karun, Glen Canyon), spillway toe erosion at hydraulic jump, scour downstream of flip bucket. The experience gained from the accidents may prove useful for stepped chute designers and operators (Table 8.2).

8.3 DESIGN SAFETY FOR STEPPED CHUTES AND SPILLWAYS

8.3.1 *Introduction*

A safe design of stepped channels must provide adequate flood discharge facilities, safe channel operation and appropriate control of the water releases. Possible material deterioration must be also taken into account.

1. Adequate spillway facilities.
The handling of flood waters can be achieved by a combination of water storage and water releases in the main channel or in auxiliary structures. A safe design procedure must include: (a) an accurate estimation of the Maximum Probable Flood event, (b) the selection of the maximum channel discharge and eventually (c) the provision for auxiliary flood releases (e.g. fuse spillways, dam overtopping).

2. Safe water release in a stepped channel.
The dynamic forces exerted by waters flowing over channels are: (a) the static load of the water, (b) the erosive (or scouring) force resulting from the high velocity pouring over the channel and the bottom shear stress associated with the flow resistance, (c) the abrasion of suspended sediment and (d) the impact force of floating ice, trees and debris of various kinds. With stepped channels, a safe water release implies a good quality of construction because the hydrodynamics of water releases induce large fluctuating efforts on the steps.

3. Appropriate control of flood release.
The crest may be gated or non-gated. Gated spillways may be operated manually or automatically. The gate must be operated properly: e.g. operator faults caused the Belci Dam failure. It should function satisfactorily: e.g. electrical failure (Tous Dam), obstruction by debris (Pinet Dam). Ungated channels do not rely upon an opening system, but they might start to operate without warning and induce downstream damage. They might also be obstructed by debris.

4. Channel material deterioration.
From the time of construction and during all the life of the structure, the chute material is subjected to natural hazards. A partial or complete deterioration of the channel material (e.g. concrete, gabion, timber) can be responsible for substantial damage during subsequent channel operations. For concrete channels, the concrete might be deteriorated by chemical reaction (e.g. alkali-aggregate reaction), freezing-thawing, settlement and cracking of the structure, vibration

of the structure caused by water surge or equipment operation, erosion of the concrete by cavitation or debris impacts.

8.3.2 *Specific problems of stepped channels*

Over twenty documented accidents and failures are listed in Table 8.1. Most cases occurred during overflow. Two dominant modes of failures were observed: basic design errors and failure processes that are specific to the stepped chute design (Table 8.3). Basic design errors cover improper hydrological assessment of the catchment. Both the Warren Dam and the Narora Weir were overtopped by flood flows significantly larger than the design discharge and this led to the failure of Narora Weir. Foundation failure is another basic design error. The complete failures of Puentes Dam, St Francis Dam and Bonshaw Weir could have been avoided with sound geological and geotechnical studies prior to construction.

Table 8.3 highlights three series of failures that are directly related to the stepped chute design in the writer's opinion. Firstly, a significant number of failures occurred during overflows at the transition flow regime: e.g. New Croton, Arizona Canal, Lahontan, Gyrandra. These flow conditions are characterised by rapid longitudinal flow variations and fluctuating flow properties. The instabilities could cause fluctuating hydrodynamic loads on the steps and unnecessary vibrations of the structure. In three well-documented cases (Arizona Canal, New Croton, Gyrandra) failures occurred during overflow conditions, very close to the transition flow conditions (Section 8.2). In another case (Lahontan), the design flow of the right spillway corresponded to a transition flow. Extensive damage to that chute was experienced to the point that it was considered to pass "flood flows over the left side only" to reduce "the cost of repairs" (Douma & Goodpasture, 1940). Basically designers must avoid the transition flow conditions and consider additional hydraulic and structural tests if they cannot avoid a flow regime transition. If the design flow regime is nappe flow, all the flow conditions up to design flow conditions will be nappe flow. If the design flow is a skimming flow, the transition flow regime is unavoidable for some flow rate below design discharge. For gated spillways, suitable gate operation must be selected to avoid the transitory flow conditions. For ungated spillways, it is recommended to select a spillway geometry (step height, width) such that the transition occurs at low flow rates.

The quality of construction is critical. Larger hydrodynamic loads are experienced on the steps compared to those on a smooth invert. In nappe flow, the impact pressure of the free-falling nappe is at least 10 times larger than the corresponding hydrostatic pressure (Section 8.3.3). In skimming flow, the boundary shear stress is about 10 times greater in average than that in a smooth chute flow, for the same flow rate and bed slope. Concentrations of shear stress exist next to the step edges. Several failures may be attributed to poor (or below standard) quality of work. At New Croton and St. Francis dams,

Table 8.3. Analysis of accident and failure cases.

Mode of failure (1)	Description (2)	Spillway/weir (3)
Design errors	Improper hydrological assessment	Narora, Warren, Debris dams in Taiwan
	Foundation failure and stability problems	Puentes, St Francis, Bonshaw, Goulburn
	Inadequate interactions between structural and hydraulic engineers	Gyrandra
Problems specific to stepped chutes	Flow instabilities associated with transition flow (between nappe and skimming flow conditions)	Arizona Canal, Gyrandra, Lahontan, New Croton
	Maintenance (lack of)	Kobila, Arizona Canal, Hilliard Creek, Binda
	Quality of construction	New Croton, St Francis (cracks) Lahontan (concrete) Bonshaw (timber) Moscovite earth dams

large cracks were observed along the spillway and within the dam masonry below the spillway[2]. In the Moscow region, two overflow earth dams failed because of poor quality drainage-filter layers. The failures of Bonshaw Weir and Lahontan Dam spillway similarly derived from lower construction standards.

Maintenance is another important issue for a safe operation. During overflow, damage might occur and repair works must be regularly conducted. Lack of maintenance may lead to partial or complete failures: e.g. Arizona Canal, Kobila and Binda weirs. Experience has shown that several stepped structures have behaved satisfactorily over more than a century with adequate maintenance: e.g. Kobila Dam in operation for 362 years before failure, Pabellon Dam built in the 1730s and still in use today, and Gold Creek dam with 110 years of operation of its concrete stepped spillway.

Another consideration is that an increase of discharge induces a significant reduction of the rate of energy dissipation for any type of flow regimes (Section 5.4). A flood discharge larger than the maximum design flow could lead to disastrous damage at the spillway toe. The selection process of the Probable Maximum Flood discharge must take into account the risks of failure and the associated costs with the underestimation of the maximum discharge capacity of the channel.

Experience of recent stepped chute overflows

A number of recent overflow events are summarised in Table 8.4. The experience provides an useful comparison with failure cases (Table 8.1). It further gives practical information on the behaviour of modern construction materials (e.g. RCC).

[2]For New Croton Dam, see Marsh (1957); for St. Francis Dam, see Noetzli (1931) and Wiley (1931).

Table 8.4. Recent examples of stepped spillway overflow events.

Dam (1)	Ref. (2)	Date / Period (Cumulative spillage duration) (3)	Overflow (4)	Remarks (5)
Concrete dam spillway				
Gold Creek Dam, Australia 1890	(CW)	1890-2000	Numerous overflows. Last in May, 1996.	Non-reinforced concrete. H_{dam} = 26 m, h = 1.5 m, l = 4 m, W = 55 m, 12 steps.
RCC dam with protective layer				
De Mist Kraal Weir, South Africa, 1986	(GO)	1986-1995 (187 days)	Maximum spillage: 2.94 m^2/s.	No damage. Design discharge: 29 m^2/s. Step height: 1 m.
Zaaihoek Dam, South Africa, 1986	(GO)	1988-1995 (44 days)	Maximum spillage: 0.17 m^2/s.	No damage. Design discharge: 15.6 m^2/s. Step height: 1 m.
Wolvedans Dam, South Africa, 1989	(GO)	1989-1995 (206 days)	Maximum spillage: 0.65 m^2/s.	No damage. Design discharge: 12.4 m^2/s. Step height: 1 m.
M'Bali, RCA, 1990	(BI)	1990-1991	q_w > 1.7 m^2/s for 3 months.	No damage. Design discharge: 16 m^2/s. Step height: 0.8 m.
Gameleira Dam, Brazil, 1991	(KU)	1991-1995	Operation every year with discharges up to 1.7 m^2/s.	No damage. Design discharge: 7.4 m^2/s. H_{dam} = 29 m, W = 150 m.
Shuidong Dam, China, 1994	(HZ)	1-3 May, 1994	Maximum discharge: 90 m^2/s.	No abnormal operation. No damage. Design discharge: 100.2 m^2/s. Step height: 0.9 m (chamfer: 0.08 m).
RCC dam without protective layer				
Kerrville Dam, USA, 1986 (refurbishment)	(SS)	1986	Maximum discharge: 3,500 m^3/s (20 m^2/s). Flood flow carrying logs, trees and other debris.	No damage. H_{dam} = 8 m, h = 0.3 m (unprotected RCC), α = 45 °.
		1986-1995	5 spillages.	No damage.
Brownwood Country Club Dam, USA, 1984	(MH)	1985-1992	6 spillages (up to 0.3 m^2/s).	No damage. Design discharge: 2.3 m^2/s. H_{dam} = 6 m, h = 0.23 m, α = 26.6° (unprotected RCC).
Ocoee Dam No. 2	(HM)	1980-1995	Regular intentional overtopping for river rafting (at least 82 times per year).	No damage. Timber crib dam with RCC buttress.
		1990	Major flood: 11.93 m^2/s	No damage.
Trigomil Dam, Mexico, 1992	(SG)	January, 1992	Major flood (1,017 m^3/s) near completion.	No damage. Unprotected RCC steps in the upper spillway section, followed by a smooth chute and stilling basin. α = 51.34°, h = 0.3 m (unprotected RCC), W = 75 m.
Camp Dyer Diversion Dam, USA, 1992	(DF)	January, 1993	Maximum discharge: 226 m^3/s.	No damage. Design discharge: 260.5 m^2/s, W = 265.8 m. H_{dam} = 21 m, h = ?? m (unprotected RCC).
Precast concrete blocks				
Dneiper Hydroplant, Ukraine	(GR)	1976 (10 hours)	Full-scale tests of concrete block systems downstream of sluice gate: V = 23 m/s q_w = 1.8 m^2/s, d = 0.5-0.7 m.	No damage. h = 0.405 m, l = 2.63 m, W = 14.2 m, α = 8.75°, 12 horizontal steps.

Table 8.4. Continued.

			$q_w = 10.7\ m^2/s$, d = 0.7-1.5 m $q_w = 17.7\ m^2/s$, d = 1-2 m $q_w = 26.5\ m^2/s$, d = 1.4-2.5 m $q_w = 35.5\ m^2/s$, d = 1.7-2.7 m $q_w = 59.4\ m^2/s$, d = 2.5-3.0 m.	
Dneister Station cofferdam	(PR)	198?	$q_w = 13\ m^2/s$.	No damage. $\alpha = 12.5°$, h < 0.25 m.
Timber crib weirs				
Cunningham Weir, Australia, 1954	(CH)	January/February, 1956 (6 weeks)	Maximum recorded stream heights: 11.13 m in January ($q_w \sim 59.6\ m^2/s$) and 8.76 m in February, 1956 ($q_w \sim 32.5\ m^2/s$). High tailwater levels.	Timber crib weir held with timber piles. H_{dam} = 4 m, W = 58.5 m, h = 0.9 m, l = 2.4 m, Design flow: 1,560 m^3/s, 4 horizontal steps.

Notes: H_{dam}: dam height; h: step height; W: chute width; a: spillway slope.
References : (BI) Bindo et al. (1993); (CH) Present study; (CW) Chanson & Whitmore (1998); (DF) Dolen & von Fay (1993); (GO) Geringer & Officer (1995); (GR) Grinchuk et al. (1977); (HM) Hansen & McLean (1995); (HZ) HE and Zeng (1995); (KU) Kuperman (1995); (MH) McLean & Hansen (1993); (PR): Pravdivets & Bramley (1989); (SG) Sanchez-Bribiesca & Gonzalez-Villareal (1996); (SS) Schrader & Stefanakos (1995).

Figure 8.9. The Dneiper Hydroplant test section, looking upstream (courtesy of Prof. Y. Pravdivets).

The Shuidong Hydropower Station experienced the largest recorded overflow (He & Zeng, 1995). The peak flow reached 90 m^2/s and it was filmed. The videotape document showed no abnormal operation and inspections after the flood event showed no sign of damage. Note that the RCC dam was protected by conventional concrete steps (h=0.9 m) with a 8-cm chamfer[3].

In North America, the Ocoee Dam No. 2 is intentionally overtopped for

[3]A chamfer is sometimes introduced for ease of construction (e.g. Shimagawa Dam, Japan).

Table 8.5. Examples of stepped spillway refurbishment associated with an increase of the maximum spillway discharge capacity.

Dam	Original maximum discharge capacity	New maximum discharge capacity	Remarks
(1)	(2)	(3)	(4)
Ternay Dam 1867, France	114 m^3/s (?)	–	Unlined rock chute. Refurbishment of spillway intake in the, 1990s.
Pas du Riot 1873, France	90 m^3/s (?)	–	Masonry construction. Enlargement and lowering of spillway crest in the, 1980s.
Gold Creek Dam, 1890, Australia	200 m^3/s	280 m^3/s (1975)[a] 280 m^3/s (1998)[a]	Concrete steps. h = 1.5 m. W = 55 m. Lowering of spillway crest in each case (broad-crest design).
La Tâche Dam 1891, France	–	–	Unlined rock chute. Enlargement and lowering of spillway crest in the, 1990s.
Cantref Dam, 1892, UK	157 m^3/s	157 m^3/s	Masonry construction. Spillway refurbishment (between, 1984 and, 1989) by addition of a reinforced grass spillway in the middle of the dam (40 m^3/s) and two siphons (73 m^3/s) for PMF outflow of 255 m^3/s.
Winterburn Dam, 1893, UK	110 m^3/s	270 m^3/s (1995)	Masonry construction. h = 2.26 m.
L'Echapre Dam, 1897, France	34 m^3/s	60 m^3/s (1989)	Unlined rock (gneiss). h = 3.2 to 4.1 m. W = 3 m. New spillway crest.
Silverleaf Weir, 1953, Australia	–	3.660 m^3/s[b]	Timber crib weir. Weir heightening by 1m concrete wall in 1995 and addition of new bottom outlet.

Notes: (–) data not available ; [a] see Section 8.4; [b]: bypass flood plain on right bank.

river rafting over 82 times per year. Large overflows were experienced during a flood in 1990 without major damage. The series of tests conducted at the Dneiper Hydroplant demonstrated the soundness of the concrete block protection systems (Fig. 8.9). Additional tests were conducted in laboratory (e.g. Baker, 2000) and in the field (e.g. Baker, 1994) in the UK.

In Queensland (Australia), the experience of the Cunningham Weir and Silverleaf Weir demonstrated the soundness of the timber crib piled weir. In 1956 the Cunningham Weir was overtopped for more than 2 months, with a maximum head above crest in excess of twice the weir height. Only minor damage was experienced and the weir is still in good condition today[4].

Refurbishment of old stepped chutes

Several old stepped chutes have been in use for over a century. These include masonry chutes[5], unlined rock cascades[6], the world's first non-reinforced concrete chute[7], a composite cribmasonry structure[8], all of which have behaved

[4]Dr. Yasuda and the writer inspected the weir in February 1998.

[5]e.g. Cantref Reservoir (UK, 1892), Goulburn Weir (Victoria, 1891) and New Croton Dam (USA, 1906).

[6]For example, the Ternay and La Tâche dams (France, 1867 and 1891).

[7]That is, Gold Creek Dam spillway, 1890, Australia (Chanson & Whitmore, 1998).

[8]That is, the Malmsbury Dam, 1870, the first large stepped spillway in Australia (Chanson, 1997d).

satisfactorily thanks to adequate maintenance. The experience learned from these successful structures highlights one critical issue, namely refurbishment (Table 8.5).

New hydrological data and risk assessment methods have led to an increase in discharge capacity of existing stepped spillways in Europe, America and Australasia. Chanson & Whitmore (1998) argued that the discharge capacity is often increased by a new spillway intake: e.g. Pas-du-Riot, Gold Creek and La Tâche dams (Table 8.5). In such a case the designer must check the new flow conditions on the spillway, the type of flow regime, and the magnitude of the hydrodynamic forces to ensure that the original stepped chute can pass the new maximum discharge without damage. The larger residual flow energy associated with the new maximum discharge might induce some damage downstream of the spillway.

A review of recent refurbishments (Table 8.5) showed that many engineers were unaware of the hydraulics of stepped cascades. They did not check the effects of a large flow rate on the stepped chute and did not assess the needs of a downstream stilling structure. One case (Gold Creek) is discussed in Section 8.4.

8.3.3 *Hydrodynamic forces on the steps (pressures and pressure fluctuations)*

The pressures and pressure fluctuations on the channel steps are important factors affecting the design and safety of stepped channels. The designer must analyse separately each flow regime, and physical modelling is strongly advised. Information on skimming and nappe flows are provided below.

Nappe flow regime

In a nappe flow situation, large hydrodynamic forces are exerted on the steps at the impact of the falling nappe, below the hydraulic jump and on the vertical face if the nappe is not adequately ventilated (Fig. 8.10). At nappe impact, large bottom pressures (i.e. larger than hydrostatic) are observed next to the impact location. The re-analysis of drop structure experiments suggests that the mean stagnation pressure P_s can be correlated by:

$$\frac{P_s - P_{atm}}{\varrho_w * g * h} = 1.253 * \left(\frac{d_c}{h}\right)^{0.349} \tag{8.1}$$

where P_s is an absolute pressure. Equation 8.1 is compared with experimental data on Figure 8.10B (Table 8.6). Although no information is available on the dynamic pressure fluctuations at drop structures, experimental results obtained with vertical plunging jets (Table 8.7), might provide an order of magnitude of the extreme maximum and minimum pressures at nappe impact:

$$P_s + 0.9 * \varrho_w * \frac{V_i^2}{2} \quad \text{instantaneous maximum pressure at nappe impact} \quad (8.2)$$

$$P_s - 0.6 * \varrho_w * \frac{V_i^2}{2} \quad \text{instantaneous minimum pressure at nappe impact} \quad (8.3)$$

where V_i is the impact velocity of the free-falling nappe (e.g. App. 3).

Downstream of the free-falling nappe, a hydraulic jump may take place (nappe flow sub-regime NA1). Although the mean pressure is quasi-hydrostatic below the jump, large pressure fluctuations are observed. The re-analysis of bottom pressure fluctuation records over long periods (Table 8.7) indicates that the instantaneous maximum and minimum pressures are about:

$$P_{hyd} + 0.6 * \varrho_w * \frac{V_1^2}{2} \quad \text{instantaneous maximum pressure below jump} \quad (8.4)$$

$$P_{hyd} - 0.4 * \varrho_w * \frac{V_1^2}{2} \quad \text{instantaneous minimum pressure below jump} \quad (8.5)$$

where P_{hyd} is the local hydrostatic pressure and V_1 is the upstream flow velocity. The velocity V_1 may deduced by combining the momentum equation and

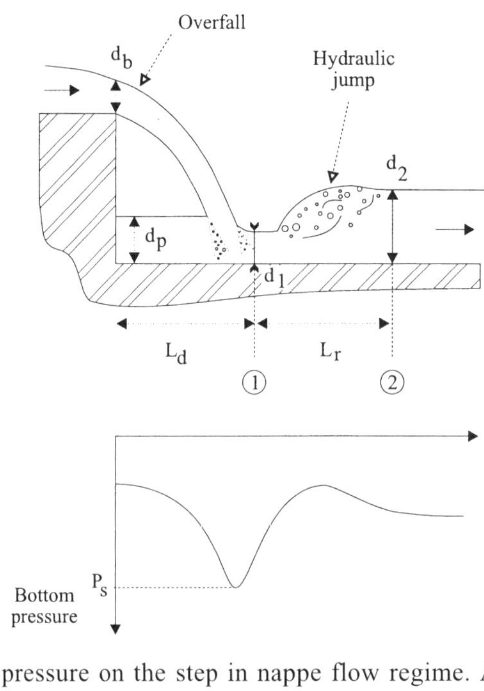

Figure 8.10. Mean pressure on the step in nappe flow regime. A. Definition sketch.

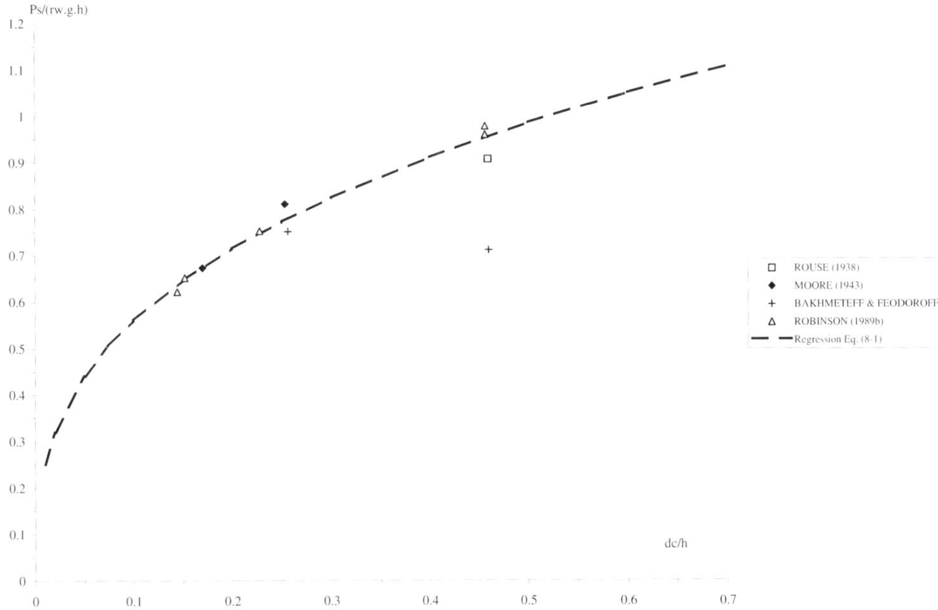

Figure 8.10. Mean pressure on the step in nappe flow regime. B. Mean stagnation pressure at the impact of a free-falling nappe. Experimental data: Rouse (1938), Moore (1943), Bakhmeteff & Feodoroff (1943), Robinson (1989).

continuity equations at nappe impact (e.g. App. 3). Underneath a hydraulic jump, the extreme minimum pressures might become negative, leading to uplift loads under the step bottoms. If the free-falling nappe is not adequately ventilated, subpressures might occur in the nappe cavity and add a suction force on the vertical face of the steps.

In summary; in a nappe flow regime, the horizontal face of the steps must be designed to sustain large positive and negative pressures for a wide range of flow conditions. As the location of the falling nappe impact and the position of the hydraulic jump on the step are functions of the discharge (Chapter 3), most of the step length must be adequately reinforced to sustain the various loading situations.

Skimming flow regime

In a skimming flow regime, the energy dissipation occurs within the flow by maintaining the eddy recirculation in the step corners beneath the pseudo-bottom formed by the step edges. The mean boundary shear stresses (i.e. $\tau_o = f*\rho_w*U_w{}^2/8$) are about 10 to 20 times greater than those on smooth channels. The steps must be carefully designed to sustain such large shear stresses including concentrations of stresses next to the step edges. The stresses on the step edges may be reduced by rounding the edges (e.g. Croton Falls Dam) or using a chamfer (e.g. Shuidong, China; Nakasujigawa, Japan). The effects of

Table 8.6. Measurements of pressures and pressure fluctuations in nappe flows and related flows.

Reference (1)	q_w m^2/s (2)	d_c/h (3)	d_c/W (4)	Remarks (5)
Drop structure				
Rouse (1938)		0.459		Aerated nappe. Sharp-crested weir.
Moore (1943)	0.0673 to 0.123	0.253 to 0.17	0.278 to 0.415	Aerated nappe. W = 0.279 m.
Robinson (1989)	0.0623 to 0.124	0.145 to 0.456	0.0803 to 0.127	Aerated nappe. W = 0.914 m. Low tailwater depths.
Bakhmeteff & Feodoroff (1943)	0.0556 to 0.134	0.256 to 0.460	0.670 to 1.201	Broad-crested weir with rounded edge. W = 0.102 m.
Vertical plunging jet				
May & Willoughby (1991)	0.225 to 0.445	N/A	0.864 to 1.36	Bi-dimensional jets (W = 0.2 m). 3.3 < V < 6.6 m/s. Scanning time: 45 minutes.
Hydraulic jump				
Vasiliev & Bukreyev (1967)	0.127	N/A		Fr_1 = 5.75. Jump downstream of a sluice gate.
Abdul Khader & Elango (1973)	0.034 to 0.126	N/A	0.082 to 0.196	4.7 < Fr_1 < 6.6. W = 0.6 m. Jump downstream of a smooth weir. Scanning rate: 100 Hz. Scanning time: 2.5 to 900 s.
Toso & Bowers (1988)		N/A		3 < Fr_1 < 10. W = 0.5 m. Jump downstream of inclined slopes (0 to 45 deg.). Scanning rate: 50 Hz. Scanning time: 600 s to 24 hours.
Fiorotto & Rinaldo (1992)	0.0484 to 0.122	N/A	0.207 to 0.384	5 < Fr_1 < 9.5. W = 0.3 m. Scanning rate: 50 to 100 Hz. Scanning time: 164 s to 12 hours.

chamfer and rounding on the flow resistance have not been investigated in details.

The efforts and the distributions of pressure and pressure fluctuations on the vertical and horizontal faces of the step are functions of the flow recirculation mechanisms. For each flow pattern (i.e. wake-step interference SK1, wake-wake interference SK2 and stable recirculation SK3 sub-regimes), different hydrodynamic loads are applied to the step faces. The recirculation mechanisms induce fluctuating pressures on both the vertical and horizontal faces of the steps. The dynamic pressure fluctuations may induce local traction and suction.

Several researchers recorded pressures on the horizontal and vertical faces of steps (Table 8.8). The results indicated consistently that: i) the mean pressures were maximum on the horizontal faces of the steps and larger than hydrostatic, ii) the mean pressures on the vertical faces were less than hydrostatic, and iii) suction (i.e. pressure below atmospheric) might eventually occur on the vertical faces, although the risks of cavitation are basically nil. Measurements also showed that the mean absolute pressures on the vertical face of the steps decreased with downward sloping steps. Although some data suggested that the pressure fluctuations were the largest next to the inception

Table 8.7. Pressures and pressure fluctuations at drop structures and below hydraulic jumps.

Experimental data (1)	Mean stagnation pressure[a] (Pa) (2)	Std of pressure fluctuations[a] (Pa) (3)	Extreme pressures[a] (Pa) (4)	Remarks (5)
Drop structure Chanson (1995a)	$P_{atm}+\varrho_w*g*h*1.25*\left(\frac{d_c}{h}\right)^{0.35}$	–	–	Low tailwater levels. Data: Moore (1943), Robinson (1989) & Rouse (1943)
Vertical plunging jet May & Willoughby (1991)	$P_{hyd}+\frac{1}{2}*\varrho_w*V_i^2$	$0.1*\frac{\varrho_w}{2}*V_i^2$	Maximum pressure: $P_{hyd}+1.9*\frac{\varrho_w}{2}*V_i^2$ Minimum pressure: $P_{hyd}+0.4*\frac{\varrho_w}{2}*V_i$	Partially developed jets.
Hydraulic jump Abdul Khader & Elango (1973)	P_{hyd}	$0.085*\frac{\varrho_w}{2}*V_1^2$	Maximum pressure: $P_{hyd}+0.26*\frac{\varrho_w}{2}*V_1^2$ Minimum pressure: $P_{hyd}-0.2*\frac{\varrho_w}{2}*V_1^2$	Partially developed U/S flow.
Toso & Bowers (1988)	P_{hyd}	$0.07*\frac{\varrho_w}{2}*V_1^2$	Maximum pressure[b]: $P_{hyd}+0.6*\frac{\varrho_w}{2}*V_1^2$ Minimum pressure[b]: $P_{hyd}-0.4*\frac{\varrho_w}{2}*V_1^2$	Partially and fully developed U/S flows.
Fiorotto & Rinaldo (1992)	P_{hyd}	$0.08*\frac{\varrho_w}{2}*V_1^{2(c)}$	Maximum pressure[d]: $P_{hyd}+0.5*\frac{\varrho_w}{2}*V_1^2$ Minimum pressure[d]: $P_{hyd}-0.4*\frac{\varrho_w}{2}*V_1^2$	Partially developed U/S flow.

Notes: [a]: absolute pressure; [b]: scanning time = 24 hours; [c]: scanning time = 164 s; [d]: scanning time = 12 hours; Std: standard deviation; V_i: impact velocity; V_1: upstream flow velocity; P_{hyd}: hydrostatic pressure.

point, measurement techniques were affected by air bubble entrainment downstream of the inception point.

It must be emphasised that only few studies were carefully detailed (Table 8.8). There is a great scatter of data, and the measurement procedures and techniques differ from one study to another. Additional experimental work is required to obtain a better understanding of the recirculation process and to predict systematically the pressure distributions and hydrodynamic efforts.

Table 8.8. Measurements of pressures and pressure fluctuations in skimming flows.

Experimental data (1)	Slope deg. (2)	Step height h m (3)	Flow conditions (4)	Remarks (5)
Houston (1987)	72 59	0.041	–	Upper Stillwater model.
BaCaRa (1991)	53.1	0.06	$0.026 \le q_w \le 0.26$ m^2/s	M'Bali model (scale 1/10).
Frizell (1991,1992)	26.6	0.051	$q_w = 0.076, 0.58$ m^2/s	USBR model. Horizontal and sloping downward steps (δ = 0, -10, -15°).
Lejeune & Lejeune (1994)	51.43	0.019 0.038	–	M'Bali model.
Hakoishi & Sumi (2000)	54.6	0.024	$q_w < 0.018$ m^2/s	Nakasujigawa model (scale 1/31.5).

8.3.4 *Safety of control structures and gates*

Channel intakes and gates are very important for a proper flood evacuation and large water releases. They must be always ready to operate and the gate opening must not fail (e.g. Belci and Tous dams). Failure risks of equipment might be reduced by diversifying the energy supply (normal and backup circuits), by providing direct gate operating mechanisms (e.g. engines, crank handle), with periodical maintenance tests and with regular training of the operators.

In any case, the gates and intakes must be accessible and the crest should not be blocked by ice and debris. The obstruction of spillways by debris was recently highlighted by the failures of the Palagnegra and Pinet dams. In Tasmania (Australia), where eucalyptus trees grow to over 70 m high, it is usual practice to provide wide spillways with ample overhead clearance (Thomas, 1976). In industrialised areas, debris can range from trees to trucks and camping-cars (e.g. Pinet Dam). The channel intakes (and the gates) must be wide and high enough to allow the passage of these debris.

8.4 DISCUSSION

Although no dramatic accident was caused directly by the selection of a stepped channel rather than a smooth channel, Goubet (1992) addressed some critical safety issues for the selection of stepped spillways. A spillway structure is designed for 20 to 50 years and might be subjected to exceptional events: e.g. heavy flooding, spillway obstruction, gate failure. How would stepped spillways behave under such extreme conditions? Prototype experience is scarce (Table 8.4; Section 7.3.2; Chapter 2).

It has already been shown that, for a given spillway geometry, the rate of energy dissipation decreases with increasing discharges (Section 5.4). In the event of an exceptional flood leading to an overflow larger than the maximum design discharge, the large residual flow energy might induce substantial damage downstream of the spillway. Furthermore, several failures occurred during water spills: e.g. Arizona Canal, Dartmouth, Minneapolis Mill, New Croton dams. In each case, the flow rates were much less than the design discharge. What would have been the damage for the design discharge?

It must be emphasized that the flow conditions in the transition flow regime are unstable and must be avoided. Some of the failure cases (e.g. Gyrandra Weir, New Croton Dam, Lahontan Dam) occurred with transition flows. The flow behaviours induced additional fluctuating loads and efforts to the structures and the resulting fatigue might have increase the rate of failure. In two cases, large cracks were observed along the spillway (New Croton Dam) or within the dam masonry below the spillway (St. Francis Dam). During spillway operation, the interactions of the fluctuating hydrodynamic load on the cracks can lead to disasterous circumstances. The examples of the New Croton and St. Francis dams highlight the requirement for a good quality of construction work. With embankment overflow spillways, the Soviet experience has shown the importance of a good-quality drainage-filter layer. Any hydraulic structure is subjected through time to deterioration and natural hazards: e.g. debris impacts, abrasion by sediments, freeze. Stepped spillways are no exceptions (e.g. Binda Weir, Kobila Dam).

Modern design errors: an attitude problem (human nature?)

Ignorance of past events and a lack of local expertise within the engineering profession can also have an impact on the safe design of stepped spillways. Many practising engineers tend to forget too easily past experience: i.e. our engineering history and heritage. In the 1970s, most expertise in stepped spillway design was lost. Well-known hydraulic textbooks did not even mention the stepped chute design (e.g. Chow, 1959; US Department of the Interior, 1965; Henderson, 1966). Leading professionals ignored local experience. For example, the late Professor Sorensen (1985) forgot the North American stepped spillways, like the New Croton Dam near Cornell University. In the UK, Essery & Horner (1978) were unaware of the Beacons and Cantref Dam spillways (Figs 2.3, 3.1 & 5.2). Interesting designs and developments are too often overlooked, and research experience is often ignored. This situation is true everywhere. Several examples illustrate the point.

In the USA, American engineers suggested that the design of stepped spillways was a new technique developed in the 1980s with the introduction of new construction materials (e.g. RCC). But there are numerous examples of stepped spillways built in North America during the 19th and early 20th centuries (Chapter 2). For example; the Ascutney Mill Dam (1834-35) is still in use today and recognised as a National Civil Engineering Landmark by the American Society of Civil Engineers since 1970; Bridgeport Dam (1886); Titicus Dam (1895); New Croton Dam (1906); Croton Falls Dam (1911) (Figs 2.2 &

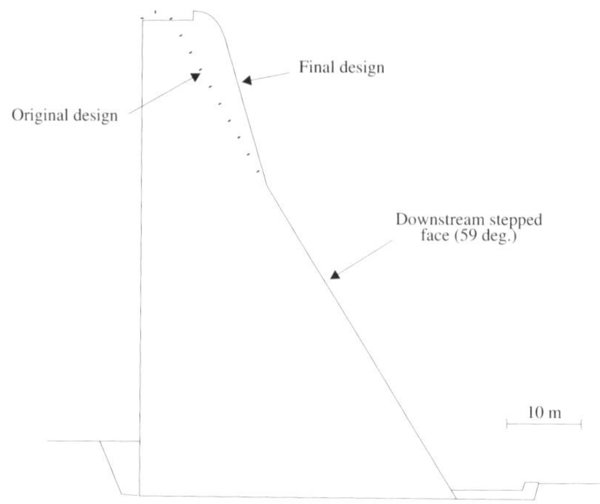

Figure 8.11. Cross-section of the Upper Stillwater Dam spillway (original design and as constructed) completed in 1987.

2.3). Another example is the spillway of the Upper Stillwater Dam, a RCC dam completed in 1987 in the USA (Fig. 8.11). The preliminary spillway design included a 4.57 m wide crest followed by a straight 59° slope chute. The final design is a 9.14 m wide crest followed by a 72° chute and then a 59° chute. "The top width was to increase to 30 feet to facilitate construction and the slope of the upper portion of the downstream face was increase to 0.32:1 so that it intersected the 0.60:1 slope at elevation 8100.0" (Houston, 1987). The break in the downstream chute slope was made to widen the crest, allowing truck access: it had no hydrodynamic or structural validity. In fact, the final design has increased the risks of jet deflection during overflows at the transition between the crest and the chute. Nevertheless, the final design was copied by other engineers: e.g. the New Victoria Dam, Australia completed in 1991 (Table 1.2).

Errors have been made twice during recent calculations for the Gold Creek Dam spillway. A leading consultant calculated incorrectly the maximum discharge capacity of the 1975 spillway to be about 400 m^3/s. However, a recent investigation showed that overtopping and associated embankment erosion would occur with spills exceeding 280 m^3/s (Chanson & Whitmore, 1998). In 1998, the Gold Creek Dam was refurbished again by raising the dam crest to 100.15 m R.L. (i.e. 1 m heightening), reinforcing the embankment downstream toe and lowering the spillway crest down to 95.75 m R.L. (i.e. 0.5 m lowering). The spillway crest was modified to pass the PMF flow of 730 m^3/s. However

[9]In January 1887, the water level was less than 1 m below Gold Creek dam crest. The spillway was discharging about 170 m^3/s (Chanson & Whitmore, 1996). In February, 1931, Gold Creek dam narrowly escaped another catastrophe when the reservoir level was barely 0.12 m below the dam crest (Cossins, 2000). At the time, the spillway crest was obstructed by a 1.2 m high concrete wall built in 1920 to increase the reservoir storage capacity.

the chute sidewalls were untouched, and sidewall overtopping and associated risk of scour would still occur with spills exceeding 280 m^3/s. Sidewall overtopping was previously experienced in January 1887, when the valve house was submerged and two men working at the valve house barely escaped (Cossins, 2000: 98)[9].

No accident has resulted from these mistakes. Nonetheless, severe accidents have been experienced in the past. It is a shame that too few engineers are willing to share their engineering failure experience with peers, for this would be of great service to the community. The teaching of stepped channel hydraulics is another issue. Of the (over two hundred) RCC dams built to date in North America, only sixty have been equipped with a stepped spillway. Why? Could it be that some researchers and engineers do not fully appreciate the needs for continuing education and undergraduate teaching of quality in hydraulic structures? It is believed that most engineers, young and senior, have never been exposed to the complexity of the stepped spillway design.

Summary

The design of stepped channels and spillways is safe if the designers are knowledgeable in stepped chute hydraulics. The hydraulics of stepped channels differs substantially from classical smooth chute calculations and it is not taught usually at student levels nor in professional courses. The author has lectured in stepped spillway hydraulics at postgraduate and undergraduate levels since 1992 in Australia (e.g. Chanson, 1999a: 313-363) and overseas (e.g. workshop lectures at Nihon University, Japan in 1998). These form a starting point for a transfer of knowledge from researchers to practising engineers. Further research investigations must be carried out to provide new information on the hydrodynamic loads on the steps and the flow behaviour. An increased knowledge of stepped channel flows will increase the safety design of stepped chutes.

CHAPTER 9

Wave phenomena and instabilities in stepped chute flows

9.1 INTRODUCTION

The design of a stepped spillway is based upon the assumption of steady flows in straight prismatic channels. However, flow instabilities and unsteady wave phenomena may be experienced. For example, shock waves, roll waves, flood waves and dam break waves. In absence of basic theory, records of wave instabilities down stepped chutes may provide useful case studies. In this section, basic wave phenomena are reviewed and stepped chute experience is documented.

9.2 SHOCK WAVES AND ASSOCIATED STANDING WAVES

In supercritical flows, any obstacle (e.g. pier, expansion, contraction) generates a surface wave moving across the flow and carried downstream. The resulting wave, called cross-wave or shock wave, is an oblique standing wave[1]. A basic example is the deflection of a vertical sidewall (Fig. 9.1). At large inflow Froude Numbers, the shock wave deflection angle θ_s equals about:

$$\theta_s \approx \beta + \frac{1}{Fr_1} \tag{9.1}$$

where β is the wall deflection angle and Fr_1 is the upstream Froude Number (Vischer & Hager, 1998).

Shock waves may be initiated by any form of obstacle: e.g. wall deflections, changes in curve slope, chute contractions and expansions, channel junctions, piers, chute bend. Cross-waves may cause standing waves when they impact on, and are reflected at, a sidewall. The supercritical flow becomes furthermore three-dimensional because shock wave development is associated with flow concentrations.

[1]Shock waves in open channel flows are analogous to the Mach shock waves in supersonic flows (e.g. Liggett, 1994; Chanson, 1999a).

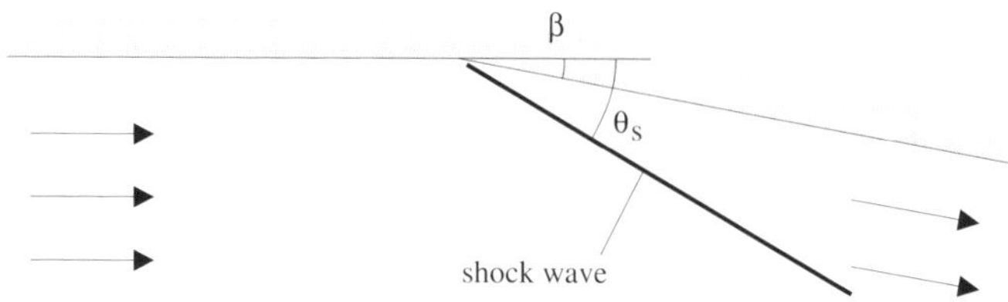

Figure 9.1. Sketch of a shock wave at a wall deflection.

Gold Creek dam spillway
Shock waves were observed at the Gold Creek Dam spillway during a spill in 1996 (Section 3.4.4). Viewed from downstream, the nappe flow exhibited some lozenges flow patterns (Fig. 3.13). At the step edges, shock wave intersections generated splashing. Standing waves and water deflections were experienced at the chute sidewalls.

Shock waves in a nappe flow without hydraulic jump (NA3)
Chanson & Toombes (1997b) observed shock waves in a 3.4° stepped chute with nappe flows without hydraulics jump (NA3) ($h = 0.143$ m, $0.35 < d_c/h < 1.0$, 10 steps). At the first drop, the inflow was smooth and two-dimensional until nappe impact. Downstream, the flow exhibited three-dimensional flow patterns with shock waves and standing waves (Fig. 9.2). Significant water splashing and jet deflection was observed associated with oblique shock waves. The cross-waves started from the sidewall near the impact of the nappe and propagated across the channel, intersecting on the channel centreline. Sidewall standing waves were seen at each step. At the first drop, the sidewall standing wave height and the cross-wave angle of deflection with the sidewall were best correlated by:

$$\frac{d_M}{E} = \frac{5.43}{Fr^{2.0}} \qquad \text{First drop (9.2)}$$

$$\theta_s = \frac{64.5}{(Fr-1)^{1.065}} \qquad \text{First drop (9.3)}$$

where E is the upstream total head measured above the downstream invert, Fr is the Froude Number at nappe impact and θ_s is in degrees (Fig. 9.2).

On the downstream steps, the dimensionless standing wave height d_M/h was nearly independent of the step number for $d_c/h < 0.84$. At greater flow rates (i.e. $d_c/h \geq 0.85$), variations in wave height were recorded from step to step (Fig. 9.2). It is believed that these fluctuations were caused by shock wave interference from the upstream step flow on the free-falling nappe and the downstream step flow. Shock wave angle data are presented in Figure 9.2C.

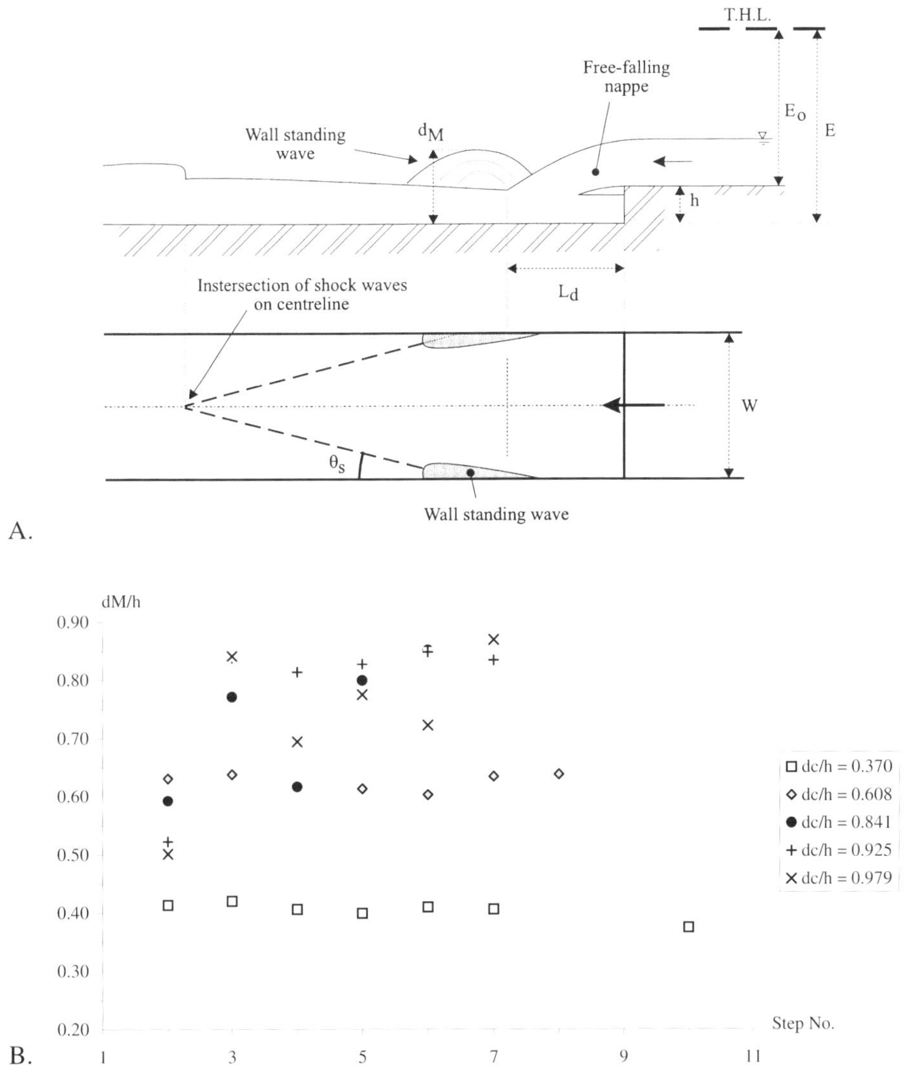

Figure 9.2. Shock wave characteristics in nappe flows without hydraulic jump (NA3) Chanson & Toombes (1997b). h = 0.143 m, W = 0.5 m, $\alpha = 3.4°$, 10 steps. A. Definition sketch of cross-waves. B. Dimensionless standing wave height d_M/h along the stepped cascade.

Effect of air entrainment on shock wave intersection

Chanson & Toombes (1997b) observed several types of shock wave intersection (Figure 9.2D). At low flow rates (i.e. $d_c/h < 0.61$), the traditional shock wave intersection was observed (Type 1(a)) for which the crosswaves intersect without significant interference[2]. Type 1(b) intersection was observed in most

[2]Type 1(a) intersection is typically observed in absence of free-surface aeration.

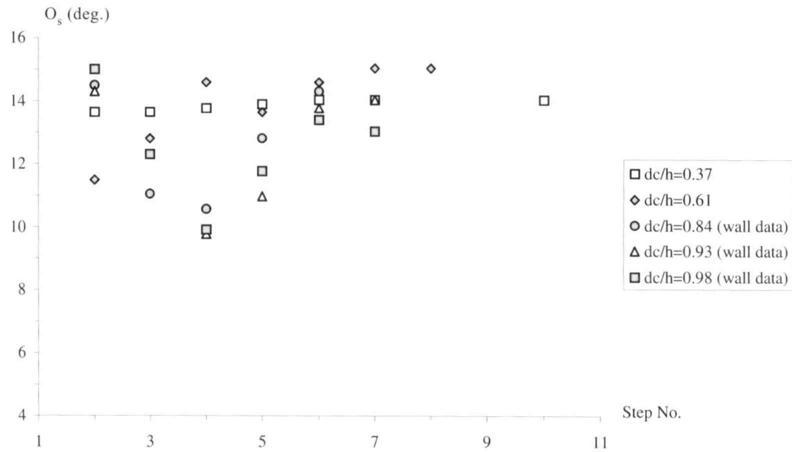

C.

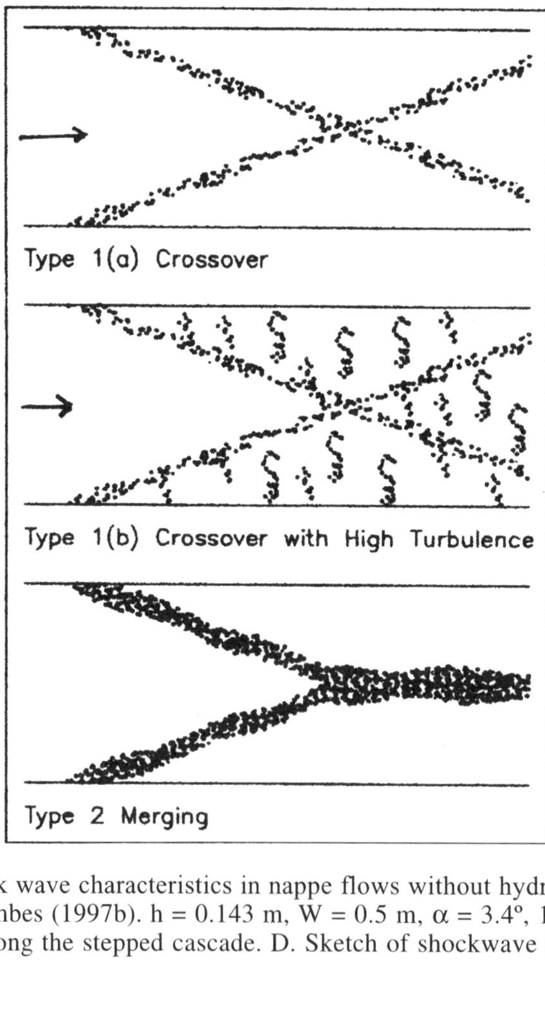

D.

Figure 9.2. Shock wave characteristics in nappe flows without hydraulic jump (NA3) Chanson & Toombes (1997b). h = 0.143 m, W = 0.5 m, α = 3.4°, 10 steps. C. Shock wave angle θ_s along the stepped cascade. D. Sketch of shockwave intersection (view in elevation).

circumstances (i.e. $d_c/h > 0.61$) downstream of the first few steps: the free surface behind the shock waves was characterised by turbulent wavelets and free-surface aeration. Downstream of the shock wave intersection, the free-surface aeration (i.e. "white waters") tended to mask the cross-waves. The free-surface aeration and turbulent wavelets appeared to increase in intensity down the channel. Downstream of the 8th step, the shock waves were completely hidden by the "white waters" at large flow rates (i.e. $d_c/h \geq 0.84$).

Type 2 shock wave intersection was observed only on the upstream steps at large flow rates (i.e. $d_c/h \geq 0.84$). The shock waves were highly turbulent and aerated. When the cross waves intersected, they merged into a straight standing wave flowing parallel to the walls. A longitudinal standing wave resulting from a shock wave intersection Type 2 extended over the next free-falling nappe.

9.3 ROLL WAVES

Roll waves are surface waves developing in steep chutes, predominantly at low flows, and caused by shear instabilities[3]. Hence, roll waves are impossible in the absence of boundary friction. Roll waves develop from small wavelets that are amplified at the upstream end. The wave growth depends primarily on the absorption of the smaller waves by faster waves. Visually, the roll waves look like a succession of small bores. Montes (1998) reviewed a number of cross-sectional shapes developed to inhibit small wave amplification, hence preventing roll waves. Unfortunately there is no information on roll wave development in stepped chutes.

A related case: Sorpe Dam spillway

Completed in 1935, the Sorpe Dam[4] spillway consists of a 100 m long crest discharging into a side channel leading to a stepped chute. The cascade is 7 m wide with 36 pooled steps. The step heights range from 0.5 to 2 m and the end sills are 0.74 m high with a rounded crest. During spillway operation on 2 November, 1998, ($Q_w = 6.9$ m^3/s), a wave phenomenon was observed in the cascade (Redeker, 2000). Surging waves rushed down the cascade, overtopping the chute sidewalls in the lower reach and causing a hazard to nearby tourists (Fig. 9.3). The event was videotaped.

[3]See Montes (1998: 585-596) for a comprehensive and modern treaty on roll waves. Liggett (1994: 317-318) developed basic equations for roll waves.

[4]The Sorpe Dam was a target of air raids during the "dam buster" campaign in World War Two. The dam was hit on 16-17 May, 1943 but it was repaired shortly after.

Figure 9.3. Photographs of Sorpe Dam spillway in operation on 2 November, 1998 (courtesy of Marq Redeker, Ruhrverband). Wave progression from right to left, top to bottom.

Discussion

It can be assumed that the wave development was a combination of several instability phenomena. The exact origin of the disturbances is not known, but it might have been flow instabilities in the side channel downstream of the crest. In the cascade, the natural period of lateral oscillation of the pools of water[5] was about 3.8 seconds on 2 November, 1998 and waves with a 3.9-s period were observed in the upper steps. Furthermore, the stepped chute operated in a transition flow regime ($0.23 \le d_c/h \le 0.92$). It is proposed that some upstream instabilities in the side channel were amplified by sloshing motion in the upper step pools and by deflected jets in the transition flow further downstream. Visually, the videotape documentary showed that the waves were amplified down the chute in a manner very similar to roll wave motion[6].

[5]That is, sloshing motion period.

[6]Visually the surging waves appear similar to huge roll waves.

9.4 FLOOD WAVES AND DAM BREAK WAVES

Flood waves resulting from dam breaks have been responsible for numerous losses. A related case was the overtopping of the Silverleaf Weir in 1995, when the upstream cofferdam was breached (Section 8.2). Stepped chutes may be subjected to similar flood waves resulting, for example, from the failure of an upstream reservoir and from some gate malfunction.

Basic theory

Considering an ideal dam break surging over a dry horizontal channel, the method of characteristics may be applied to solve the wave profile as first proposed by Ritter in 1892[7]. For a horizontal channel, the shape of the ideal surge free-surface satisfies:

$$\frac{x}{t} = 3 * \sqrt{g * d} - 2 * \sqrt{g * d_o} \tag{9.4}$$

where the longitudinal origin ($x = 0$) is the dam location, the time origin ($t = 0$) is the instantaneous dam break, d is the flow depth and d_o is the initial reservoir water depth (Fig. 9.4) (e.g. Henderson, 1966; Montes, 1998). After dam break, the flow depth and discharge at the origin $x = 0$ are constants:

$$d_{(x=0)} = \frac{4}{9} * d_o \tag{9.5}$$

$$Q_{(x=0)} = \frac{8}{27} * d_o * \sqrt{g * d_o} * B \tag{9.6}$$

and the celerity of the wave front equals:

$$C_s = 2 * \sqrt{g * d_o} \tag{9.7}$$

Although Equations 9.4 to 9.7 assume no boundary friction, model and prototype experiments showed good agreement with the theory, but for the leading edge of the wave. Bottom friction affects significantly the propagation of the leading tip and Whitham (1955) proposed an analytical solution of the wave front celerity that is best correlated by:

[7]Ritter's work was initiated by the South Fork Dam (Johnstown) catastrophe (Table 8.2).

$$\frac{C_s}{\sqrt{g*d_o}} = \frac{2}{1.0004 + 2.907*\left(\frac{f}{8}*\sqrt{\frac{g*t^2}{d_o}}\right)^{0.4255}} \quad (9.8)$$

where f is the Darcy Friction Factor. Whitham (1955) commented that his work was applicable only for $C_s/\sqrt{g*d_o} > 2/3$.

Escande et al. (1961) investigated specifically the effects of bottom roughness on dam break wave in a natural valley. They showed that, with a very-rough bottom, the wave celerity could be about 20 to 30% lower than for a smooth bed. Hunt (1988) developed an asymptotic method for sloping channels (smooth chutes). His results implied: $x_s \propto (\sin\alpha)^{2/3}$ where x_s is the distance between the wave front and the initial dam location.

Flood wave on stepped chutes

A series of laboratory experiments was conducted in a 3.4° stepped chute (h = 0.143 m, W = 0.5 m) (Fig. 9.4) and a sudden discharge release was videotaped down the Brushes Clough Dam spillway[8]. A summary of the flow conditions and results is given in Table 9.1. Visual observations showed that the flood wave propagated as a nappe flow and it was highly aerated, particularly at Brushes Clough. Experimental observations of the wave front propagation are presented in Figure 9.4C. For these data, the results are best correlated by:

$$\frac{\frac{x_s}{t}}{\sqrt{g*d_o}} = \left(12.6*(\sin\alpha)^{2/3} + 0.2\right)*\left(1.09 - \frac{0.14}{d_o/h}\right) \quad (9.9)$$

where x_s is the distance between the wave front and the start of the stepped chute, and $d_o = \frac{9}{4}*\sqrt[3]{\frac{(Q_w(t=0+)/W)^2}{g}}$ is the initial water level in an ideal dam break. Note that Equation 9.9 should not be applied outside of $3° < \alpha <, 19°$ and $0.8 < d_o/h < 2.2$.

Dam break flood wave run-up downstream of a drop structure

A series of experiments were conducted downstream of a vertical overfall (W = 0.75 m) discharging into a 20 m long horizontal flume (Chanson et al., 2000b). The initial upstream head above channel invert ranged from 1.65 m up to 2.1 m. For an initial dry channel, the data showed consistently a greater

[8]The test at Brushes Clough was conducted in 1994. The video footage was taken by Dr. R. Baker and analysed by the writer.

wave front celerity than for the classical dam break analysis for $x/d_o < 25$, where d_o is a function of the initial discharge (Eq. 9.6). Close to the nappe impact, the wave celerity was found to be nearly twice that of the dam break wave (Whitham's solution) (Fig. 9.5). The experimental data highlighted a large initial momentum of the wave despite energy loss at nappe impact. The results imply further some fluid deceleration up to $x/d_o = 25$ to 30. Practically, the experimental data for an initially dry channel were best correlated by:

$$\frac{C_s}{V_i} = \frac{0.598}{1 - 0.0336 * \frac{x}{d_o} + 0.00237 * \left(\frac{x}{d_o}\right)^2} \qquad 4 < x/d_o < 30 \quad (9.9)$$

Table 9.1. Flood wave propagation down a 3.4° chute (h = 0.143 m, W = 0.5 m).

Reference	Slope ?	Run	Q_w (t=0) m^3/s	d_o/h	$\frac{x_s/t}{\sqrt{g*d_o}}$	Remarks
(1)	(2)	(3)	(4)	(5)	(6)	
University of Queensland	3.4°					Rectangular channel: W = 0.5 m, h = 0.143 m.
		CT1	0.019	0.83	1.96	
		CT2	0.030	1.13	2.21	
		CT3	0.040	1.37	2.18	
		CT4	0.075	2.08	2.05	
		CT5	0.019	0.83	1.87	
		CT6	0.030	1.13	1.99	
		CT7	0.040	1.37	2.09	
		CT8	0.075	2.08	2.20	
Brushes Clough	18.4°					Trapezoidal channel: 2m bottom width.
		BC1	0.5	2.2	6.1	

Note: $d_o = \frac{9}{4} * \sqrt[3]{\frac{(Q_w(t=0+)/W)^2}{g}}$

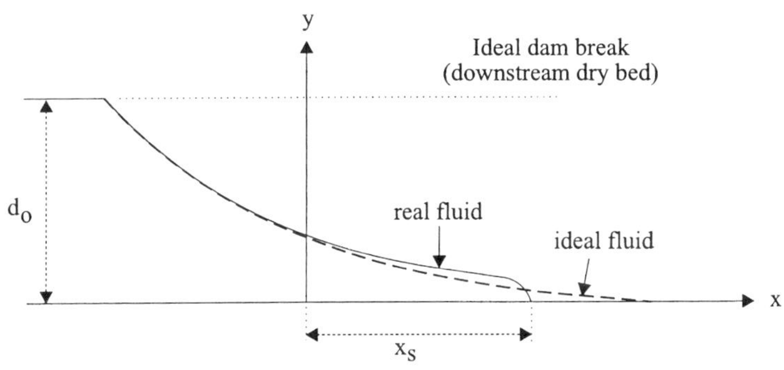

Figure 9.4. Dam break flood wave on stepped chutes. A. Idealised dam break problem.

B.

C.

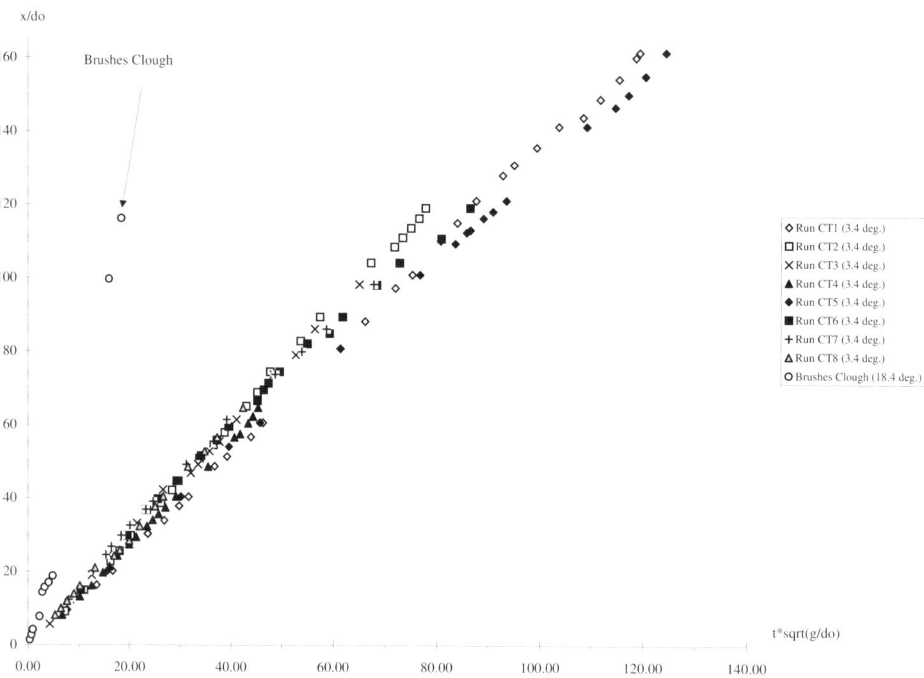

Figure 9.4. Dam break flood wave on stepped chutes. B. Flood wave propagation down a 3.4° slope stepped cascade (h = 0.143 m, W = 0.5 m). Left: Q_w = 0.030 m^3/s, t = 7.6 s, Step 7, x/L = 0.7, Run CT6. Right: Q_w = 0.075 m^3/s, t = 4.72 s, Step 5, x/L = 0.5, Run CT8. C. Experimental observations of wave front propagation down stepped chutes.

where V_i is the initial jet impact velocity. For $x/d_o > 30$, the agreement between the data and Whitham's solution was fair.

For a channel is initially filled with water, the propagation of the flood wave is a different situation because the dam break wave is led by a positive surge (e.g. Henderson, 1966). The data of Chanson et al. (2000b) showed a slightly greater wave front celerity than for the theoretical analysis for $x/d_o <$ 20 to 25. As a first estimate, the data were best fitted by:

$$\frac{C_s}{V_i} = \frac{0.598}{1 + 0.0263 * \frac{x}{d_o} + 0.00117 * \left(\frac{x}{d_o} \right)^2} \tag{9.10}$$

Equation 9.10 was obtained for $0.1 < d_1/d_o < 0.8$ and $10 < x/d_o < 25$, where d_1 is the initial water depth in the channel. Note that the result is little affected by the initial water depth within the range $0.1 < d_1/d_o < 0.8$.

In summary, the study consistently showed a larger wave celerity downstream of nappe impact for $x/d_o < 30$ (Chanson et al., 2000b). Downstream, the effect of the initial flow momentum is expected to be reduced because of flow resistance and turbulence losses. The results imply that the warning time of flood wave arrival is significantly smaller than that predicted by a classical dam break wave theory.

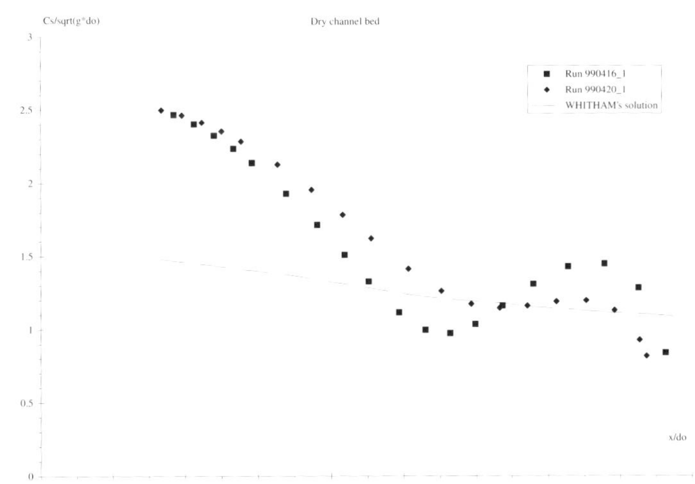

Figure 9.5. Propagation of a flood wave on a dry horizontal bed downstream of an overfall $C_s/\sqrt{g * d_o}$ (dry channel bed). Comparison with Whitham's (1955) theory.

CHAPTER 10

Summary, conclusions and recommendations

The stepped channel design has been used for more than 3,500 years. The Greek and Minoan engineers were probably the first to design an overflow stepped weir and stepped storm waterways respectively. Later, Roman, Moslem, Mughal and Spanish designers used a similar technique. At the end of the 19th century, a significant number of dams were built with overflow stepped spillways (e.g. Schuyler in 1909, Wegmann in 1911 and Kelen in 1933). Since the beginning of the 20th century, stepped chutes have been designed more specifically to dissipate flow energy (e.g. New Croton Dam). The steps increase significantly the rate of energy dissipation taking place on the channel face, reducing the size of the required downstream energy dissipation and the risks of scouring. Recently, new construction materials (e.g. RCC, strengthened gabions) have increased the interest for stepped channels and spillways. The construction of stepped chutes is compatible with the slipforming and placing methods of roller compacted concrete, with the construction techniques of gabion dams and with the precast concrete block system technique.

10.1 PRACTICAL CONSIDERATIONS FOR STEPPED CHUTE DESIGN

Of the over two hundred RCC dams built to date in North America, only sixty were equipped with a stepped spillway. Why? It is believed that many professional engineers have never been exposed to the complexity of the stepped spillway design. Professionals and students have many simple, practical questions.

Is it safe to design a stepped spillway?
Yes, but ... A number of stepped spillways have been in operation for several decades although prototype experience of overflow is scarce (Table 8.4; Section 7.3.2; Chapter 2; Table 1.1). Past failures and accidents suggest that a safe operation relies upon a sound design, a good quality of construction, suitable flow conditions, and regular maintenance (Section 8.3.2). A related issue is the education of professional engineers (Section 8.4).

Who designed the first stepped spillway?
Stepped channels have been used for over 3,500 years. The first stepped chutes were probably a series of stepped culverts in Crete and an overflow stepped

weir in Greece (Chapter 2). The recent renewal of interest in stepped spillways has been linked with the introduction of new construction techniques (e.g. RCC; Fig. 10.1) and the need to increase the discharge capacity of embankment dams (i.e. overtopping protection system; Fig. 10.2) (Sections 7.3 to 7.5).

Is there a simple equation to calculate the stepped channel flow properties?
No. Stepped chute overflows may be either a nappe flow regime, a transition flow regime or a skimming flow regime depending upon the flow rate per unit width, step height and step length (Section 1.3). Each flow regime must be analysed in its context. Detailed informations are presented in Chapters 3, 4 and 5 for nappe, transition and skimming flows respectively.

What is the maximum discharge capacity achievable in a stepped spillway?
Practically the maximum discharge per unit width is achieved with a skimming flow regime. Design guidelines are proposed in Section 7.3 and Table 7.2. It must be understood that the largest discharge per unit width is not the most efficient flow situation in terms of energy dissipation.

What are the most efficient flow conditions in terms of energy dissipation?
Experimental results suggest that greater energy dissipation is achieved in nappe flow situations (Section 5.4). In practical terms, a nappe flow regime implies large step heights and flat slopes that might not be achievable. In any case, experimental observations indicate that the residual energy increases significantly with increasing flow rates for a given stepped chute design (Chapters 3 to 5). The residual head is a more important design parameter than the total energy loss. For small weirs and dams, a nappe flow regime is recommended.

In skimming flow, does it matter if the flow is not fully developed at the downstream end of the chute?
Yes. Fully-developed flow conditions may not be achieved at the end of the chute for very large flow rates. In such cases, the flow consists of a developing boundary layer with an ideal fluid flow above. The kinetic energy of the ideal fluid flow region is considerable and the residual energy is very large. (Note that free-surface aeration is basically nil.)

In skimming flow, what is an order of magnitude of the flow resistance?
The review of a large number of experimental data showed a scatter of results. A careful re-analysis of detailed studies, taking into account free-surface aeration, suggest that the equivalent Darcy Friction Factor characterising the form drag on steep slopes may be estimated as: $f_e \sim 0.2$ in first approximation (Section 5.3.3). An analytical model of the form drag mechanism is developed in Appendix 5. The results are in agreement with experimental observations.

In skimming flow, how can the free-surface profile be computed accurately?
Skimming flows are characterised by significant free-surface aeration. The free-surface profile must be defined by two parameters: the clear-water depth d and the mean air concentration C_{mean} (or the characteristic depth Y_{90}). Complete calculations are difficult and inaccurate because of the high levels of turbulence and the rapidly-varied flow region next to the inception point. In practice a simple method was developed to estimate the flow properties at any position along the chute (App. 7). The results are summarised in a design chart (Fig. 5.22, Section 5.3.4).

What are the pressure forces acting on the step faces?
In a nappe flow, the invert pressure is maximum at nappe impact on the horizontal step face. Extreme pressure fluctuations are observed at nappe impact and below the hydraulic jump (nappe flow regime NA1). In a skimming flow, shear stress concentrations occur at the step edges. Rounded corner and chamfer may assist in a reduction of the stresses (Section 8.3.3) although there are needs for further research. Maximum invert pressures are observed on the horizontal faces of the steps and minimum pressures on the vertical faces[1].

What are the risks of cavitation damage?
Nil. The risk of cavitation damage is negligible because the slower flow velocities and greater water depths imply a cavitation index 10 to 100 times larger than on smooth-invert chutes for identical flow rates. Further strong free-surface aeration, and entrained bubbles within the step cavities, prevent cavitation damage[2].

What are the effects of free-surface aeration?
Free-surface aeration contributes to a flow bulking, and a greater air-water flow depth Y_{90}, and some substantial air-water mass transfer. In skimming flows, air entrainment assists in preventing cavitation damage and it induces some drag reduction (Section 5.5.3).

Is the stepped cascade design an efficient means for mass transfer?
Yes. Basically the stepped cascade design is characterised by greater residence times and air-water interface areas than in smooth chute flows. Both effects combine to yield greater aeration efficiencies. For example, a stepped chute is about 4 to 10 times more efficient in terms of oxygen transfer than a smooth chute for the same flow rate and total drop height (Sections 6.3 to 6.5).

[1]In skimming flows, pressure on the vertical step faces may become sub-atmospheric.
[2]Small quantities of undissolved air produce a large reduction in cavitation damage. For concrete chute spillways, 4 to 8% of air in the flow layers close to the invert prevent cavitation damage for flow velocities up to 45 m/s (e.g. Chanson, 1997a: 200-204).

Figure 10.1. Santa Cruz Dam, New Mexico USA in 1991 (courtesy of the US Bureau of Reclamation). H_{dam} = 46 m. The 1929 concrete arch dam was reinforced by RCC and conventional concrete buttresses in 1987, with a new stepped spillway. Design flow: 56.3 m^3/s (W = 22.9 m), with any larger flood being passed over the dam crest. Note the splitter piers to aerate the undernappe.

Figure 10.2. Zaraysk Dam spillway (Russia) (courtesy of Prof. Y. Pravdivets).

What is the most efficient flow regime for mass transfer?
Both nappe and skimming flow regimes are efficient flow situations for air-water mass transfer. Air-water gas transfer is maximum for small discharges with skimming flows and for relatively large discharges in nappe flow situations. This apparent contrast may be explained by the different mechanisms of air entrainment and air-water gas transfer between the two flow regimes (Section 6.5.1).

Is there some design guideline for inclined steps and pooled steps?
No.

How to design a stepped chute with a non-rectangular cross-section?
There is little information on the hydraulics of stepped channels with non-rectangular cross-sections. Some work has been done with flat trapezoidal cross-sections for pre-cast concrete block systems (Section 7.5) (Figure 10.2).

What is the best construction material?
There is no simple answer. For a temporary structure, a timber, timber crib or gabion structure may be appropriate even for large overflows. For embankment overtopping systems, unprotected roller compacted concrete (RCC) may be appropriate. Concrete stepped spillways are designed to last and they can pass large flow rates. For example, Gold Creek Dam spillway (1890), is still in use, and the Shuidong Dam spillway operated with up to 90 m^2/s in May 1994. For aesthetical purposes, a cut-stone masonry construction can provide good results, such as the marble stone steps at Peterhof's Grande Cascade and at Marly's La Rivière.

Is the use of roller compacted concrete (RCC) restricted to new structures?
No. RCC has been used for the refurbishment of a number of ancient embankment dams (Section 7.4). It may also be used to strengthen old gravity and arch dams. Figure 10.1 shows a concrete arch dam reinforced with conventional concrete buttresses and RCC, with provision for a stepped spillway.

How should the chute sidewall be sized?
The design of chute sidewalls must take into account the flow bulking resulting from the free-surface aeration. If some splashing is acceptable, the training wall height may be sized to contain the characteristic air-water depth Y_{90} for all flow rates up to design flows. If the surroundings (e.g. embankment dam) are at risk of erosion, the sidewall height must be designed for $1.4*Y_{90}$. When the developing spray can lead to fog or ice on surrounding roads or settlements, a greater safety margin must be considered (e.g. 1.6 to 1.8 times Y_{90}). The calculations of sidewall heights depend upon the type of flow regime. Detailed information is given in Chapters 3, 4 and 5 for nappe, transition and skimming flows respectively.

Does the superposition of small and large steps improve the design efficiency?
Yes possibly, but the experience is limited. Two Australian stepped spillways (Malmsbury, 1870 and Upper Coliban, 1903) were designed with a combination of large drops (ΔH = 2 to 4 m) and stepped chutes (h = 0.305 m) (Figs 2.2 & 2.10). At design flows, the drops operate with a nappe flow and the chutes in the skimming flow regime. Their successful operation for over a century suggests that the design is sound. On the model of the Kennedy's Vale Dam (α = 45°), Stephenson (1988) superimposed large steps to smaller steps. For this model, occasional large steps induced a 10% increase in specific energy dissipation. Additional experimental investigations are required to verify this finding. In any case, occasional large steps might induce shock waves and

flow disturbances that are not acceptable. Designers should not consider the introduction of occasional large steps without extensive physical modelling.

For a physical model, what is the minimum model size to avoid scale effects?
Physical modelling of open channel flows is based upon a Froude similitude (e.g. Henderson, 1966; Chanson, 1999a). In skimming flows, scale effects with a Froude similitude are small as long as the ratio of prototype to model lengths (geometric similarity) is less than 10 to 12. For larger values, the air entrainment and the flow resistance are not properly scaled (Section 5.3.3). Although no systematic study was conducted with nappe and transition flows, the same advice is believed to apply.

Is it possible to experience wave instabilities in a stepped chute?
Yes. The Sorpe Dam spillway experienced surging waves during an operation in 1998 (Chapter 9). The successive waves look like huge roll waves. It is believed that the waves were initiated by transition flow instabilities and sloshing motion in the step pools at the upper steps. In 1996, shock waves are observed at the Gold Creek dam spillway (Chapter 3). The cross waves impacted onto the sidewalls generating water splashing over the chute walls. A detailed study of shock wave propagation in a nappe flow without hydraulic jump (NA3) is described in Section 9.2.

Stepped chutes may also experience overtopping waves resulting from an upstream dam break (e.g. Silverleaf Weir in 1995) or from the sudden opening of the gates (Section 9.4).

What are typical applications of the stepped chute design?
There are numerous applications of stepped chutes (Table 1.1; Chapter 7). With roller compacted concrete (RCC) construction, the stepped spillway design is more economical and more rapid to place than conventional concrete smooth chutes. It contributes further to some energy dissipation over the chute reducing the size of the downstream stilling structure. The construction of a spillway is possible above an embankment dam, with precast concrete blocks and RCC (Sections 7.4 & 7.5). The former technique was developed in Russia while the latter is preferred is USA.

Stepped storm waterways are similarly efficient designs (Figs 1.2 & 10.3). Figure 10.3 shows parts of a modern storm waterway system which includes 3 stepped chutes. In Figure 10.3, the chute is equipped with pooled steps and a 5 m high wall at the downstream end to prevent water splashing onto the nearby road[3]. It is believed that the pooled step design was selected to maximise energy dissipation in the small available space. A related application is river training (e.g. Fig. 1.2) and debris dams (Sections 7.6 & 7.9).

[3]At the downstream end of the chute, the storm waterway turns 90° to the left, following the road for a further 500 m.

Figure 10.3. Stepped storm waterway at Villeneuve-Lès-Avignons, France (photographed by the author on 20 September 2000). Design characteristics: 7 pooled steps (h = 0.7 to 1 m, d_p = 0.4 m), α = 45°. Construction between 1999 and 2000.

Is it efficient to use a stepped chute design in a sewer system?
Yes. In sewers, the stepped chute design is particularly efficient because of the relative ease of construction and maintenance access, the high rate of energy dissipation taking place above the stepped cascade and the strong aeration performances. The latter is induced by the larger amounts of entrained air and slower residence times than on smooth chutes for identical flow conditions (Chapter 6). Aeration contributes to the re-oxygenation of the waters which in turn favours oxidation processes. A related design is the dropshaft cascade, or series of dropshafts. The Romans used large dropshaft cascades along some aqueducts: e.g. Valdepuentes, Yzeron, Montjeu (Section 2.4). Such a design is also used in underground drainage systems.

10.2 SUMMARY

The main hydraulic feature of stepped channel flows is the three different flow regimes: that is, either nappe flow regime for small discharges and flat channel slopes, transition flow regime, or skimming flow regime at large flow rates (Table 10.1, Section 1.3).

In a *nappe flow regime*, the water proceeds in a series of plunges from one step to another. The flow from each step hits the step below as a free-falling jet, eventually followed by a hydraulic jump on the horizontal step face (sub-regimes NA1 and NA2). The air cavity beneath the free-falling nappe must be adequately ventilated to prevent oscillations of the free-falling nappe and the associated noise. Stepped spillways with nappe flows with hydraulic jumps may be analysed as a succession of drop structures. The head loss at any intermediary step equals the step height. The energy dissipation occurs by jet breakup in air, by jet mixing on the step and with the formation of a hydraulic jump on the step. In a nappe flow without hydraulic jump, three-dimensional flow patterns are observed and air entrainment is significant. The occurrence of nappe flow regime implies relatively large steps and flat slopes. This situation is not often practical, but it may apply to relatively flat spillways, streams, creeks, river training and storm waterways.

In a *transition flow regime*, the cascading waters have a chaotic appearance. The dominant flow feature is the lower nappe stagnation at each horizontal step face and the downstream splashing. Water droplets and spray are observed to reach heights in excess of five to seven times the step height. There are significant longitudinal variations of the air-water flow properties within a step and from one step to the next one. In a series of experiments, a deflecting nappe was observed few steps downstream of the inception point of free-surface aeration, with the upper nappe reaching heights of up to $1.5*d_c$. The flow aeration is very rapid downstream of the inception point of free-surface aeration, and very strong aeration takes place. Mean air concentrations were measured in excess of the maximum acknowledged in skimming flows (i.e. uniform equilibrium mean air content). Despite these early results, there is a need for further systematic experimental data sets.

In the *skimming flow regime*, the water flows down the stepped invert as a coherent stream, skimming over the steps. The external edges of the steps form a pseudo-bottom over which the flow passes. Beneath this, recirculating vortices develop and are maintained through the transmission of shear stress from the waters flowing past the step edges. Form loss is a dominant process.

For flat slopes ($\alpha < 15$ to 25°), large stable recirculation vortices cannot develop. The recirculating vortices do not fill the entire cavity between the edges, and the wake from each step edge interferes with the next step. The flow pattern is characterised by the impact of the wake on the next step, a three-dimensional unstable recirculation in the wake and some friction drag (skin friction) on the step downstream of the wake impact. For an increase of slope, the tail of the wake starts interfering with the next wake and the friction drag component disappears. For steep slopes ($\alpha > 25^o$), a quasi-stable recirculation in the cavities between adjacent steps is observed. The recirculating eddies are large three-dimensional vortices with some irregular fluid ejection

Table 10.1. Summary of flow regimes above stepped chutes.

Regime (1)	Description (2)	Flow conditions (3)	Remark (4)
NA1	Nappe flow with fully-developed hydraulic jump	$\frac{d_c}{h} < 0.0916 * \left(\frac{h}{l}\right)^{-1.276}$	
NA2	Nappe flow with partially-developed hydraulic jump	$\frac{d_c}{h} > 0.0916 * \left(\frac{h}{l}\right)^{-1.276}$ and $\frac{d_c}{h} < 0.89 - 0.4 * \frac{h}{l}$	
NA3	Nappe flow without hydraulic jump	$\frac{d_c}{h} < 0.89 - 0.4 * \frac{h}{l}$	Steep slope, inclined downward step
TRA	Transition flow	$0.89 - 0.4 * \frac{h}{l} < \frac{d_c}{h} < 1.2 - 0.325 * \frac{h}{l}$	
SK1	Skimming flow with wake-step interference	$\frac{d_c}{h} > 1.2 - 0.325 * \frac{h}{l}$ and $\frac{h}{l} < 0.3 \text{ to } 0.5$	Flat slope ($\alpha < 15$ to 25°)
SK2	Skimming flow with wake-wake interference	$\frac{d_c}{h} > 1.2 - 0.325 * \frac{h}{l}$ and $\frac{h}{l} \sim 0.3 \text{ to } 0.5$	Moderate slope ($\alpha \sim 15$ to 25°)
SK3	Skimming flow with recirculating cavity flow	$\frac{d_c}{h} > 1.2 - 0.325 * \frac{h}{l}$ and $\frac{h}{l} > 0.3 \text{ to } 0.5$ $f = \frac{2}{\sqrt{\pi}} * \frac{1}{K}$ $\frac{f_e}{f} = 0.5 * \left(1 + \tanh\left(2.5 * \frac{0.5 - C_{mean}}{C_{mean} * (1 - C_{mean})}\right)\right)$	Steep slope ($\alpha > 15$ to 25°) Analytical flow resistance calculation ($K = 4.5$ to 6)

process. The flow resistance is form drag predominantly and it can be modelled analytically (App. 5).

At the upstream end of a stepped chute, the flow is smooth and glassy. Next to the bottom, turbulence is generated and a boundary layer grows until the outer edge of the boundary layer reaches the free-surface. At that location, called the inception point of free-surface aeration, the turbulence next to the free surface becomes large enough to overcome surface tension and buoyancy effects and free-surface aeration occurs. The inception point of air entrainment is a rapidly-varied flow region characterised by a rapid flow aeration. Downstream of the inception point, the flow becomes fully aerated and the air bubble entrainment may be modelled with an advective diffusion model (App. 6). The air entrainment affects the flow properties, particularly by inducing a flow bulking and reducing the flow resistance.

Maximum *energy dissipation* is achieved in nappe flows. Yet this may not be practical as a nappe flow regime requires flat slope and large step height. Nappe flows are advisable, in terms of energy dissipation, for small dams. For short channels, skimming flows do not reach fully-developed flow conditions: i.e. the flow consists of a boundary layer and an ideal-fluid flow region above with large kinetic energy. It must be emphasised that, with both nappe, transition and skimming flows, an increase of water discharge generates a sharp increase in residual energy.

In both nappe and skimming flow regimes, the *air-water mass transfer* taking place above the chute is enhanced by the large residence time and by the aeration of the flow compared to smooth chute flows for identical discharges. Both effects combine to contribute to greater re-oxygenation above stepped cascades. Direct applications include polluted streams and rivers which can explain the high fish mortality downstream of large hydraulic structures.

The flow patterns of stepped channel flows are very different from smooth chute flow situations. Designers must analyse carefully stepped chutes and engineers must select the appropriate flow regime. Next, hydraulic calculations should be developed for that flow regime; i.e. either nappe, transition or skimming flow. Designers must take into account the hydraulic characteristics, the flow aeration and the hydrodynamic interactions between the flow and the steps. In the past, several accidents involved stepped chute structures. The experience suggests that sound design, good quality of construction and regular maintenance are essential. Moreover, a possible transition from skimming flow to nappe flow (and opposite) must be avoided. Transition flow are characterised by flow instabilities that might induce fluctuating hydrodynamic loads and possible vibrations to the hydraulic structure. For both nappe and skimming flow regimes, an increase of discharge induces a significant increase in residual energy. Such a consideration must be seriously examined in the design stages of a stepped chute. A flood release larger than the maximum design discharge could lead to disastrous damage at the downstream end of the structure. For stepped spillways, the selection of the maximum design discharge must take into account the risks of failure and the costs associated with the underestimation of the maximum discharge capacity.

10.3 FUTURE RESEARCH

Further research investigations must be carried out to provide new information on the flow characteristics of stepped channels. An increased knowledge of stepped channel flows will increase the safety of stepped chutes. This study has highlighted the lack of expertise and knowledge on:

- the hydraulic characteristics of *skimming flows past flat slopes*,
- the hydraulics of the *transition flow regime* and the air-water flow properties,

- the mechanisms of *flow recirculation* in skimming flow regime,
- the *air-water gas transfer process* in nappe flow regime and in skimming flow regime,
- *prototype experience*, and
- the hydraulics of stepped chutes with non-rectangular cross-sections, inclined steps and pooled steps.

With regard to skimming flows, new experimental work is required to provide a better understanding of the flow recirculation in the cavities beneath the pseudo-bottom formed by the step edges. Indeed, most of the energy is dissipated in maintaining the recirculation of the cavity. Little attention has been paid to nappe flows without hydraulic jump. New experimental studies are required to gain new information on the air-water flow properties, and more work required to gain a better understanding of the transition flow regime. Furthermore, this work has characterised the hydraulic properties of stepped chute flow for prismatic rectangular channels with flat horizontal steps. Little information is available for non-rectangular channels and three-dimensional flows and only a few studies have analysed inclined- or pooled-step geometries.

CORRECTIONS AND UPDATES

For completeness, corrections and updates of the manuscript will be posted on the Internet at:

http://www.uq.edu.au/~e2hchans/reprints/book4.htm

APPENDIX 1

Physical and chemical properties of fluids

A1.1 ACCELERATION OF GRAVITY

Standard acceleration of gravity

The standard acceleration of gravity is:

$$g = 9.80665\,\text{m/s}^2 \tag{A1.1}$$

The standard gravity corresponds roughly to that at sea level and 45-degree latitude. The gravitational acceleration varies with latitude and elevation owing to the form and rotation of the earth. Rouse (1938) proposed the empirical correlation:

$$g = 9.806056 - 0.025027*\cos(2*\,\text{Latitude}) - 3\text{E}{-}6*z \text{ (in m/s}^2\text{)} \tag{A1.2}$$

where z is the altitude positive upwards (in metres) with the sea level as the origin.

Geographic altitude z m (1)	Standard acceleration of gravity g m/s^2 (2)
-1000	9.810
0	9.807
1000	9.804
2000	9.801
3000	9.797
4000	9.794
5000	9.791
6000	9.788
7000	9.785
10000	9.776

Absolute gravity values

Location	g m/s^2	Location	g m/s^2	Location	g m/s^2
(1)	(2)	(1)	(2)	(1)	(2)
Addis Ababa, Ethiopia	9.7743	Helsinki, Finland	9.81090	Québec, Canada	9.80726
Algiers, Algeria	9.79896	Kuala Lumpur, Malaysia	9.78034	Quito, Ecuador	9.7726
Anchorage, USA	9.81925	La Paz, Bolivia	9.7745	Sapporo, Japan	9.80476
Ankara, Turkey	9.79925	Lisbon, Portugal	9.8007	Reykjavik, Iceland	9.82265
Aswan, Egypt	9.78854	Manilla, Philippines	9.78382	Taipei, Taiwan	9.7895
Bangkok, Thailand	9.7830	Mexico city, Mexico	9.77927	Teheran, Iran	9.7939
Bogota, Colombia	9.7739	Nairobi, Kenya	9.77526	Thule, Greenland	9.82914
Brisbane, Australia	9.794	New Delhi, India	9.79122	Tokyo, Japan	9.79787
Buenos Aires, Argentina	9.7949	Paris, France	9.80926	Vancouver, Canada	9.80921
Christchurch, NZ	9.8050	Perth, Australia	9.794	Ushuaia, Argentina	9.81465
Denver, USA	9.79598	Port-Moresby, P.N.G.	9.782		
Guatemala, Guatemala	9.77967	Pretoria, South Africa	9.78615		

Source: Morelli (1971)

A1.2 PROPERTIES OF WATER

Temperature	Density	Dynamic viscosity	Surface tension	Vapour pressure
Celsius	ρ_w kg/m^3	μ_w Pa.s	σ N/m	P_v Pa
(1)	(2)	(3)	(4)	(5)
0	999.9	1.792E-3	0.0762	0.6E+3
5	1000.0	1.519E-3	0.0754	0.9E+3
10	999.7	1.308E-3	0.0748	1.2E+3
15	999.1	1.140E-3	0.0741	1.7E+3
20	998.2	1.005E-3	0.0736	2.5E+3
25	997.1	0.894E-3	0.0726	3.2E+3
30	995.7	0.801E-3	0.0718	4.3E+3
35	994.1	0.723E-3	0.0710	5.7E+3
40	992.2	0.656E-3	0.0701	7.5E+3

Source: Streeter & Wylie (1981)

A1.3 SALINITY AND CHLORINITY

The salinity characterises the amount of dissolved salts in the water: e.g, carbonates, bromides, organic matter. The definition of the salinity is based on the electrical conductivity of water relative to a specified solution of KCl and H_2O (Bowie et al., 1985). The scale is dimensionless but the salinity is defined often in parts per thousand (ppt).

The chlorinity Chl is defined in relation to salinity Sal as:

$$\mathrm{Sal} = 1.80655 * \mathrm{Chl} \tag{A1.3}$$

References: Bowie et al. (1985)

A1.4 SOLUBILITY OF NITROGEN, OXYGEN AND ARGON IN WATER

Dissolved gas concentrations in water are usually expressed in mass of dissolved chemical per unit volume or kg/m^3. Another unit is the ppm (parts per million). The conversion is: 1 ppm = 1 mg/L.

Solubility of oxygen

The solubility of oxygen in water at equilibrium with water saturated air C_s(Pstd) at standard pressure (i.e. Pstd = 1 atm = 1.01325 E+5 Pa) is calculated as:

$$\begin{aligned} \mathrm{LN}(1000 * C_s(\mathrm{Pstd})) &= -139.24411 + \frac{1.575701\mathrm{E}+5}{\mathrm{TK}} \\ &- \frac{6.642308\mathrm{E}+7}{\mathrm{TK}^2} + \frac{1.243800\mathrm{E}+10}{\mathrm{TK}^3} - \frac{8.621949\mathrm{E}+11}{\mathrm{TK}^4} \\ &- \mathrm{Chl} * \left(3.1929\mathrm{E}-2 - \frac{19.428}{\mathrm{TK}} + \frac{3.8673\mathrm{E}+3}{\mathrm{TK}^2}\right) \end{aligned} \tag{A1.4}$$

where C_s(Pstd) is the solubility of oxygen at standard pressure in kg/m^3, TK is the temperature in Kelvin and Chl is the chlorinity in ppt. The saturation concentration of dissolved oxygen at non-standard pressure is:

$$C_s(P) = C_s(\mathrm{Pstd}) * P * \left(\frac{\left(1 - \frac{P_v}{P}\right) * (1 - \mathrm{Teta} * P)}{(1 - P_v) * (1 - \mathrm{Teta})} \right) \tag{A1.5}$$

where P is the absolute pressure in atm (within 0 to 2 atm), P_v is the partial pressure of water vapour in atm and TK is the temperature in Kelvin. In Equation A1.5, the expression of Teta is:

$$\text{Teta} = 0.000975 - 1.426\text{E}-5*\text{TC} + 6.436\text{E}-8*\text{TC}^2 \tag{A1.6}$$

The partial pressure of water vapour may be computed as:

$$\text{LN}(P_v) = 11.8571 - \frac{3840.70}{\text{TK}} - \frac{216961}{\text{TK}^2} \tag{A1.7}$$

where P_v is in atm.

Sources: Apha (1985, 1989); Bowie et al. (1985)

Temperature	C_s(Pstd) Oxygen		
Celsius (1)	[Chl = 0] kg/m^3 (2)	[Chl = 10 ppt] kg/m^3 (3)	[Chl = 20 ppt] kg/m^3 (4)
0	14.621E-3	12.388E-3	11.355E-3
5	12.770E-3	11.320E-3	10.031E-3
10	11.238E-3	10.058E-3	8.959E-3
15	10.084E-3	9.027E-3	8.079E-3
20	9.0982E-3	8.174E-3	7.346E-3
25	8.263E-3	7.457E-3	6.728E-3
30	7.559E-3	5.845E-3	6.197E-3
35	6.950E-3	6.314E-3	5.734E-3
40	6.412E-3	5.842E-3	5.321E-3

Source: Bowie et al. (1985)

Volumetric solubility of nitrogen, oxygen and argon

Weiss (1970) proposed an expression of the volumetric solubility of nitrogen, oxygen and argon at one atmosphere total pressure (i.e. standard pressure):

$$\text{LN}(C_s(\text{Pstd})) = \text{A}1 + \text{A}2*\frac{100}{\text{TK}} + \text{A}3*\text{LN}\left(\frac{\text{TK}}{100}\right) + \text{A}4*\frac{\text{TK}}{100}$$

$$+\text{Sal}*\left(\text{B}1 + \text{B}2*\frac{\text{TK}}{100} + \text{B}3*\left(\frac{\text{TK}}{100}\right)^2\right) \tag{A1.8}$$

where C_s(Pstd) is the solubility in mL/L, TK is the temperature in Kelvin and Sal is the salinity in ppt. The constants A1, A2, A3, A4, B1, B2 and B3 are summarised in the next table for nitrogen (N_2), oxygen (O_2) and argon (Ar).

Gas (1)	A1 (2)	A2 (3)	A3 (4)	A4 (5)	B1 (6)	B2 (7)	B3 (8)
Nitrogen	-172.4965	248.4262	143.0738	-21.7120	-0.049781	0.025018	-0.0034861
Oxygen	-173.4292	249.6339	143.3483	-21.8492	-0.033096	0.014259	-0.0017000
Argon	-173.5146	245.4510	141.8222	-21.8020	-0.034474	0.014934	-0.0017729

Source: Weiss (1970)

A1.5 DIFFUSION COEFFICIENTS

The gas-liquid diffusivity of oxygen and nitrogen in water is:

Temperature Celsius (1)	$D(O_2)$ Oxygen m^2/s (2)	$D(N_2)$ Nitrogen m^2/s (3)
10	1.54E-9	1.29E-9
25	2.20E-9	2.01E-9
40	3.33E-9	2.83E-9
55	4.50E-9	3.80E-9

Source: Ferrell & Himmelblau (1967)

The data of Ferrell & Himmelblau (1967) can be correlated as:

$$D(O_2) = 1.16793E-27 * TK^{7.3892} \qquad \text{Oxygen} \quad (A1.9)$$

$$D(N_2) = 5.567E-11 * TK - 1.453E-8 \qquad \text{Nitrogen} \quad (A1.10)$$

where D is the molecular diffusivity in m^2/s and TK is the temperature in Kelvin.

A1.6 COEFFICIENT OF MASS TRANSFER K_L

Kawase & Moo-Young (1992) reviewed some correlations for the mass transfer coefficient (K_L) calculations in turbulent gas-liquid flows. They showed that K_L is almost constant regardless of the bubble size and the flow situations. The transfer coefficient of gas bubbles affected by surface active impurities can be correlated as:

$$K_L = 0.28 * D_{gas}^{2/3} * \left(\frac{\mu_w}{\rho_w} \right)^{-1/3} * \sqrt[3]{g} \qquad (d_{ab} < 0.25 \text{ mm}) \quad \text{(A1.11)}$$

$$K_L = 0.47 * \sqrt{D_{gas}} * \left(\frac{\mu_w}{\rho_w} \right)^{-1/6} * \sqrt[3]{g} \qquad (d_{ab} > 0.25 \text{ mm}) \quad \text{(A1.12)}$$

where D_{gas} is the molecular diffusivity (e.g. Section A1.5), μ_w and ρ_w are the dynamic viscosity and density of the liquid, d_{ab} is the gas bubble diameter and g is the gravity constant. All variables are expressed in SI units. Equations (A1.11) and (A1.12) were compared successfully with more than a dozen of experimental studies. Typical values for oxygen and nitrogen transfer in water are summarised below.

Temperature	K_L (O_2) in water		K_L (N_2) in water	
Celsius	Small bubbles m/s Eq. (A1.11)	Large bubbles m/s Eq. (A1.12)	Small bubbles m/s Eq. (A1.11)	Large bubbles m/s Eq. (A1.12)
(1)	(2)	(3)	(4)	(5)
5	6.35E-5	3.44E-4	5.06E-5	2.90E-4
10	7.31E-5	3.77E-4	6.49E-5	3.45E-4
15	8.32E-5	4.11E-4	7.55E-5	3.83E-4
20	9.44E-5	4.47E-4	8.82E-5	4.25E-4
25	1.05E-4	4.80E-4	9.90E-5	4.59E-4

Note: calculations performed assuming $g = 9.807$ m/s^2.

APPENDIX 2

Unit conversions

A2.1 INTRODUCTION

The systems of units derived from the metric system have gradually given way to a single system, called the Système International d'Unités (SI) and the present monograph presents results expressed in SI Units.

Since a large number of countries continue to use British and American units, Section A2.2 gives their equivalents against the SI units.
In the last section (A2.3), a table of conversion is provided for ancient (and often discussed) units.

A2.2 SI UNITS AND CONVERSION FACTORS

Quantity (1)	Unit (symbol) (2)	Conversion (3)	Comments (4)
Length:	1 inch (in)	= 25.4 10^{-3} m	Exactly.
	1 foot (ft)	= 0.3048 m	Exactly.
	1 yard (yd)	= 0.9144 m	Exactly.
	1 mile	= 1.609.344 m	Exactly.
Area:	1 square inch (in^2)	= 6.4516 10^{-4} m^2	Exactly.
	1 square foot (ft^2)	= 0.09290306 m^2	Exactly.
Volume:	1 cubic inch (in^3)	= 16.387064 10^{-6} m^3	Exactly.
	1 cubic foot (ft^3)	= 28.3168 10^{-3} m^3	Exactly.
	1 gallon UK (gal UK)	= 4.54609 10^{-3} m^3	
	1 gallon US (gal US)	= 3.78541 10^{-3} m^3	
	1 barrel US	= 158.987 10^{-3} m^3	Petroleum,...
Velocity:	1 foot per second (ft/s)	= 0.3048 m/s	Exactly.
	1 mile per hour (mph)	= 0.44704 m/s	Exactly.
Acceleration:	1 foot per second squared (ft/s^2)	= 0.3048 m/s^2	Exactly.
Mass:	1 pound (lb or lbm)	= 0.45359237 kg	Exactly.

	1 ton UK	= 1016.05 kg	
	1 ton US	= 907.185 kg	
Density:	1 pound per cubic foot (lb/ft^3)	= 16.0185 kg/m^3	
Force:	1 kilogram-force (kgf)	= 9.80665 N (exactly)	Exactly.
	1 pound force (lbf)	= 4.4482216152605 N	
Moment of force:	1 foot pound force (ft.lbf)	= 1.35582 N.m	
Pressure:	1 Pascal (Pa)	= 1 N/m^2	
	1 standard atmosphere (atm)	= 101325 Pa	Exactly.
	1 bar	= 10^5 Pa	Exactly.
	1 Torr 1 conventional metre of water (m of H_2O)	= 133.322 Pa = 9.80665 10^3 Pa	Exactly.
	1 conventional metre of Mercury (m of Hg)	= 1.33322 10^5 Pa	
	1 Pound per Square Inch (PSI)	= 6.8947572 10^3 Pa	
Temperature:	T (Celsius)	= T (Kelvin) - 273.16	0 Celsius is 0.01 K below the temperature of the triple point of water.
	T (Fahrenheit)	$= T(Celsius) * \frac{9}{5} + 32$	
	T (Rankine)	$= \frac{9}{5} * T(Kelvin)$	
Dynamic viscosity:	1 Pa.s	= 0.006720 lbm/ft/s	
	1 Pa.s	= 10 Poises	
Kinematic viscosity:	1 square foot per second (ft^2/s)	= 0.0929030 m^2/s	
	1 m^2/s	= 10.7639 ft^2/s	
	1 m^2/s	= 10^4 Stokes	
Work energy:	1 Joule (J)	= 1 N.m	
	1 Joule (J)	= 1 W.s	
	1 Watt hour (W.h)	= 3.600 10^3 J	Exactly.
	1 electronvolt (eV)	= 1.60219 10^{-19} J	
	1 foot pound force (ft.lbf)	= 1.35582 J	
Power:	1 Watt (W)	= 1 J/s	
	1 foot pound force per second (ft.lbf/s)	= 1.35582 W	
	1 horsepower (hp)	= 745.700 W	
Dissolved gas concentration:	1 parts per million (ppm)	= 1 mg/L	Exactly.

Sources: Degremont (1979); International Organization for Standardization (1979)

A2.3 ANCIENT UNITS AND CONVERSION TABLES

Quantity (1)	Ancient unit (symbol) (2)	Conversion (3)	Comments (4)
Length:			
	1 arshin	= 16 vershoks = 711.2 E-3 m	Ancient Russian imperial unit.
	1 braccio	= 0.6096 m	Equal 2 feet exactly. From the Italian "braccia".
	1 ell	= 45 inches = 1.143 m	Old English unit.
	1 foot	= 0.3048 m	Ancient Russian imperial unit. Equal 1 English foot.
	1 foot	≈ 0.3 m.	Roman unit.
	1 palmo	= 8.25 inches = 0.20955 m	Old Spanish unit.
	1 sazhen	= 3 arshins = 2.336 m	Ancient Russian imperial unit. Equal 7 feet.
	1 vara	= 33 inches = 0.8382 m	Old Spanish unit.
	1 vershok	= 44.45 E-3 m	Ancient Russian imperial unit.
	1 versta (or verst)	= 500 sazhens =1,066.8 m	Ancient Russian imperial unit.
Mass/Weight:			
	1 dolya	= 44.43 E-6 kg	Ancient Russian imperial unit.
	1 funt	= 96 zolotniks = 409.5 E-3 kg	Ancient Russian imperial unit. Russian pound.
	1 pud (or pood)	= 40 funts = 16.38 kg	Ancient Russian imperial unit.
	1 scruple	= 1.295 E-3 kg	Ancient Roman unit of weight (and also unit of coinage weight).
	1 zolotnik	= 96 dolyas = 4.266 E-3 kg	Ancient Russian imperial unit.

APPENDIX 3

Nappe flow trajectory at a drop structure

A3.1 INTRODUCTION

In a nappe flow regime, the flow from each step hits the step below as a falling jet. Nappe flows with hydraulic jumps are analysed as a succession of drop structures. For a horizontal step, a simple expression of the nappe trajectory can be deduced from the motion equation assuming that the velocity of the flow at the brink of the step is nearly horizontal. Application of the momentum equation at the base of the overfall provides most flow properties at the jet impact. Most calculations are developed for aerated nappes. The particular case of un-aerated nappes is detailed in Sections A3.3 and A3.4.

A3.2 NAPPE TRAJECTORY OF AERATED NAPPE

Considering an aerated nappe, the flow direction at the brink of the step is nearly horizontal. Once the fluid leaves the step (Fig. A3.1), the horizontal acceleration is zero and the vertical acceleration equals minus the gravity acceleration. The nappe velocity components in the x- and y-directions are:

$$V_x = V_b \tag{A3.1a}$$

$$V_y = -g*t \tag{A3.1b}$$

where V_b is the flow velocity at the brink of the step and t is the time. The trajectory equations of the nappe centreline are:

$$x = V_b*t \tag{A3.2a}$$

$$y = h + \frac{d_b}{2} - \frac{1}{2}*g*t^2 \tag{A3.2b}$$

where h is the step height and d_b is the flow depth at the brink.

Drop length

When the fluid leaves the step, the time t', taken to hit the next step is given by:

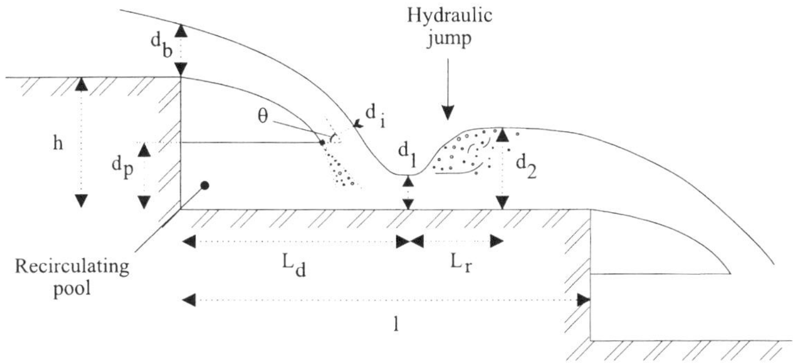

Figure A3.1. Nappe flow.

$$0 = -\frac{1}{2} * g * t'^2 + \left(h + \frac{d_b}{2}\right) \tag{A3.3}$$

The length of the drop L_d equals:

$$L_d = V_b * t' \tag{A3.4}$$

It yields:

$$\frac{L_d}{h} = \left(\frac{d_c}{h}\right)^{3/2} * \sqrt{\frac{h}{d_b}} * \sqrt{1 + 2 * \frac{h}{d_b}} \tag{A3.5}$$

Impact flow conditions
When the fluid leaves the step, the time t" taken to reach the pool free-surface is given by:

$$d_p = -\frac{1}{2} * g * t''^2 + \left(h + \frac{d_b}{2}\right) \tag{A3.6}$$

where d_p is the height of water in the pool behind the overfalling jet. The nappe thickness d_i and the flow velocity V_i at the intersection of the falling nappe with the receiving pool are then:

$$\frac{d_i}{d_c} = \left(\left(\frac{d_c}{d_b}\right)^2 + 2 * \frac{h + \frac{d_b}{2} - d_p}{d_c}\right)^{-1/2} \tag{A3.7}$$

$$\frac{V_i}{V_c} = \sqrt{\left(\frac{d_c}{d_b}\right)^2 + 2 * \frac{h + \frac{d_b}{2} - d_p}{d_c}} \tag{A3.8}$$

The angle of the falling nappe with the horizontal (Fig. A3.1) is given by:

$$\tan\theta = \sqrt{2} * \sqrt{\frac{d_b}{d_c}} * \sqrt{\frac{h + \frac{d_b}{2} - d_p}{d_c}} \tag{A3.9}$$

Pool height

The pool of water behind the overfalling jet (Fig. A3.1) is important as its weight provides the force parallel to the step surface which is required to change the jet from an angle θ to the step bottom to parallel to the horizontal face of the step. For a ventilated nappe, the momentum equation resolved along the step surface is:

$$\frac{1}{2} * \varrho_w * g * d_p^{\,2} - \frac{1}{2} * \varrho_w * g * d_1^{\,2} = \varrho_w * q_w * (V_1 - V_i * \cos\theta) \tag{A3.10}$$

assuming that the edges of the jet fluid have not disintegrated into spray and neglecting the shear forces on the surfaces. Assuming that the velocity entering the control volume is approximately the same as leaving it, it yields:

$$\frac{d_p}{d_i} = \sqrt{1 + 2 * \frac{V_i^{\,2}}{g + d_{i_t}} * (1 - \cos\theta)} \tag{A3.11}$$

Flow depth downstream of the jet impact

White (1943) developed the momentum equation at the base of the overfall to estimate the energy loss. His results gave the flow depth downstream of the jet impact (Fig. A3.1):

$$\frac{d_1}{d_c} = \frac{\sqrt{2}}{\frac{1.5}{\sqrt{2}} + \sqrt{\frac{h}{d_c} + \frac{3}{2}}} \tag{A3.12}$$

Note that the flow immediately downstream of the jet impact is supercritical: i.e., $d_1/d_c < 1$. Rajaratnam & Chamani (1995) argued the validity of the assumptions underlying Equation A3.12.

Flow depth downstream of the hydraulic jump

The flow conditions downstream of the hydraulic jump are deduced from the momentum and continuity equations. For a horizontal step, constant channel width, and neglecting the bed and wall friction, it yields:

$$\frac{d_2}{d_1} = \frac{1}{2} * \left(\sqrt{1 + 8 * \left(\frac{d_c}{d_1} \right)^3} - 1 \right) \tag{A3.13}$$

where the subscripts 1 and 2 refer respectively to the upstream and downstream flow conditions of the hydraulic jump (Fig. A3.1) (e.g. Chanson, 1999a). Equation A3.13 implies hydrostatic pressure distributions upstream and downstream of the jump. This is likely to be incorrect if the hydraulic jump is located immediately downstream of the nappe impact.

Discussion

Equations A3.5, A3.7, A3.9, A3.11, A3.12 and A3.13 form a system of six non-linear equations in terms of the drop length L_d, the impact flow conditions d_i and θ, the pool height d_p, the flow depths d_1 and d_2. Assuming that the flow upstream of the step edge is subcritical and that $d_b/d_c = 0.715$, Chanson (1995a) solved numerically the system of equations. The results are best fitted by the following correlations:

$$\frac{L_d}{h} = 2.171 * \left(\frac{d_c}{h} \right)^{0.525} \tag{A3.14}$$

$$\frac{d_i}{h} = 0.688 * \left(\frac{d_c}{h} \right)^{1.483} \tag{A3.15}$$

$$\tan q = 0.855 * \left(\frac{d_c}{h} \right)^{-0.582} \tag{A3.16}$$

$$\frac{d_p}{h} = 0.998 * \left(\frac{d_c}{h} \right)^{0.675} \tag{A3.17}$$

$$\frac{d_1}{h} = 0.625 * \left(\frac{d_c}{h} \right)^{1.326} \tag{A3.18}$$

$$\frac{d_2}{h} = 1.565 * \left(\frac{d_c}{h} \right)^{0.809} \tag{A3.19}$$

Practical calculations of the nappe geometry with subcritical inflow
For aerated nappes, Table A3.1 reviews nappe geometry calculations and experimental observations. All results were obtained in rectangular channels with horizontal overfall crest and assuming subcritical flow conditions upstream of the drop. For engineering applications, the flow depth at the brink d_b is computed as Rouse (1936):

$$d_b = 0.715 * d_c \tag{A3.20}$$

The following correlations were successfully verified with experimental data (e.g. Fig. A3.2). They are commonly used to estimate the drop length, pool height and the flow depths d_1 and d_2:

$$\frac{L_d}{h} = 4.30 * \left(\frac{d_c}{h}\right)^{0.81} \tag{A3.21}$$

$$\frac{d_p}{h} = \left(\frac{d_c}{h}\right)^{0.66} \tag{A3.22}$$

$$\frac{d_1}{h} = 0.54 * \left(\frac{d_c}{h}\right)^{1.275} \tag{A3.23}$$

$$\frac{d_2}{h} = 1.66 * \left(\frac{d_c}{h}\right)^{0.81} \tag{A3.24}$$

$$\frac{d_i}{d_c} = 0.687 * \left(\frac{d_c}{h}\right)^{0.483} \tag{A3.25}$$

$$\frac{V_i}{V_c} = 1.455 * \left(\frac{d_c}{h}\right)^{-0.483} \tag{A3.26}$$

$$\tan q = 0.838 * \left(\frac{d_c}{h}\right)^{-0.586} \tag{A3.27}$$

$$\frac{\Delta H}{H_1} = 1 - \left(\frac{0.54 * \left(\frac{d_c}{h}\right)^{0.275} + \frac{3.43}{2} * \left(\frac{d_c}{h}\right)^{-0.55}}{\frac{3}{2} + \frac{h}{d_c}}\right) \tag{A3.28}$$

On stepped channels with nappe flow regime, the experiments of Peyras et al. (1991) indicated that the correlations of Rouse (1936) and Rand (1955) are reasonably accurate for nappe flows with fully-developed jumps. For nappe flows with partially-developed hydraulic jump, the correlations of Rand (1955) (i.e. Eqs A3.21 to A3.23) provide reasonable estimate of the flow characteristics at the jet impact (Peyras et al., 1991).

Table A3.1. Calculations of the nappe geometry for aerated nappes.

Variable (1)	Formula (2)	Ref. (3)	Remarks (4)
d_b ([a])	$d_b = 0.715 * d_c$	[RO]	Experimental data.
	$d_b = 0.65 * d_c$	[CR]	Analytical solution.
	$d_b = 0.781 * d_c$	[BA]	Model data. $d_c < 0.124$ m. W = 0.46 m.
	$d_b = 0.714 * d_c$	[KR]	Model data. $0.021 < d_c < 0.068$ m. W = 0.5 m.
	$0.72 < d_b/d_c < 0.93$	[GI]	Model data. $0.075 < d_c/h < 0.45$.
	$d_b = 0.76 * d_c$	[FE]	Model data. $0.15 < d_c/h < 0.93$.
L_d	$\frac{L_d}{h} = 2.17 * \left(\frac{d_c}{h}\right)^{0.525}$	[CH]	Best fit of Equation A3.5.
	$\frac{L_d}{h} = 4.30 * \left(\frac{d_c}{h}\right)^{0.81}$	[RA]	Model data. $0.045 < d_c/h < 1$.
d_i	$\frac{d_i}{h} = 0.688 * \left(\frac{d_c}{h}\right)^{1.483}$	[CH]	Best fit of Equation A3.7.
$\tan\theta$	$\tan\theta = 0.855 * \left(\frac{d_c}{h}\right)^{-0.582}$	[CH]	Best fit of Equation A3.9.
d_p	$\frac{d_p}{d_c} = \sqrt{\left(\frac{d_1}{d_c}\right)^2 + 2 * \frac{d_c}{d_1} - 3}$	[MO]	Analytical solution.
	$\frac{d_p}{h} = \left(\frac{d_c}{h}\right)^{0.66}$	[RA]	Model data. $0.045 < d_c/h < 1$.
	$\frac{d_p}{h} = 1.067 * \left(\frac{d_c}{h} - 0.0016\right)^{0.697}$	[GI]	Model data. $0.075 < d_c/h < 0.45$.
	$\frac{d_p}{h} = 1.107 * \left(\frac{d_c}{h}\right)^{0.719}$	[RC]	Model data. $0.07 < d_c/h < 1$.
	$\frac{d_p}{h} = 0.998 * \left(\frac{d_c}{h}\right)^{0.675}$	[CH]	Best fit of Equation A3.11

Table A3.1. Continued.

Variable (1)	Formula (2)	Ref. (3)	Remarks (4)
d_1	$\frac{d_1}{d_c} = \frac{\sqrt{2}}{\frac{1.5}{\sqrt{2}} + \sqrt{\frac{h}{d_c} + \frac{3}{2}}}$	[WH]	Analytical solution.
	$\frac{d_1}{h} = 0.54 * \left(\frac{d_c}{h}\right)^{1.275}$	[RA]	Model data. $0.045 < d_c/h < 1$.
	$\frac{d_1}{h} = 0.625 * \left(\frac{d_c}{h}\right)^{1.326}$	[CH]	Best fit of Equation A3.12.
d_2	$\frac{d_2}{d_1} = \frac{1}{2} * \left(\sqrt{1 + 8 * Fr_1^2} - 1\right)$		Analytical solution.
	$\frac{d_2}{h} = 1.66 * \left(\frac{d_c}{h}\right)^{0.81}$	[RA]	Model data. $0.045 < d_c/h < 1$.
	$\frac{d_2}{h} = 1.565 * \left(\frac{d_c}{h}\right)^{0.809}$	[CH]	Best fit of Equation A3.13.
ΔH	$\frac{\Delta H}{H_1} = \left(\frac{0.54 * \left(\frac{d_c}{h}\right)^{0.275} + \frac{3.43}{2} * \left(\frac{d_c}{h}\right)^{-0.55}}{\frac{3}{2} + \frac{h}{d_c}}\right)$	[CH]	Semi-analytical calculation.
	$\frac{\Delta H}{H_1} = 0.0896 * \left(\frac{d_c}{h}\right)^{-0.766}$	[RC]	Correlation.

Notes:
([a]): subcritical flow in rectangular channel
H_1: total head upstream of the drop
[BA] Bauer & Graf (1971); [CH] Chanson (1995a); [CR] Craya (1948); [FE] Ferro (1992); [GI] Gill (1979); [KR] Kraijenhoff & Dommerholt (1977); [MO] Moore (1943); [RC] Rajaratnam & Chamani (1995); [RA] Rand (1955); [RO] Rouse (1936); [WH] White (1943).

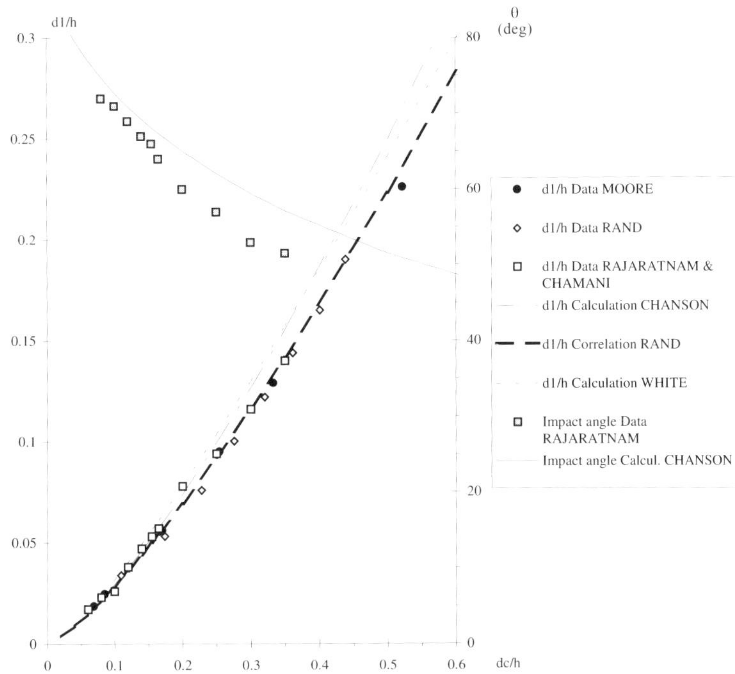

A.

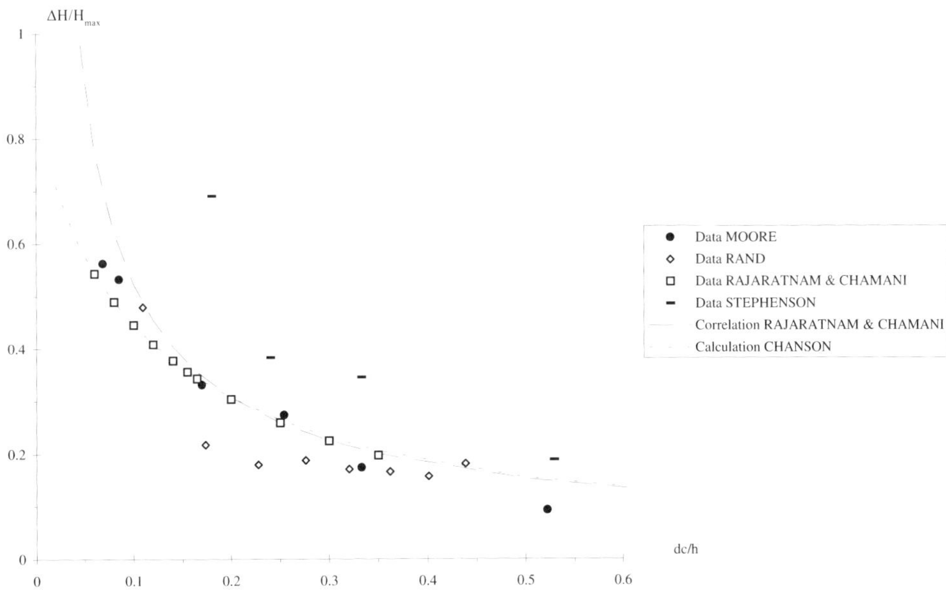

B.

Figure A3.2. Flow properties. Comparison between experimental data (Moore, 1943; Rand, 1955; Stephenson, 1979a; Chamani & Rajaratnam, 1995), analytical calculations (White, 1943; Chanson, 1995a) and empirical correlations (Rand, 1955). A. Dimensionless supercritical flow depth d_1/h and impact angle θ. B. Dimensionless energy dissipation $\Delta H/H_{max}$.

A3.3 NAPPE TRAJECTORY OF UN-AERATED NAPPE

Considering an un-aerated nappe, the pressure in the cavity between the nappe and the step is $P_{atm} - \Delta P$, where P_{atm} is the atmospheric pressure. At the brink of the step, the vertical acceleration of the flow Γ_y equals:

$$\Gamma_y = -g * \left(1 + \frac{\Delta P}{\rho_\omega * g * d_b}\right) \tag{A3.29}$$

The complete jet trajectory can be computed as a function of the flow properties at the brink of the step and the cavity subpressure using the finite element method (Wei & De Fazio, 1982), analytical methods (Schwartz & Nutt, 1963; Chanson, 1988) or a numerical method (Yuan, 1990).

Chanson (1988) developed a two-dimensional jet calculation based on the continuity and momentum equations. In the general case, the solution of the trajectory equations, is:

$$x' = \frac{Fr_b^2}{P_{Nb}} * \sqrt{-\frac{P_{Nb}^2}{Fr_b^4} * y'^2 + 2 * \frac{1 + P_{Nb} * \cos\theta_b}{Fr_b^2} * y' + (\sin\theta_b)^2}$$

$$+ \frac{Fr_b^2}{P_{Nb}^2} * \text{ArcSin}\left(\frac{(1 + P_{Nb} * \cos\theta_b) - \frac{P_{Nb}^2}{Fr_b^2} * y'}{\sqrt{1 + 2 * P_{Nb}^2 * \cos\theta_b + P_{Nb}^2}}\right) \tag{A3.30}$$

$$- \frac{Fr_b^2}{P_{Nb}^2} * \left(\sin\theta_b + \frac{1}{P_{Nb}} * \text{ArcSin}\left(\frac{1 + P_{Nb} * \cos\theta_b}{\sqrt{1 + 2 * P_{Nb} * \cos\theta_b + P_{Nb}^2}}\right)\right)$$

assuming that ΔP is non zero, and where $y' = y_1/d_b$, $x' = x_1/d_b$, $P_{N_b} = \Delta P/(\rho_w * g * d_b)$, $Fr_b = V_b/g * d_b$ and θ_b is the initial angle of the streamlines (Fig. A3.3). If the pressure difference across the jet is zero (i.e. $\Delta P = 0$), the analytical solution is the same as for the method of Schwartz & Nutt (1963):

$$x' = Fr_b^2 * \cos\theta_b * \left(\sqrt{\frac{2}{Fr_b^2} * y' + (\sin\theta_b)^2} - \sin\theta_b\right) \tag{A3.31}$$

Equations A3.29 and A3.30 apply to both subcritical and supercritical flow upstream of the step edge. The effects of the initial velocity distribution can be taken into account in these calculations. Complete calculations were detailed in Chanson (1988, 1992).

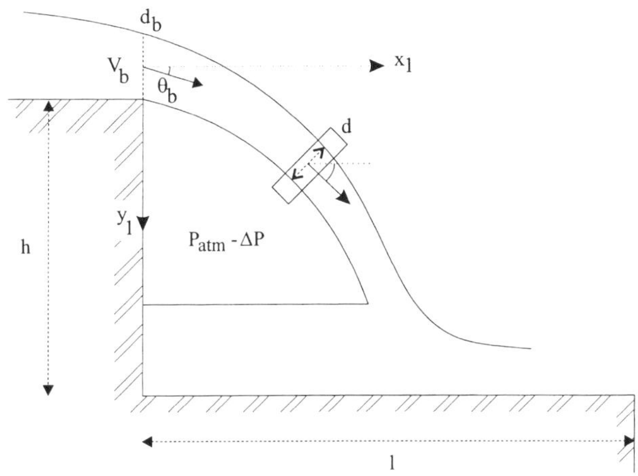

Figure A3.3. Jet trajectory.

A3.4 FREQUENCY OF NAPPE OSCILLATIONS

Thin sheets of free-falling water are subjected to fluttering instabilities. The instability is related to the shear flow instability of surfaces separating fluids moving at different velocities: they are referred as Kelvin-Helmholtz instabilities (Casperson, 1993).

For a nappe leaving a horizontal step, the possible frequencies of nappe oscillation can be determined as (Schwartz, 1964b; Casperson, 1993):

$$\frac{F*d_b}{\sqrt{g*d_b}} = \frac{I+\frac{1}{4}}{Fr_b*\left(-1+\sqrt{1+\frac{2}{Fr_b^{\ 2}}*\frac{h}{d_b}}\right)} \tag{A3.32}$$

where I is an integer: I = 1, 2, 3, 4 or more. I (or [I + 1/4]) is approximately the number of wavelengths in the nappe. Estimating the flow depth at the brink as Rouse (1936), it yields:

$$F*\sqrt{\frac{d_c}{g}} = 0.715*\left(\frac{I+\frac{1}{4}}{-1+\sqrt{1+1.022*\frac{h}{d_c}}}\right) \tag{A3.33}$$

APPENDIX 4

Bubble rise velocity

Air bubbles submerged in water are subjected to a net upward force (i.e. buoyancy). Buoyancy is a vertical force caused by the pressure difference between the upper and lower surfaces of the bubble (e.g. Chanson, 1997a: 274-277). The buoyant force counteracts the pressure force and equals the weight of displaced liquid.

Small air bubbles (i.e. $d_{ab} < 1$ mm) act as rigid spheres. Surface tension imposes the shape of small bubbles but the motion of these bubbles is dominated by the balance between the viscous drag force and the buoyant force. For very small bubbles (i.e. $d_{ab} < 0.1$ mm) the bubble rise velocity u_r is given by the Stokes' Law:

$$u_r = \frac{2}{9} * \frac{g * (\rho_w - \rho_{gas})}{\mu_w} * d_{ab}^2 \qquad (d_{ab} < 0.1 \text{ mm}) \quad \text{(A4.1)}$$

where ρ_w is the liquid density, ρ_{gas} is the gas density, μ_w is the liquid dynamic viscosity, u_r is the rise velocity and d_{ab} is the bubble diameter.

For small rigid spherical bubbles (i.e. $0.1 < d_{ab} < 1$ mm), the rise velocity is best fitted by (Comolet, 1979a):

$$u_r = \frac{g * \rho_\omega}{18 * \mu_\omega} * d_{ab}^2 \qquad (0.1 < d_{ab} < 1 \text{ mm}) \quad \text{(A4.2)}$$

As far as the fluid viscosity is neglected (i.e. $d_{ab} > 1$ mm) Comolet (1979b) showed that the bubble rise velocity can be estimated as:

$$u_r = \sqrt{\frac{2.14 * \sigma}{\rho_w * d_{ab}} + 0.52 * g * d_{ab}} \qquad (d_{ab} > 1 \text{ mm}) \quad \text{(A4.3)}$$

Figure A4.1 summarises these results.

Discussion: effects of turbulence

It is well known that solid particles settling in a liquid are affected by turbulence. Numerous researchers have discussed possible effects. Nielsen argued that the particle fall velocity may increase or decrease depending upon turbu-

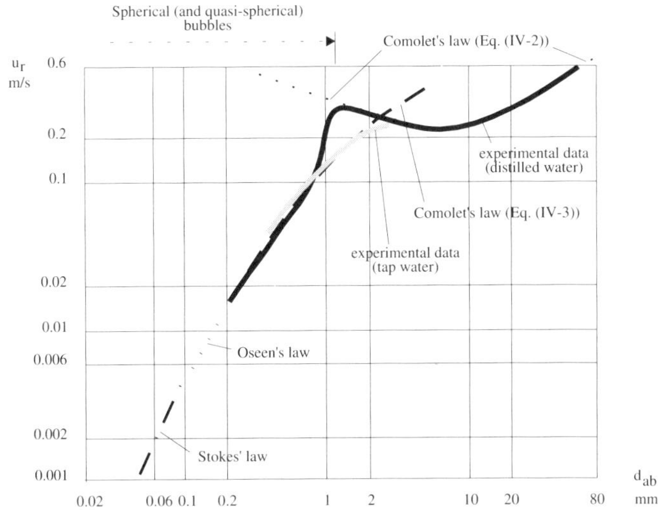

Figure A4.1. Bubble rise velocity in still water (after Comolet, 1979a).

lence intensity, the particle density, and the characteristic length scale and time scale of the turbulence (Nielsen, 1993; Nielsen & Zeng, 1998). In turbulent air-water flows, several researchers (e.g. Falvey, 1980; Chanson, 1988, 1989b, 1993a) discussed the effects of turbulence upon the bubble rise velocity. Most discussions suggested that high level of turbulence would retard the bubble rise, resulting in smaller rise velocities than in still water. The problem, however, is still subject to discussion. No study has proved conclusively either an increase or a decrease of rise velocity. It is thought that the effects of turbulence on bubble rise velocity might be dependent upon the bubble size, characteristic turbulent length-scale and time-scale, and the turbulence intensity.

APPENDIX 5

Modelling form drag and flow resistance in skimming flows

Skimming flows are characterised by unsteady momentum exchanges between the main stream and cavity flows. The recirculating fluid will, at irregular time intervals, flow outward into the main flow and be replaced by fresh fluid. The ejection mechanism appears sequential. Once one cavity outflow occurs, it induces a sequence of outflows at the downstream cavities. The process was observed visually for slopes ranging from 5.7° to 55° but the sequential mechanism was possibly most effective on the steep chute geometries (Chanson et al., 2000). This similar pattern is observed with skimming flows past rectangular and triangular strip roughness (Djenedi et al., 1994; Tantirige et al., 1994; Elavarasan et al., 1995), while sequential fluid ejection process was earlier observed on the M'Bali stepped spillway model by Professor Lejeune.

For flat slope ($\alpha < 20°$), the flow resistance includes the form drag of the recirculating cavity plus the skin friction on the downstream end of each step. The free-surface is not parallel to the pseudo-bottom but it exhibits an "undular" pattern in phase with the step geometry. Between adjacent steps, the cavity of recirculating fluid is affected by the interference of the mixing layer onto the downstream step.

For steep slope ($\alpha > 20°$), there is no skin friction between the main flow and the step faces, and the form drag associated with the recirculation is preponderant. The cavity flow is three-dimensional as visually observed by Professor Lejeune on the M'Bali model and by Chanson et al. (2000). The flow resistance is a function of the recirculation process and energy dissipation is dominated by the transfer of momentum between the cavity flow and the main stream.

Estimating the boundary shear stress along a recirculating cavity

At each step, the cavity flow is driven by the developing shear layer and the transfer of momentum across (Fig. A5.1). The mixing layer is basically a free shear layer. The equivalent boundary shear stress of the cavity flow equals the maximum shear stress in the shear layer that may be modelled by a mixing length model:

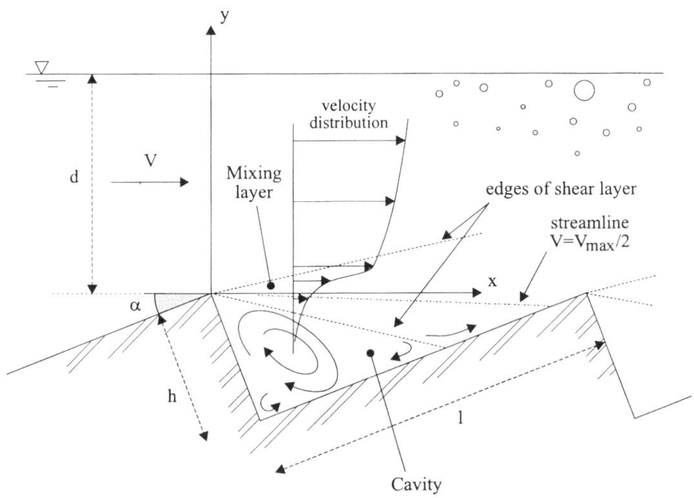

Figure A5.1. Skimming flow: definition sketch.

$$\frac{v}{V_o} = \frac{1}{2} * \left(1 + \operatorname{erf}\left(\frac{K*(y - y_{50})}{x}\right)\right) \tag{A5.1}$$

assuming a constant eddy viscosity n_t across the shear layer:

$$v_t = \frac{1}{4*K^2} * x * V_o \tag{A5.2}$$

where 1/K is the dimensionless rate of expansion of the shear layer, y_{50} is the location of the streamline $v = V_o/2$ and V_o is the free-stream velocity (Goertler, 1942). The boundary shear stress between the outer flow and the cavity flow equals the maximum shear stress in the mixing layer:

$$\tau_o = \tau_{max} = \rho * v_t * \left(\frac{\partial v}{\partial y}\right)_{y=y_{50}} \tag{A5.3}$$

where v_t is the momentum exchange coefficient (Eq. A1.2). For Goertler's approximation, the dimensionless pseudo-boundary shear stress equals:

$$f = \frac{8*\tau_{max}}{\rho * V_o^2} = \frac{2}{\sqrt{\pi} * K} \tag{A5.4}$$

where the left hand side term is basically a Darcy Friction Factor (Chanson et al., 2000). In air-water mixing layers of plane plunging jets, Brattberg & Chanson (1998) observed K ~ 6 for velocities ranging from 2 to 8 m/s[1]. Equation A5.4

[1]For monophase flows, K ~ 12.

predicts a form drag: f ≈ 0.2 for skimming flow, a result close to friction factor data in steep stepped flows. Brattberg & Chanson (1998) showed further that $y_{50}/d_o = -0.094*(x/d_o+5.3)$, implying the presence of a stagnation point associated with maximum pressures on the horizontal face of the step. This is consistent with the pressure measurements of Houston (1987); Frizell (1991, 1992) and Lejeune & Lejeune (1994) on the Upper Stillwater model ($\alpha = 72$ & 59°, h = 0.041 m), USBR model ($\alpha = 26.6°$, h = 0.051 m) and M'Bali model ($\alpha = 51.43°$, h = 0.019 & 0.038 m) respectively.

In monophase flows, Wygnanski & Fiedler (1970) observed that the maximum shear stress is independent from the distance from the singularity and their data yield f = 0.18. For cavity flows, Haugen & Dhanak (1966) and Kistler & Tan (1967) observed similar results.

Discussion

It must be emphasised that the above development (i.e. Eq. A5.4) does not take into account the interactions between the developing mixing layer and the downstream step, and the associated losses. Pearson et al. (1997) demonstrated that turbulent flows past a recirculating square cavity are associated with a sharp rise of boundary shear stress downstream of the cavity for up to 6 cavity lengths. Their data showed the magnitude of the shear stress peak being about 3 times the equivalent skin friction stress. They attributed the sharp rise to local intense favourable pressure gradient that is associated with the downstream stagnation edge of the cavity. This is in fact a kind of form drag.

APPENDIX 6

Void fraction distributions in chute flows

In supercritical open channel flows, free-surface aeration is often observed when the turbulence acting next to the free-surface is large enough to overcome both surface tension for the entrainment of air bubbles and buoyancy to carry downwards the bubbles (Section 1.4.1). Within the air-water flow, the diffusion of air bubbles can be approximated by a simple diffusion model. At uniform equilibrium, the buoyancy counteracts exactly the turbulent diffusion. The diffusion equation for air bubbles yields:

$$0 = \frac{\partial}{\partial y}\left(D_t * \frac{\partial C}{\partial y} \right) - \cos\alpha * \frac{\partial}{\partial y}(u_r * C) \tag{A6.1}$$

where y is the direction normal to the invert, α is the invert slope, D_t is the turbulent diffusivity in the direction normal to the flow direction and u_r is the bubble rise velocity. The latter characterises the buoyant force acting on the submerged bubbles.

By definition, the buoyant force acting on a submerged body is the difference between the vertical components of the pressure force on its underside and on its upper side. For a single bubble rising at a constant velocity in a quiescent surrounding fluid, the drag force counterbalances the resultant of the weight force and the buoyant force. Neglecting the weight of the air bubble, the rise velocity squared is proportional to the pressure gradient:

$$u_r^2 \sim -\frac{\partial P}{\partial y} \tag{A6.2}$$

In air-water flow, the bubble rise velocity in a fluid of density ρ_w*(1 - C) equals:

$$u_r^2 = [(u_r)_{Hyd}]^2 * (1 - C) \tag{A6.3}$$

where C is the void fraction and $(u_r)_{Hyd}$ is the rise velocity in hydrostatic pressure gradient (App. 4). Equation A6.3 gives the rise velocity in a two-phase flow mixture of void fraction C as a function of the rise velocity in hydrostatic pressure gradient. It demonstrates that the buoyant force is smaller in aerated waters than in clear-water. Replacing u_r by its expression (Eq. A6.3), the diffusion equation for air in the equilibrium flow region becomes:

$$\frac{\partial}{\partial y}\left(D_t * \frac{\partial C}{\partial y}\right) = (u_r)_{Hyd} * \cos\alpha * \frac{\partial}{\partial y}(C * \sqrt{1-C}) \tag{A6.4}$$

where the rise velocity in hydrostatic pressure gradient $(u_r)_{Hyd}$ is assumed a constant. Assuming a homogeneous turbulence across the flow, the integration yields:

$$C = 1 - \tanh^2\left(K' - \frac{y'}{2 * D'}\right) \tag{A6.5}$$

where $y' = y/Y_{90}$, $D' = \frac{D_t}{(u_r)_{Hyd} * \cos\alpha * Y_{90}}$, K' is an integration constant, tanh(x)

= (exp(+x) - exp(-x))/(exp(+x) + exp(-x)) (Chanson, 1995b, 1997a).

The air concentration distribution is a function of two constant parameters: the dimensionless diffusivity D' and the dimensionless constant K'. A relationship between D' and K' is deduced at the boundary condition C = 0.9 for y' = 1:

$$K' = K^* + \frac{1}{2 * D'} \tag{A6.6}$$

where $K^* = \tanh^{-1}(\sqrt{0.1}) = 0.32745015...$
If the diffusivity is unknown, it can deduced from the mean air content:

$$C_{mean} = \int_0^1 C * dy'$$

where C_{mean} is the mean air concentration defined in terms of Y_{90} (Table A6.1) Equation A6.5 was successfully validated with smooth invert model and prototype data in both the uniform equilibrium and gradually-varied flow regions. It was also validated in the gradually-varied flow region of skimming flows on stepped chutes (Chapter 5, Section 5.3.3). Chanson (1997a: 115-122, 293-303) discussed in further details some limitation of the air bubble diffusion model.

Table A6.1. Relationship between C_{mean}, D' and K' (after Chanson, 1997a).

C_{mean} (1)	D' (2)	K' (3)
0.01	0.007312	68.70445
0.05	0.036562	14.0029
0.10	0.073124	7.16516
0.15	0.109704	4.88517
0.20	0.146489	3.74068
0.30	0.223191	2.567688
0.40	0.3111	1.93465
0.50	0.423441	1.508251
0.60	0.587217	1.178924
0.70	0.878462	0.896627

Notes: For $C_{mean} \leq 0.4$, the dimensionless turbulent diffusivity and mean air content are best correlated by: $D' = 0.75699 * C_{mean}^{1.010401}$ with a normalised coefficient of correlation of 0.9998894.

APPENDIX 7

Stepped spillway chute flow calculations (skimming flow)

On an uncontrolled chute spillway, the flow is accelerated by the gravity force component in the flow direction. For an ideal fluid flow, the velocity at the downstream chute end can be deduced from the Bernoulli Equation:

$$V_{max} = \sqrt{2 * g * (H_{max} - d * \cos\alpha)} \qquad \text{Ideal fluid flow} \quad \text{(A7.1)}$$

where H_{max} is the upstream total head, α is the channel slope and d is the downstream flow depth (Fig. A7.1). In practice, however, friction losses occur and the flow velocity at the downstream end of the chute is less than the ideal-fluid velocity (called the maximum flow velocity). At the upstream end of the chute, a bottom boundary layer is generated by bottom friction and develops in the flow direction. When the outer edge of the boundary layer reaches the free-surface, the flow becomes fully developed.

Developing flow region

In the developing flow region, the flow consists of a turbulent boundary layer next to the invert and an ideal-fluid flow region above. In the ideal-fluid region ($\delta_{BL} < y < d$), the velocity, called the free-stream velocity, may be deduced from the Bernoulli Equation (Eq. A7.1). In the boundary layer, experimental data indicate that the velocity distribution follows closely a power law:

$$\frac{V}{V_{max}} = \left(\frac{y}{\delta_{BL}}\right)^{1/N} \qquad 0 < y/\delta < 1 \quad \text{(A7.2)}$$

where δ_{BL} is the boundary layer thickness and y is the distance normal to the channel bed. The velocity distribution exponent equals about N = 5 for stepped chutes (Section 5.2.1). Combining Equations A7.1 and A7.2, the continuity equation gives:

$$q_w = V_{max} * \left(d - \frac{\delta_{BL}}{N+1}\right) \qquad \text{(A7.3)}$$

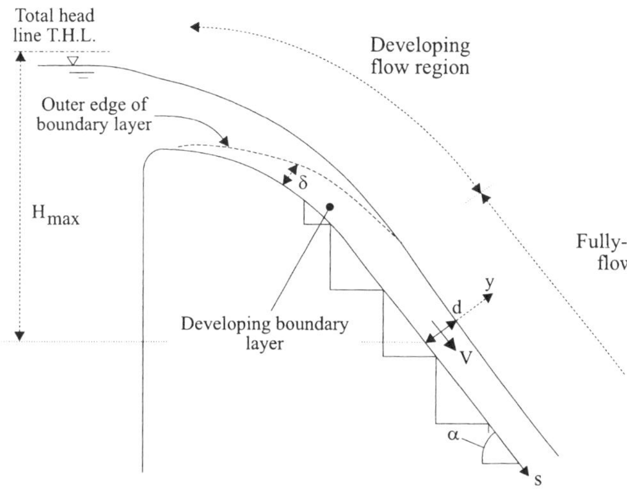

Figure A7.1. Sketch of a stepped spillway flow.

The boundary growth in skimming flow is enhanced by the turbulence generated by the steps. It may be estimated in first approximation:

$$\frac{\delta_{BL}}{s} = 0.06106 * (\sin\alpha)^{0.133} * \left(\frac{s}{h * \cos\alpha}\right)^{-0.17} \quad \text{Stepped chute (skimming flow)} \tag{A7.4}$$

where h is the step height and s is the curvi-linear coordinate along the flow direction. Equation A7.4 was checked with model and prototype data (e.g. Chanson, 1995a). It applies only to steep chutes (i.e. $\alpha > 30°$). Note that it is nearly independent of the crest shape.

Fully-developed flow region

In the fully-developed flow region, the flow is accelerated until the boundary friction counterbalances exactly the gravity force component in the flow direction (i.e. uniform equilibrium flow or normal flow). At equilibrium, the normal flow conditions can be deduced from the momentum equation. For wide channel, the momentum equation yields:

$$U_w = \sqrt{\frac{8 * g}{f}} * \sqrt{d * \sin\alpha} \tag{A7.5a}$$

where U_w is the flow velocity, f is the Darcy Friction Factor, α is the channel slope and d is the normal depth (e.g. Henderson, 1966; Chanson, 1999a). Combining with the continuity equation, it becomes:

$$U_w = \sqrt[3]{\frac{8*g}{f}*q_w*\sin\alpha} \tag{A7.5b}$$

On a long chute, uniform equilibrium flow conditions are reached before the downstream end of the chute, and the dimensionless downstream flow velocity U_w equals:

$$\frac{U_w}{V_{max}} = \sqrt{\frac{4*\sin\alpha}{f*\left(\frac{H_{max}}{d_c}*\sqrt[3]{\frac{8*\sin\alpha}{f}}-\cos\alpha\right)}} \tag{A7.5c}$$

where V_{max} is the ideal flow velocity (Eq. A7.1) and d_c is the critical flow depth. Equation A7.5c is valid for large values of H_{max}/d_c and for wide rectangular channels.

Gradually-varied flow
In the fully-developed flow region, the flow is gradually varied in the upstream part until becoming uniform equilibrium. In the gradually-varied flow region, the flow properties can be deduced from a differential form of the energy equation:

$$\frac{\partial H}{\partial s} = -S_f \tag{A7.6}$$

where H is the mean total energy, s is the direction along the channel bed and S_f is the friction defined as:

$$S_f = f*\frac{1}{D_H}*\frac{U_w^2}{2*g} \tag{A7.7}$$

and D_H is the hydraulic diameter.

Practical applications
For steep chutes (typically $\alpha \sim 45$ to 55°), both the acceleration and boundary layer development affect the flow properties significantly. In addition free-surface aeration[1] may take place downstream of the intersection of the outer edge of the developing boundary layer with the free-surface. Air entrainment can be clearly identified by the "white water" appearance of the free-surface flow.

The complete calculations of the flow properties may be tedious. In prac-

[1]Air entrainment in open channels is called free-surface aeration, self-aeration, insufflation or white waters.

A.

B.

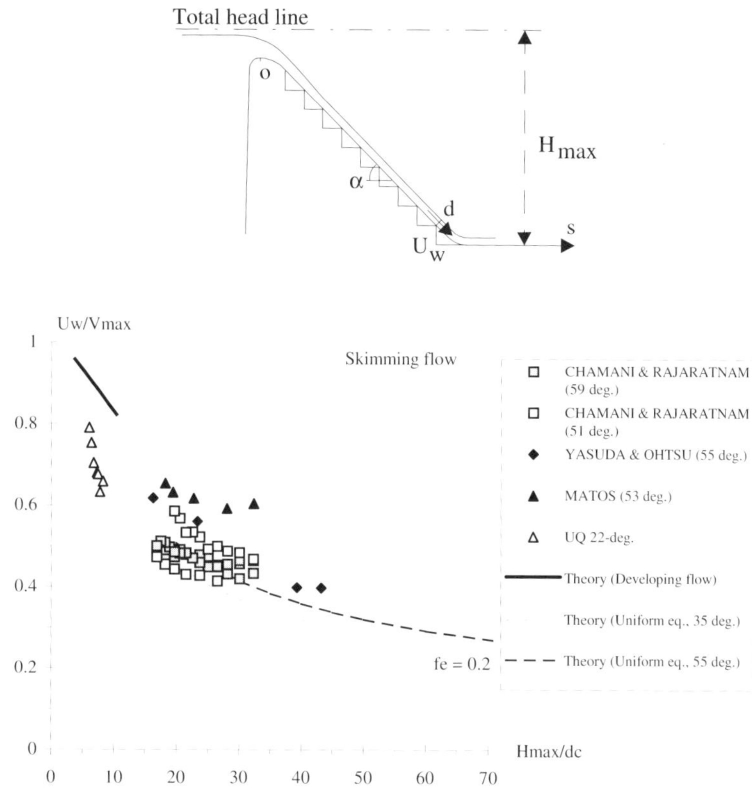

Figure A7.2. Flow velocity at the spillway toe as a function of the upstream head. Comparison with experimental data. A. Definition sketch. B. Results.

tice, the combination of the flow calculations in developing flow and in uniform equilibrium flow give a general trend which may be used for a preliminary design. Figure A7.2 combines these results with smooth transition curves. It provides some information on the mean flow velocity at the end of the chute as a function of the theoretical velocity V_{max} (Eq. A7.1), the upstream total head (above spillway toe) and the critical depth. The results are shown for slopes ranging from 35 to 55° (i.e. 1V:1.4H to 1V:0.7H) assuming f = 0.2 (Section 5.3.3). Experimental results[2] are also shown.

In the developing region, the trend indicates that the mean velocity departs from the theoretical velocity with increasing upstream total head. Indeed, for a given discharge, an increase in upstream total head brings an increase in dam height, in distance s from the crest, in boundary layer thickness and hence in bottom friction effect. In the fully-developed flow region, the flow resistance

[2]The date are clear-water flow velocities deduced from air concentration distribution measurements.

becomes dominant and the mean flow velocity tends to the uniform equilibrium flow conditions.

Simplified results

At uniform equilibrium in a wide channel, the residual flow velocity equals:

$$\frac{U_w}{\sqrt{g*d}} = \sqrt{\frac{8*\sin\alpha}{f}} \qquad \text{Uniform equilibrium} \quad (A7.5)$$

In gradually-varied flows, a smooth transition curve between developing and uniform equilibrium flow conditions may be correlated as:

$$\frac{U_w}{V_{max}} \approx \frac{1}{0.924 + 0.046*\frac{H_{dam}}{d_c} - 9.671E{-}05*\left(\frac{H_{dam}}{d_c}\right)^2} \qquad \alpha = 55°,\ H_{dam}/d_c < 200 \quad (A7.9)$$

Figure A7.2 and Equations A7.5 and A7.9 may be used for pre-design calculations.

APPENDIX 8

Modelling air-water gas transfer in skimming flows

A8.1 INTRODUCTION

Fick's Law states that the mass transfer rate of a chemical across an interface varies directly as the coefficient of molecular diffusion D_{gas} and the gradient of gas concentration. For the gas transfer of a dissolved gas across an air-water interface, it is usual to rewrite Fick's Law as:

$$\frac{d}{dt} C_{gas} = K_L * a * (C_s - C_{gas}) \tag{A8.1}$$

where C_{gas} is the concentration of dissolved gas, K_L is the coefficient of mass transfer, a is the specific surface area defined as the air-water interface area per unit volume of air and water and C_s is the gas saturation concentration. In free-surface flows along a chute, the coefficient of transfer K_L and the saturation concentration C_s can be assumed constant (App. 1). Along the chute Equation A8.1 may be numerically integrated between two cross-sections as:

$$\frac{\Delta C_{gas}}{C_s - C_{gas}} = \frac{K_L * a}{U_w} * \Delta s \tag{A8.2a}$$

where a and U_w are the mean specific interface area and flow velocity over the interval Δs between cross-sections, and the transfer coefficient k_L is a function of the physical and chemical fluid properties only. It yields:

$$\Delta C_{gas} = (C_s - C_{gas})_1 * \left(1 - \exp\left(-\frac{K_L * a}{U_w} * \Delta s\right)\right) \tag{A8.2b}$$

where the subscribe 1 refers to the upstream cross-section.

A8.2 AIR-WATER FLOW CHARACTERISTICS

Downstream of the inception point of air entrainment, the distribution of void fraction may be described by a diffusion model:

$$C = 1 - \tanh^2\left(K' - \frac{y}{2 * D' * Y_{90}}\right) \tag{A8.3}$$

where tanh is the hyperbolic tangent function, y is the distance normal to the invert, Y_{90} is the distance where C = 90%, D' is a dimensionless turbulent diffusivity and K' is an integration constant (App. 6). D' and K' are functions of the mean air content C_{mean} only (Table A8.1).

The velocity distribution follows a power law:

$$\frac{V}{V_{90}} = \left(\frac{y}{Y_{90}}\right)^{1/N} \tag{A8.4}$$

where V_{90} is the characteristic velocity at $y = Y_{90}$ and N = 3.7 (Section 5.3.2).

In open channel flow, the relationship between the bubble frequency and air content may be approximated by a parabolic shape:

$$\frac{F_{ab}}{(F_{ab})_{max}} = 4 * C * (1 - C) \tag{A8.5}$$

where the maximum bubble frequency $(F_{ab})_{max}$ is always observed for C = 50% (Chanson, 1997b). The parabolic shape was observed first in open channel flows, in high-velocity water jets and in hydraulic jumps (e.g. Brattberg et al., 1998; Chanson & Brattberg, 2000). Equation A8.5 holds also in skimming flows.

Table A8.1. Air concentration distribution coefficients (after Chanson, 1997a)

C_{mean} (1)	D' (2)	K' (3)
0.0	0.0	$+\infty$
0.01	0.007312	68.70445
0.05	0.036562	14.0029
0.10	0.073124	7.16516
0.15	0.109704	4.88517
0.20	0.146489	3.74068
0.30	0.223191	2.567688
0.40	0.3111	1.93465
0.50	0.423441	1.508251
0.60	0.587217	1.178924
0.70	0.878462	0.896627

8.3 AIR-WATER INTERFACE AREA

Downstream of the inception point, the air-water interface area may be estimated from the air-water flow properties. For any bubble shape, bubble size distribution and chord length distribution, the mean chord length size[1] equals $C*V/F_{ab}$ where V is local air-water velocity and F_{ab} is the bubble count rate. The specific air-water interface area derives from the mass conservation for air:

$$a = \frac{4*F_{ab}}{V} \tag{A8.6}$$

Equation A8.6 is valid in bubbly flows (i.e. C < 0.3 to 0.5). In high air content regions (C > 0.3 to 0.5), the flow structure is more complex and the result is not exactly the true specific interface area. It becomes simply proportional to the number of air-water interfaces per unit length of air-water mixture (i.e. $a \propto 2*F_{ab}/V$) (Chanson, 1997c). The depth-averaged mean specific interface area is defined as:

$$a_{mean} = \frac{1}{Y_{90}} * \int_0^{Y_{90}} a*dy \tag{A8.7}$$

The dimensionless specific interface area $a_{mean}*V_{90}/(F_{ab})_{max}$ is a function of the mean air concentration only. It may be correlated as:

$$\frac{a_{mean}*V_{90}}{(F_{ab})_{max}} = 8.54*C_{mean} \qquad C_{mean} < 0.45 \tag{A8.8a}$$

$$\frac{a_{mean}*V_{90}}{(F_{ab})_{max}} = 7.2*C_{mean} + 9.43*C_{mean}^2 - 16.24*C_{mean}^3 \qquad C_{mean} < 0.7 \tag{A8.8b}$$

A8.4 AERATION EFFICIENCY OF STEPPED CHUTES WITH SKIMMING FLOW

In the fully-developed flow region (i.e. downstream of the inception point), the mass transfer equation (Eq. A8.1) may be numerically integrated (Eq. A8.2) using the mean specific interface area (Eq. A8.7). The mean air concentration may deduced from the continuity equation for the air phase (Eq. 5.29, Section 5.3.4). In skimming flows the boundary conditions may be set as: $s_* = L_I$, $C_* = 0.20$ and $u_r = 0.4$ m/s. The flow velocity may be calculated as in Appendix 7. The specific interface area calculations require an estimate of the maximum

[1]That is, the number mean size.

bubble count rate $(F_{ab})_{max}$. Experimental results are summarised in Table A8.2. Despite the different flow conditions, the data may be correlated by:

$$(F_{ab})_{max} = 20.65 * U_w^{1.74} \qquad 2 \le U_w \le 8 \quad (A8.9)$$

where F_{ab} and U_w are in SI units (i.e. Hz and m/s respectively).

Numerical modelling

The numerical integration of the mass transfer equation was conducted for slopes ranging from 22 to 53°, d_c/h between 1.5 and 10, q_w between 1 and 10 m^2/s, and assuming $u_r = 0.4$ m/s. The maximum bubble count rate $(F_{ab})_{max}$ was estimated using Equation A8.9. Typical results are presented in Figure A7.1 in terms of oxygen aeration efficiency at standard pressure and 15 Celsius as a function of the ratio of the dam height H_{dam} to the critical depth d_c. The calculations are compared with DO measurements in skimming flow in a laboratory (Essery et al., 1978) and DO data on prototype smooth chutes (Rindels & Gulliver, 1989). The latter were re-analysed by Chanson (1995c). The results (Fig. A8.1) highlight greater aeration efficiencies on stepped chutes with skimming flow (white symbols) than those on smooth-invert chutes (black symbols). With a stepped cascade, a greater aeration per metre drop in elevation is achieved with flat stepped chutes ($\alpha < 30°$). For a given geometry, the largest aeration is obtained at low flow rates.

Table A8.2. Experimental values of maximum air bubble frequency in turbulent shear flows.

Flow conditions (1)	q_w m2/s (2)	U_w m/s (3)	$(F_{ab})_{max}$ Hz (4)	Comments (5)
Open channel flow	0.150	4.1	133	Chanson (1997b). Smooth invert.
	0.150	3.7	78	
	0.150	3.1	68	
Hydraulic jump	0.0328	2.6	133	Chanson & Brattberg (2000). Partially-developed inflow conditions.
	0.0440	3.5	243	
Plunging jet flow	0.0172	2.0	29	Brattberg & Chanson (1998).
	0.0318	3.0	138	
	0.0450	4.0	250	
	0.0576	5.0	390	
	0.06996	6.0	565	
	0.0823	7.0	767	
	0.09456	8.0	935	

Table A8.2. Continued.

Flow conditions	q_w m2/s	U_w m/s	$(F_{ab})_{max}$ Hz	Comments
(1)	(2)	(3)	(4)	(5)
Stepped chute flow	0.1052	3.55	290	Toombes & Chanson (2000). Nappe flow without hydraulic jump. Downstream of nappe impact
	0.124	2.75	171	University of Queensland (a = 22°).
	0.103	2.65	182	Skimming flow: d/s end of chute.

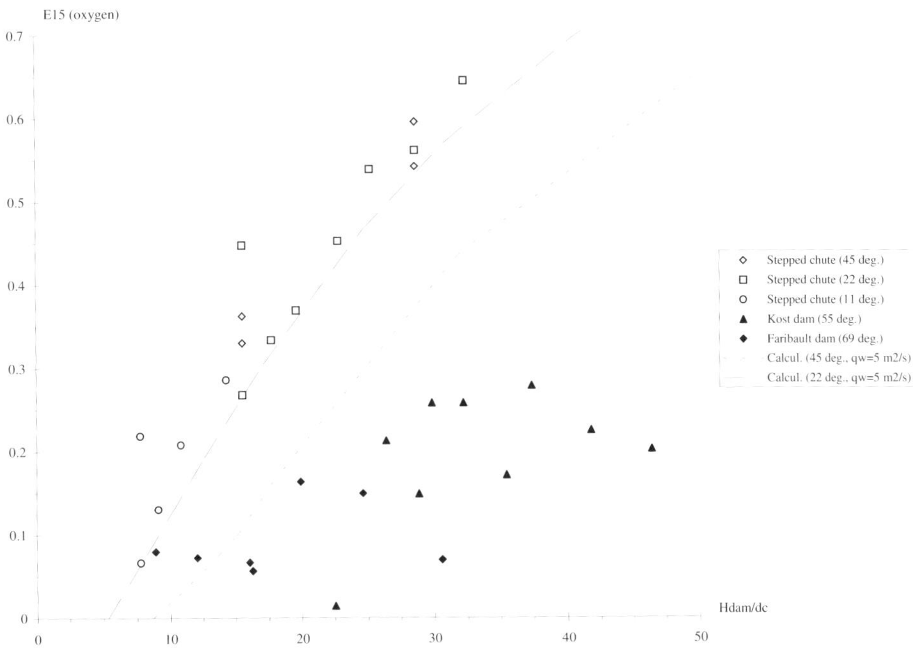

Figure A8.1. Aeration efficiency in terms of dissolved oxygen at standard pressure and 15 Celsius. Comparison between calculations (Eq. A7.2), stepped chute data (Essery et al., 1978) and prototype smooth chute data (Rindels & Gulliver, 1989). Black symbols: smooth-invert chute (prototype). White symbols: stepped chute (laboratory).

APPENDIX 9

Corrections

If you find a mistake or an error, your help and assistance would be most appreciated to improve the next edition of this book. Please record the errors on this page, and send a copy to the author:

Dr. H. Chanson
Dept. of Civil Engineering, The University of Queensland, Brisbane QLD 4072, Australia
Fax.: (61 7) 33 65 45 99   Email: h.chanson@mailbox.uq.edu.au
URL: http://www.uq.edu.au/~e2hchans   http://www.uq.edu.au/~e2hchans/photo.html

Corrections and updates will be posted on the Internet at:
http://www.uq.edu.au/~e2hchans/reprints/book4.htm

CORRECTION FORM

Contact
Name: ....

Address: ....

....

....

Tel.: ....

Fax: ....

Email: ....

Description of the error

Page number: Line number: Equation number:

....

....

....

....

....

Proposed correction

....

....

....

....

ABOUT THE AUTHOR

Hubert Chanson received a degree of Ingénieur Hydraulicien from the Ecole Nationale Supérieure d'Hydraulique et de Mécanique de Grenoble (France) in 1983 and a degree of Ingénieur Génie Atomique from the Institut National des Sciences et Techniques Nucléaires in 1984. He worked for the industry in France as a R&D engineer at the Atomic Energy Commission from 1984 to 1986, and as a computer professional in fluid mechanics for Thomson-CSF between 1989 and 1990. From 1986 to 1988, he studied at the University of Canterbury (New Zealand) as part of a Ph.D. project. He was awarded a Doctor of Engineering from the University of Queensland in 1999 for outstanding research achievements in gas-liquid flows.

Since 1990, Hubert Chanson has been a Reader in Environmental Fluid Mechanics and Water Engineering at the University of Queensland. His research interests include design of hydraulic structures, experimental investigations of two-phase flows, coastal hydrodynamics, water quality modelling, environmental management and natural resources. He has authored three books: *Hydraulic Design of Stepped Cascades, Channels, Weirs and Spillways* (Pergamon, 1995), *Air Bubble Entrainment in Free-Surface Turbulent Shear Flows* (Academic Press, 1997) and *The Hydraulics of Open Channel Flows: An Introduction* (Butterworth-Heinemann, 1999). His publication record includes more than 180 internationally-refereed papers and his work was cited over 550 times during the period 1990-2000. Hubert Chanson has also been active as a consultant for both governmental agencies and private organisations.

He has been awarded five fellowships from the Australian Academy of Science. In 1995 he was a Visiting Associate Professor at National Cheng Kung University (Taiwan R.O.C.) and he was a Visiting Research Fellow at Toyohashi University of Technology (Japan) in 1999.

Hubert Chanson was the keynote lecturer at the 1998 ASME Fluids Engineering Symposium on Flow Aeration (Washington DC), at the Workshop on Flow Characteristics around Hydraulic Structures (Nihon University, Japan, 1998), and at the 1st International Conference of the International Federation for Environmental Management System IFEMS'01 (Tsurugi, Japan, 2001). He gave an invited lecture at the International Workshop on Hydraulics of Stepped Spillways (ETH-Zürich, 2000). He has also lectured several short courses in Australia and overseas (e.g. Taiwan).

His Internet home page is {http://www.uq.edu.au/~e2hchans}. He has created a number of Internet-based resources, including a gallery of photographs that has received over 1,250 hits per month since inception:
{http://www.uq.edu.au/~e2hchans/photo.html}

References

Abdul Khader, M.H., & Elango, K. 1973. Turbulent Pressure Field beneath a Hydraulic Jump. *Jl of Hyd. Res.*, IAHR, Vol. 12, No. 4: 469-489.

Abt, S.R., & Johnson, T.L. 1991. Riprap Design for Overtopping Flow. *Jl of Hyd. Engrg.*, ASCE, Vol. 117, No. 8: 959-972.

Adachi, S. 1964. On the Artificial Strip Roughness. *Disaster Prevention Research Institute Bulletin*, No. 69, Kyoto University, Japan, March.

Agostini, R. et al. 1987. Flexible Gabion and Reno Mattress Structures in River and Stream Training Works. Section One: Weirs. *Officine Maccaferri*, Bologna, 2nd edition.

Aivazyan, O.M. 1986. Stabilized Aeration on Chutes. *Gidrotekhnicheskoe Stroitel'stvo*, No. 12: 33-40 (in Russian). (Translated in *Hydrotechnical Construction*, Plenum, 1987, 713-722).

Allen, P. 1984. *Modelling Flow Over and Through Overtopped Rockfill Embankments.* M.Eng. thesis, Dept. of Civil Eng., University of Queensland, Brisbane, Australia.

Apelt, C.J. 1983. Hydraulics of Minimum Energy Culverts and Bridge Waterways. *Australian Civil Engrg Trans.*, I.E.Aust., Vol. CE25, No. 2: 89-95.

APHA, AWWA, & WPCF 1985. Standard Methods for the Examination of Water and Wastewater. *American Public Health Association Publ.*, 6th edition.

APHA, AWWA, & WPCF 1989. Standard Methods for the Examination of Water and Wastewater. *American Public Health Association Publ.*, 7th edition.

Armanini, A., Dellagiacoma, F., & Ferrari, L. 1991. From the Check Dam to the Development of Functional Check Dams. In Arminini & Di Dilvio (ed.) *International Work. Fluvial Hydraulics of Mountain Regions.* IAHR: 331-344. Berlin: Springer Verlag.

ASCE Task Committee 1994. Alternatives for Overtopping Protection of Dams. *ASCE*, Task Committee on Overtopping Protection, New York.

Ashby, T. 1935. *The Aqueducts of Ancient Rome.* I.A. Richmond (ed.) Oxford: Clarendon Press.

Avery, S.T., & Novak, P. 1978. Oxygen Transfer at Hydraulic Structures. *Jl of Hyd. Div.*, ASCE, Vol. 104, No. HY11: 1521-1540.

Babbitt, H.E. 1922. *Sewerage and Sewage Treatment.* New York: John Wiley & Sons.

BaCaRa 1991. Etude de la Dissipation d'Energie sur les Evacuateurs à Marches. (Study of the Energy Dissipation on Stepped Spillways.) *Rapport d'Essais*, Projet National BaCaRa, CEMAGREF-SCP, Aix-en-Provence, France, Oct., 111 pages (in French).

Baker, R. 1990. Precast Concrete Blocks for High Velocity Flow Applications. *J. IWEM*, Vol. 4, Dec: 552-557. Discussion: Vol. 4: 557-558.

Baker, R. 1994. Brushes Clough Wedge Block Spillway - Progress Report No. 3. *SCEL Project Report No. SJ542-4*, University of Salford, UK.

Baker, R. 2000. The CIRIA Guide for the Design of Stepped-Block Spillways. In H.E. Minor & W.H. Hager (ed.) *International Workshop on Hydraulics of Stepped Spillways*, Zürich, Switzerland: 155-161. Balkema.

Baker, R., & Gardiner, K. 1994. The Construction and Performance of a Wedge Block Spillway at Brushes Clough Reservoir. *Proc. 8th Conf. British Dam Society*, Exeter University, UK: 214-223. Thomas Telford.

Baker, R., Pravdivets, Y., & Hewlett, H. 1994. Design Considerations for the Use of

Wedge-Shaped Precast Concrete Blocks for Dam Spillways. *Proc. Instn Civ. Engrs Wat. Marit. & Energy*, Vol. 106, Dec: 317-323.

Bakhmeteff, B.A., & Fedoroff, N.V. 1943. Energy Loss at the Base of a Free Overfall - Discussion. *Transactions,* ASCE, Vol. 108: 1364-1373.

Bakhmeteff, B.A., & Matzke, A.E. 1936. The Hydraulic Jump in Terms of Dynamic Similarity. *Transactions,* ASCE, Vol. 101: 630-647. Discussion: Vol. 101: 648-680.

Bass, R. 1998. Learning Lessons on RCC Design. *Intl Water and Dam Construction*, Vol. 50, No. 9: 14-18.

Bathurst, J.C. 1978. Flow Resistance of Large-Scale Roughness. *Jl of Hyd. Div.*, ASCE, Vol. 104, No. HY12: 1587-1603.

Bathurst, J.C. 1988. Velocity Profile in High-Gradient, Boulder-Bed Channels. *Proc. Intl Conf. on Fluvial Hydraulics*, IAHR, May, Budapest, Hungary: 29-34.

Battison, E.A. 1975. Ascutney Gravity-Arch Mill Dam - Windsor, Vermont 1834. *IA Jl of the Soc. of Industrial Archeology*, Summer, Vol. 1, No. 1: 53-58.

Bauer, S.W., & Graf, W.H. 1971. Free Overfall as Flow Measuring Device. *Jl of Irrig. and Drain.*, ASCE, Vol. 97, No. IR1: 73-83.

Bayat, H.O. 1991. Stepped Spillway Feasibility Investigation. *Proc. 17th ICOLD Congress*, Vienna, Austria, Q. 66, R. 98: 1803-1817.

Bazin, H. 1865. Recherches Expérimentales sur l'Ecoulement de l'Eau dans les Canaux Découverts. (Experimental Research on Water Flow in Open Channels) *Mémoires présentés par divers savants à l'Académie des Sciences*, Paris, France, Vol. 19: 1-494 (in French).

Beitz, E., & Lawless, M. 1992. Hydraulic Model Study for dam on GHFL 3791 Isaac River at Burton Gorge. *Water Resources Commission Report*, Ref. No. REP/24.1, Sept., Brisbane, Australia.

Bélanger, J.B. 1828. *Essai sur la Solution Numérique de quelques Problèmes Relatifs au Mouvement Permanent des Eaux Courantes.* (Essay on the Numerical Solution of Some Problems relative to Steady Flow of Water) Paris: Carilian-Goeury (in French).

Bélidor, B.F. de 1737-1753. *Architecture Hydraulique.* (Hydraulic Architecture). Paris: Charles-Antoine Jombert, 4 volumes (in French).

Bindo, M., Gautier, J., & Lacroix, F. 1993. The Stepped Spillway of M'Bali Dam. *Intl Water Power and Dam Construction*, Vol. 45, No. 1: 35-36.

Binnie, A.R. 1913. *Rainfall Reservoirs and Water Supply.* London: Constable & Co.

Binnie, G.M. 1981. *Early Victorian Water Engineers.* London: Thomas Telford.

Binnie, G.M. 1987. *Early Dam Builders in Britain.* London: Thomas Telford.

Boes, R.M. 1999. Physical Model Study on Two-Phase Cascade Flow. *Proc. 28th IAHR Congress*, Graz, Austria, Session S1 (CD-ROM).

Boes, R.M. 2000. *Zweiphasenströmung und Energieumsetzung auf Grosskaskaden.* (Two-Phase Flow and Energy Dissipation on Cascades) Ph.D. thesis, VAW-ETH, Zürich, Switzerland (in German).

Bos, M.G. 1976. Discharge Measurement Structures. *Publication No. 161*, Delft Hydraulic Laboratory, Delft, The Netherlands (also Publication No. 20, ILRI, Wageningen, The Netherlands).

Bossut, Abbé C. (1772. *Traité Elémentaire d'Hydrodynamique.* (Elementary Treaty on Hydrodynamics) Paris: Imprimerie Chardon, 1st edition (in French). (2nd edition: 1786. 3rd edition: 1796.

Boussinesq, J.V. 1877. Essai sur la Théorie des Eaux Courantes. (Essay on the Theory of Water Flow) *Mémoires présentés par divers savants à l'Académie des Sciences*, Paris, France, Vol. 23, ser. 3, No. 1, supplément 24: 1-680 (in French).

Boussinesq, J.V. (1896. Théorie de l'Ecoulement Tourbillonnant et Tumultueux des

Liquides dans les Lits Rectilignes à Grande Section (Tuyaux de Conduite et Canaux Découverts) quand cet Ecoulement s'est régularisé en un Régime Uniforme, c'est-à-dire, moyennement pareil à travers toutes les Sections Normales du Lit. (Theory of Turbulent and Tumultuous Flow of Liquids in prismatic Channels of Large Cross-sections (Pipes and Open Channels) when the Flow is Uniform, i.e., Constant in Average at each Cross-section along the Flow Direction) *Comptes Rendus des séances de l'Académie des Sciences*, Paris, France, Vol. 122: 1290-1295 (in French).

Bouyge, B. et al. 1988. Construction et Contrôle d'un Barrage en Béton Compacté au Rouleau (BCR): un Travail d'Equipe. (Roller Compacted Concrete Dam Construction and Works Supervision: a Team Job) *Proc. 16th ICOLD Congress*, San Francisco, USA, Q. 62, R. 34: 589-612 (in French).

Bowie, G. et al. 1985. Rates, Constants, & Kinetics Formulations in Surface Water Quality Modeling. *Tetra Tech. Report*, No. EPA/600/3-85-040, 2nd edition.

Boyer, P.B. 1971. Gas Supersaturation Problem in the Columbia River. *Proc. Intl Symp. on Man-Made Lakes*, American Geophysical Union, Knoxville TN, USA: 701-705.

Brattberg, T., & Chanson, H. 1998. Air Entrapment and Air Bubble Dispersion at Two-Dimensional Plunging Water Jets. *Chemical Engineering Science*, Vol. 53, No. 24, Dec.: 4113-4127. Errata: 1999, Vol. 54, No. 12: 1925.

Brattberg, T., Chanson, H., & Toombes, L. 1998. Experimental Investigations of Free-Surface Aeration in the Developing Flow of Two-Dimensional Water Jets. *Jl of Fluids Eng.*, Trans. ASME, Vol. 120, No. 4: 738-744.

Bresse, J.A. 1860. *Cours de Mécanique Appliquée Professé à l'Ecole des Ponts et Chaussées*. (Course in Applied Mechanics lectured at the Pont-et-Chaussées Engineering School). Paris: Mallet Bachelier (in French).

Breznik, M. 1984. The Safety and Endurance of the Old Dams of Idrija. *Proc. Intl Conf. on Safety of Dams*, Portugal: 133-139. Rotterdam: Balkema.

Bruschin, J. et al. 1982. The Overtopping of the Palagnedra dam. *Intl Water and Dam Construction*, Vol. 34, No. 1, Jan.: 13-19.

Buckley, R.B. 1905. *The Irrigation Works in India*. London: E.&F.N. Spon. 2nd edition.

Bushnell, D.M., & Hefner, J.N. 1990. Viscous Drag Reduction in Boundary Layers. In *Progress in Astronautics and Aeronautics*, Vol. 123. Washington DC: AIAA Publications.

Cargill, G.D. 1994. Water Quality in the Windy City. *Water Environ. & Technol.*, Vol. 6, No. 9: 52-56.

Casperson, L.W. 1993. Fluttering Fountains. *Jl Sound and Vibration*, Vol. 162, No. 2: 251-262.

Chamani, M.R. 2000. Air inception in Skimming Flow regime over Stepped Spillways. In H.E. Minor & W.H. Hager (ed.) *International Workshop on Hydraulics of Stepped Spillways, Zürich, Switzerland*: 61-67. Balkema.

Chamani, M.R., & Rajaratnam, N. 1994. Jet Flow on Stepped Spillways. *Jl of Hyd. Engrg.*, ASCE, Vol. 120, No. 2: 254-259. Discussion: *Jl of Hyd. Engrg.*, ASCE, Vol. 121, No. 5: 441-448.

Chamani, M.R., & Rajaratnam, N. 1999a. Characteristics of Skimming Flow over Stepped Spillways. *Jl of Hyd. Engrg.*, ASCE, Vol. 125, No. 4: 361-368.

Chamani, M.R., & Rajaratnam, N. 1999b. Onset of Skimming Flow on Stepped Spillways. *Jl of Hyd. Engrg.*, ASCE, Vol. 125, No. 9: 969-971.

Chanson, H. 1988. *A Study of Air Entrainment and Aeration Devices on a Spillway Model*. Ph.D. thesis, Ref. 88-8, Dept. of Civil Engineering, University of Canterbury, New Zealand.

Chanson, H. 1989a. Study of Air Entrainment and Aeration Devices. *Jl of Hyd. Research*, IAHR, Vol. 27, No. 3: 301-319.

Chanson, H. 1989b. Flow downstream of an Aerator. Aerator Spacing. *Jl of Hyd. Research*, IAHR, Vol. 27, No. 4: 519-536.

Chanson, H. 1992. Air Entrainment in Chutes and Spillways. *Research Report No. CE 133*, Dept. of Civil Engineering, University of Queensland, Australia, Feb., 85 pages.

Chanson, H. 1993a. Self-Aerated Flows on Chutes and Spillways. *Jl of Hyd. Engrg.*, ASCE, Vol. 119, No. 2: 220-243. Discussion: Vol. 120, No. 6: 778-782.

Chanson, H. 1993b. Velocity Measurements within High Velocity Air-Water Jets. *Jl of Hyd. Res.*, IAHR, Vol. 31, No. 3: 365-382 & No. 6: 858.

Chanson, H. 1993c. Stepped Spillway Flows and Air Entrainment. *Can. Jl of Civil Eng.*, Vol. 20, No. 3, June: 422-435.

Chanson, H. 1994a. Drag Reduction in Open Channel Flow by Aeration and Suspended Load. *Jl of Hyd. Res.*, IAHR, Vol. 32, No. 1: 87-101.

Chanson, H. 1994b. Hydraulics of Nappe Flow Regime above Stepped Chutes and Spillways. *Aust. Civil Engrg Trans.*, I.E.Aust., Vol. CE36, No. 1, Jan.: 69-76.

Chanson, H. 1994c. Comparison of Energy Dissipation between Nappe and Skimming Flow Regimes on Stepped Chutes. *Jl of Hyd. Res.*, IAHR, Vol. 32, No. 2: 213-218. Errata: Vol. 33, No. 1: 113. Discussion: Vol. 33, No. 1: 114-143.

Chanson, H. 1994d. Hydraulics of Skimming Flows over Stepped Channels and Spillways. *Jl of Hyd. Res.*, IAHR, Vol. 32, No. 3: 445-460. Discussion: Vol. 33, No. 3: 414-419.

Chanson, H. 1995a. *Hydraulic Design of Stepped Cascades, Channels, Weirs and Spillways*. Pergamon, Oxford.

Chanson, H. 1995b. Air Bubble Diffusion in Supercritical Open Channel Flow. In R.W. Bilger (ed.) *Proc. 12th Australasian Fluid Mechanics Conference AFMC, Sydney, Australia*, Vol. 2: 707-710.

Chanson, H. 1995c. Predicting Oxygen Content Downstream of Weirs, Spillways and Waterways. *Proc. Instn Civ. Engrs Wat. Marit. & Energy*, UK, Vol. 112, Mar.: 20-30. Discussion: Vol. 130, No. 2: 115-116.

Chanson, H. 1996. Prediction of the Transition Nappe/Skimming Flow on a Stepped Channel. *Jl of Hyd. Res.*, IAHR, Vol. 34, No. 3: 421-429.

Chanson, H. 1997a. *Air Bubble Entrainment in Free-Surface Turbulent Shear Flows*. London: Academic Press.

Chanson, H. 1997b. Air Bubble Entrainment in Open Channels. Flow Structure and Bubble Size Distributions. *Intl Jl of Multiphase Flow*, Vol. 23, No. 1: 193-203.

Chanson, H. 1997c. Measuring Air-Water Interface Area in Supercritical Open Channel Flow. *Water Res.*, IAWPRC, Vol. 31, No. 6: 1414-1420.

Chanson, H. 1997d. A Short History of Stepped Cascades in Australia. *ANCOLD Bulletin*, No. 106, Aug.: 101-111.

Chanson, H. 1998a. Le Développement Historique des Cascades et Fontaines en Gradins. (Historical Development of Stepped Cascades and Fountains) *Jl La Houille Blanche*, No. 7/8: 76-84 (in French).

Chanson, H. 1998b. Extreme Reservoir Sedimentation in Australia: a Review. *Intl Jl of Sediment Research*, UNESCO-IRTCES, Vol. 13, No. 3: 55-63.

Chanson, H. 1999a. *The Hydraulics of Open Channel Flows: An Introduction*. Oxford: Butterworth-Heinemann.

Chanson, H. 2000. Hydraulics of Roman Aqueducts: Steep Chutes, Cascades and Dropshafts. *American Jl of Archaeology*, Vol. 104, No. 1, Jan.: 47-72.

Chanson, H., & Brattberg, T. 2000. Experimental Study of the Air-Water Shear Flow in a Hydraulic Jump. *Intl Jl of Multiphase Flow*, Vol. 26, No. 4: 583-607.

Chanson, H., & Montes, J.S. 1998. Overflow Characteristics of Circular Weirs: Effect of Inflow Conditions. *Jl of Irrigation and Drainage Engrg.*, ASCE, Vol. 124, No. 3: 152-162.

Chanson, H., & Toombes, L. 1997a. Energy Dissipation in Stepped Waterway. In F.M. Holly Jr. and A. Alsaffar (ed.) *Proc. 27th IAHR Congress, San Francisco, USA*, Vol. D: 595-600.

Chanson, H., & Toombes, L. 1997b. Flow Aeration at Stepped Cascades. *Research Report No. CE155*, Dept. of Civil Engineering, University of Queensland, Australia, June, 110 pages.

Chanson, H., & Toombes, L. 1998. Supercritical Flow at an Abrupt Drop: Flow Patterns and Aeration. *Can. Jl of Civil Eng.*, Vol. 25, No. 5, Oct.: 956-966.

Chanson, H., & Toombes, L. 2000. Stream Reaeration in Nonuniform Flow: Macroroughness Enhancement. Discussion. *Jl of Hyd. Engrg.*, ASCE, Vol. 126, No. 3: 222-224.

Chanson, H., & Whitmore, R.L. 1996. Investigation of the Gold Creek Dam Spillway, Australia. *Research Report No. CE153*, Dept. of Civil Engineering, University of Queensland, Australia.

Chanson, H., & Whitmore, R.L. 1998. Gold Creek Dam and its Unusual Waste Waterway (1890-1997): Design, Operation, Maintenance. *Can. Jl of Civil Eng.*, Vol. 25, No. 4, Aug.: 755-768 & Front Cover.

Chanson, H., Yasuda,Y., & Ohtsu, I. 2000. Flow Resistance in Skimming Flow: a Critical Review. In H.E. Minor & W.H. Hager (ed.) *International Workshop on Hydraulics of Stepped Spillways, Zürich, Switzerland*: 95-102. Balkema.

Chanson, H., Aoki, S., & Maruyama, M. 2000b. Experimental Investigations of Wave Runup Downstream of Nappe Impact. Applications to Flood Wave Resulting from Dam Overtopping and Tsunami Wave Runup. *Coastal/Ocean Engineering Report*, No. COE00-2, Dept. of Architecture and Civil Eng., Toyohashi University of Technology, Japan.

Chen, C.L. 1990. Unified Theory on Power Laws for Flow Resistance. *Jl of Hyd. Engrg.*, ASCE, Vol. 117, No. 3: 371-389.

Chow, V.T. 1959. *Open Channel Hydraulics*. New York: McGraw-Hill.

Christodoulou, G. C. 1993. Energy Dissipation on Stepped Spillways. *Jl of Hyd. Engrg.*, ASCE, Vol. 119, No. 5: 644-650. Discussion: Vol. 121, No. 1: 80-87.

Colas des Francs, E. 1975. Sealing and Drainage Facilities for the Ste Croix and Quinson Dams. *Intl Water Power & Dam Construction*, Vol. 27, Oct.: 356-361.

Collett, K.O. 1975. Unusual Surfaces for Large Spillways. *ANCOLD Bulletin*, No. 42, July: 3-10.

Comolet, R. 1979a. Sur le Mouvement d'une bulle de gaz dans un liquide. (Gas bubble motion in a liquid medium) *Jl La Houille Blanche*, 1979, No. 1: 31-42 (in French).

Comolet, R. 1979b. Vitesse d'ascension d'une bulle de gaz isolée dans un liquide peu visqueux. (The terminal Velocity of a Gas Bubble in a Liquid of very low Viscosity) *Jl de Mécanique Appliquée*, Vol. 3, No. 2: 145-171 (in French).

Concrete 1993. Clywedog Dam - Mid-Wales. *Concrete*, Vol. 27, No. 5: 27-29.

Cook, O.F. 1916. Staircase Farms of the Ancients. *National Geographic Magazine*, Vol. 29: 474-534

Coriolis, G.G. 1836. Sur l'établissement de la formule qui donne la figure des remous et sur la correction qu'on doit introduire pour tenir compte des différences de vitesses dans les divers points d'une même section d'un courant. (On the establishment of the formula giving the backwater curves and on the correction to be introduced to take into account the velocity differences at various points in a cross-section of a stream) *Annales des Ponts et Chaussées*, 1st Semester, Series 1, Vol. 11 (in French).

Corsi, R.L. et al. 1992. Oxygen Transfer and VOC Emissions from Sewer Drop Structures. *Proc. 1992 Nat. Conf. on Hydraulic Engrg.* Water Forum '92, ASCE, Baltimore, USA: 305-310.

Cossins, G. 2000. *The Gold Creek Dam Story*. Brisbane: Inst. of Engineers, Australia; Queensland Division.

Craya, A. 1948. Hauteur d'eau à l'Extrémité d'un Long Déversoir. (Flow Depth at the Downstream End of a Long Spillway) *Jl La Houille Blanche*, Mar.-Apr.: 185-186 (in French).

Creager, W.P. 1917. *Engineering of Masonry Dams*. New York: John Wiley & Sons.

Creager, W.P., Justin, J.D., & Hinds, J. 1945. *Engineering for Dams*. New York: John Wiley & Sons. 3 volumes.

Crocker, N. 1987. Fountains. *Landscape Australia*, No. 2: 137-154, No 3: 201-206, No. 4: 324-332.

Crouch, D.P. 1991. Spanish Water Technology in New Spain - Transfer and Alterations. *8th Intl Symp. on History of Hydraulics*, Merida, Spain (Mitteilungen, Inst. für Wasserbau der Technische Univ. Braunschweig, Germany, Col. 117, 1992: 180-227).

Cummins, P.J., Smith, B.R., & Evans, R.C. 1985. Rehabilitation of Goulburn Weir Foundations and Superstructures. *Proc. 15th ICOLD Congress*, Lausanne, Switzerland, Q. 59, R. 38: 603-618.

Curtis, R.P., & Lawson, J.D. 1967. Flow over and through Rockfill Banks. *Jl of Hyd. Div.*, ASCE, Vol. 93, No. HY5: 1-21.

Daniil, E.I., & Gulliver J. 1988. Temperature Dependence of Liquid Film Coefficient for Gas Transfer. *Jl of Envir. Engrg.*, ASCE, Vol. 114, No. 5: 1224-1229.

Daniil, E.I., Gulliver J., & Thene, J.R. 1991. Water-Quality Assessment for Hydropower. *Jl of Envir. Engrg.*, ASCE, Vol. 117, No. 2: 179-193.

Danilevslkii, V.V. 1940. History of Hydroengineering in Russia before the Nineteenth Century. *Gosudarstvennoe Energeticheskoe Izdatel'stvo*, Leningrad, USSR (in Russian) (English translation: *Israel Program for Scientific Translation*, IPST No. 1896, Jerusalem, Israel, 1968).

Darcy, H.P.G. 1856. *Les Fontaines Publiques de la Ville de Dijon*. (The Public Fountains of the City of Dijon). Paris: Victor Dalmont (in French).

Darcy, H.P.G. 1858. Recherches Expérimentales relatives aux Mouvements de l'Eau dans les Tuyaux. (Experimental Research on the Motion of Water in Pipes) *Mémoires Présentés à l'Académie des Sciences de l'Institut de France*, Vol. 14: 141 (in French).

Darcy, H.P.G., & Bazin, H. 1865. Recherches Hydrauliques. (Hydraulic Research) *Imprimerie Impériales*, Paris, France, Parties 1ère et 2ème (in French).

Degremont 1979. Water Treatment Handbook. *Halsted Press Book*. New York: John Wiley & Sons, 5th edition.

Delagrave, A. et al. 1994. Résistance au Gel-Dégel du Béton Compacté au Rouleau pour les Barrages à Base de liant Rolac. (Freeze-thaw Durability of Roller-compacted Concrete made with Rolac Binder) *Materials and Structures*, Vol. 27, No. 165: 26-32 (in French).

Department of the Environment 1973. Aeration at Weirs. *Notes on Water Pollution*, No. 61, Water Res. Lab., Elder Way, Stevenage, Herts, England.

Diacon, A., Strmatiu, D., & Mircea, N. 1992. An Analysis of the Belci Dam Failure. *Intl Water Power & Dam Construction*, Vol. 44, No. 9, Sept.: 67-72.

Diez-Cascon et al. 1991. Studies on the Hydraulic Behaviour of Stepped Spillways. *Intl Water Power and Dam Construction*, Vol. 43, No. 9, Sept.: 22-26.

Ditchey, E.J., & Campbell, D.B. 2000. Roller Compacted Concrete and Stepped Spillways. In H.E. Minor & W.H. Hager (ed.) *International Workshop on Hydraulics of Stepped Spillways, Zürich, Switzerland*: 171-178. Balkema.

Djenedi, L, Anselmet, F., & Antonia, R.A. 1994. LDA Measurements in a Turbulent Boundary Layer over a D-Type Rough Wall. *Experiments in Fluids*, Vol. 16: 323-329.

DNR 1989. Private Communication.

Dolen, T.P., & von Fay, K.F. 1993. The Development of Roller Compacted Concrete Mixtures for Bureau of Reclamation Mass Concrete Construction. In Liu & Hoff (ed.) *American Concrete Institute*, SP-141, USA: 133-163.

Douma, J.H., & Goodpasture, R.A. 1940. Hydraulic Model Studies for Reconstruction of the Lahontan Dam Spillway. *Hydraulic Laboratory Report No. HYD-71*, US Bureau of Reclamation, Dept. of the Interior, Denver, 13 Jan.

Dupuit, A.J.E. (1848. *Etudes Théoriques et Pratiques sur le Mouvement des Eaux Courantes*. (Theoretical and Practical Studies on Flow of Water). Paris: Dunod (in French).

Dussart, J. et al. 1993. Le Béton Compacté au Rouleau (BCR) du Barrage de Petit-Saut (Guyane). (The Roller Compacted Concrete of the Petit-Saut Dam (Guyana)) *Travaux*, No. 688, June: 14-21 (in French).

Elavarasan, R., Pearson, B.R., & Antonia, R.E. 1995. Visualization of Near Wall Region in a Turbulent Boundary Layer over Transverse Square Cavities with Different Spacing. *Proc. 12th Australasian Fluid Mech. Conf. AFMC*, Sydney, Australia, Vol. 1: 485-488.

Ellis, J. 1989. Guide to Analysis of Open-Channel Spillway Flow. *CIRIA Technical Note No. 134*. London, 2nd edition.

Elmore, H.L., & West, W.F. 1961. Effect of Water Temperature on Stream Reaeration. *Jl of Sanitary Engrg.*, ASCE, Vol. 87, No. SA6: 59-71.

Elviro, V., & Mateos, C. 1995. Spanish Research into Stepped Spillways. *Intl Jl Hydropower & Dams*, Vol. 2, No. 5: 61-65.

Engels, H. 1914. *Handbuch des Wasserbaues*. (Handbook of Hydraulic Structures) Leipzig: Wilhelm Engelmann. 2 volumes (in German).

The Engineer 1939. The Ladybower Reservoir. *The Engineer*, Vol. 168: 440-442.

Engineering 1966. Britain's Highest Dam. *Engineering*, Vol. 202: 755-759.

Engineering News. 1905a. Construction, Repairs and Subsequent Partial Destruction of the Arizona Canal Dam, near Phoenix, Arizona. *Engineering News*, Vol. 53, No. 17, 27 April: 450-451.

Engineering News. 1905b. Debris Barrier No. 1, Yuba River, California. *Engineering News*, Vol. 53, No. 24, 15 June: 609-610.

Erbe, J. 1974. *Handbook for Designing Decorative Fountains*. Albuquerque (NM): Roman Fountains Inc.

Ervine, D.A., & Falvey, H.T. 1987. Behaviour of Turbulent Water Jets in the Atmosphere and in Plunge Pools. *Proc. Instn Civ. Engrs.*, Part 2, Mar. 1987, 83: 295-314. Discussion: Part 2, Mar.-June 1988, 85: 359-363.

Escande, L. et al. 1961. Influence de Quelques Paramètres sur une Onde de Crue Subite à l'Aval d'un Barrage. (The Influence of certain Parameters on a Sudden Flood Wave Downstream from a Dam) *Jl La Houille Blanche*, No. 5: 565-575.

Essery, I.T.S., & Horner, M.W. 1978. The Hydraulic Design of Stepped Spillways. *CIRIA Report No. 33*, Jan., London, 2nd edition.

Essery, I.T.S., Tebbutt, T.H.Y., & Rasaratnam, S.K. 1978. Design of Spillways for Re-aeration of Polluted Waters. *CIRIA Report No. 72*, Jan., London, UK, 36 pages.

Etcheverry, B.A. 1916. *Irrigation Practice and Engineering*. New York: McGraw Hill. 3 Volumes.

Ethembabaoglu, S. 1978. Some Characteristics of Unstable Flow past Slots. *Jl o Hyd. Div.*, Proc. ASCE, Vol. 104, No. HY5: 649-666.

Evans, A.H. 1928. *The Palace of Minos: a comparative account of the successive stages of the early Cretan civilization as illustrated by the discoveries at Knossos*. Vol. II, Part 1. London: Macmillan.

Evans, R.C. 1984. Goulburn Weir Remodelling. *ANCOLD Bulletin*, No. 70, Dec.: 36-38.

Fahlbusch, H. 1987. Alte Talsperren im Gebiet des Königreichs Saudi Arabien. (Ancient Dams in the region of the Saudi Arabia Kingdom). In *Historische Talsperren*, Vol. 1, K. Wittwert, Stuttgart: 199-220 (in German).

Falvey, H.T. 1980. *Air-Water Flow in Hydraulic Structures*. USBR Engrg. Monograph, No. 41, Denver, Colorado.

Farrington, I.S. 1980. The Archaeology of Irrigation Canals with Special Reference to Peru. *World Archaeology*, Vol. 11, No. 3: 287-305.

Farrington, I.S., & Park, C.C. 1978. Hydraulic Engineering and Irrigation Agriculture in the Moche Valley, Peru: c. A.D. 1250-1532. *Jl of Archaeological Science*, Vol. 5: 255-268.

Ferrell, R.T., & Himmelblau, D.M. 1967. Diffusion Coefficients of Nitrogen and Oxygen in Water. *Jl of Chem. Eng. Data*, Vol. 12, No. 1: 111-115.

Ferro, V. 1992. Flow Measurement with Rectangular Free Overfall. *Jl of Irrig. and Drain.*, ASCE, Vol. 118, No. 6: 956-964.

Ferro, V., & Baiamonte, G. 1994. Flow Velocity Profiles in Gravel-Bed Rivers. *Jl of Hyd. Engrg.*, ASCE, Vol. 120, No. 1: 60-80.

Fick, A.E. 1855. On Liquid Diffusion. *Philos. Mag.*, Vol. 4, No. 10: 30-39.

Fiorotto, V., & Rinaldo, A. 1992. Turbulent Pressure Fluctuations under Hydraulic Jumps. *Jl of Hyd. Res.*, IAHR, Vol. 30, No. 4: 499-520.

Forbes, B.A. 1995. Australian RCC Practice: Nine Dams Each Different. *Proc. Intl Symp. on Roller Compacted Concrete Dams, Santander, Spain, IECA and CNEGP*, Vol. 3: 1407-1435.

Forbes, R.J. 1955. *Studies in Ancient Technology*. Leiden: E.J. Brill, 9 volumes.

Ford, A.C. 1957. Repair of Spillway leaks at New Croton, N.Y.. *Jl American Waterworks Association*, Feb.: 198-204.

Foree, E.G. 1976. Reaeration and Velocity Prediction for Small Streams. *Jl of Envir. Engrg.*, ASCE, Vol. 102, No. EE5: 937-952.

Fourier, J.B.J. 1822. *Théorie Analytique de la Chaleur*. (Analytical Theory of Heat) Paris: Didot (in French).

Franc, J.P. et al. 1995. *La Cavitation. Mécanismes Physiques et Aspects Industriels*. (The Cavitation. Physical Mechanisms and Industrial Aspects) Presses Universitaires de Grenoble, Collection Grenoble Sciences (in French).

Frizell, K.H. 1992. Hydraulics of Stepped Spillways for RCC Dams and Dam Rehabilitations. *Proc. 3rd Specialty Conf. on Roller Compacted Concrete*, ASCE, San Diego CA: 423-439.

Frizell, K.H., & Mefford, B.W. 1991. Designing Spillways to Prevent Cavitation Damage. *Concrete International*, Vol. 13, No. 5: 58-64.

Furuya, Y., Miyata, M., & Fujita, H. 1976. Turbulent Boundary Layer and Flow Resistance on Plates Roughened by Wires. *Jl of Fluids Eng.*, Trans. ASME, Vol. 98: 635-644.

Gameson, A.L.H., Vandyke, K.G., & Ogden, C.G. 1958. The Effect of Temperature on Aeration at Weirs. *Water and Water Engrg.*, Nov.: 489-492.

Garbrecht, G. 1987. *Hydraulics and Hydraulic Research: a Historical Review*. Rotterdam: Balkema.

Garcia-Diego, J.A. 1976. The Chapter on Weirs in the Codex of Juanelo Turriano. *Technology and Culture*: 217-234.

Garcia-Diego, J.A. 1977. Old Dams in Extramadura. *History of Technology*, Vol. 2: 95-124.

Gasparotto, R. 1991. Urban Waterfalls Combine Aesthetics with Water Quality Improvement. *Water Quality Intl.*, No. 4: 36-39.

Gasparotto, R. 1992. Waterfall Aeration Works. *Civil Engineering*, ASCE, Vol. 62, No. 10: 52-54.

Gauckler, P.G. 1867. Etudes Théoriques et Pratiques sur l'Ecoulement et le Mouvement des Eaux. (Theoretical and Practical Studies of the Flow and Motion of Waters) *Comptes Rendues de l'Académie des Sciences*, Paris, Tome 64: 818-822 (in French).

Gausmann, R.W., & Madden, C.M. 1923. Experiments with Models of the Gilboa Dam and Spillway. *Transactions*, ASCE, Vol. 86: 280-305.

Geringer, J.J. 1995. The Design and Construction of RCC Dams in Southern Africa. *Proc. Intl Symp on RCC Dams*, Santander, Spain, IECA-CNEGP, Vol. 3: 1459-1495.

Geringer, J.J., & Officer, A.K. 1995. Stepped Spillway Hydraulic Research for RCC Dams - Quo Vadis ? *Proc. Intl Symp on RCC Dams*, Santander, Spain, IECA-CNEGP, Vol. 1: 549-563.

Gill, M.A. 1979. Hydraulics of Rectangular Vertical Drop Structures. *Jl of Hyd. Res.*, IAHR, Vol. 17, No. 4: 289-302.

Goertler, H. 1942. Berechnung von Aufgaben der freien Turbulenz auf Grund eines neuen Näherungsansatzes. *Z.A.M.M.*, 22: 244-254 (in German).

Gomez-Perez, F. 1942. Mexican Irrigation in the Sixteenth Century. *Civil Engineering*, ASCE, Vol. 12, No. 1: 24-27.

Gordienko, P.I. 1944. Rockfill Overflow Dams. *Gidrotekhnicheskoe Stroitel'stvo*, No. 3 (in Russian).

Gordienko, P.I. 1978. Reinforced-Concrete-Earth Overflow Dams. *Dams & Spillways*, Collection of Works No. 61, Issue 2, MISI, Moscow: 3-17 (in Russian).

Gosse, P., & Gregoire, A. 1997. Dispositif de Réoxygénation Artificielle du Sinnamary à l'Aval du Barrage de Petit-Saut (Guyane). (Artificial Re-Oxygenation of the Sinnamary, Downstream of Petit-Saut Dam (French Guyana)) *Hydroécol. Appl.*, Tome 9, No. 1-2: 23-56 (in French).

Goubet, A. 1992. Evacuateurs de Crues en marches d'Escalier. (Stepped Spillways) *Jl La Houille Blanche*, No. 2/3: 159-162. Discussion: No. 2/3: 247-248 (in French).

Greenhalgh, F. 1974. *Bath Flood Protection Scheme*. Wessex Water Authority, Bath, UK.

Grinchuk, A.S., Pravdivets, Y.P., & Shekhtman, N.V. 1977. Test of Earth Slope Revetments Permitting Flow of Water at Large Specific Discharges. *Gidrotekhnicheskoe Stroitel'stvo*, No. 4: 22-26 (in Russian). (Translated in Hydrotechnical Construction, 1978, Plenum: 367-373).

Gulliver, J.S., Thene, J.R., & Rindels, A.J. 1990. Indexing Gas Transfer in Self-Aerated Flows. *Jl of Environm. Engrg.*, ASCE, Vol. 116, No. 3: 503-523. Discussion, Vol. 117: 866-869.

Gyr, A. 1989. Structure of Turbulence and Drag Reduction. *IUTAM Symposium*, Zürich. Berlin: Springer-Verlag.

Haddad, A.A. 1998. *Water Flow Over Stepped Spillway*. Masters Thesis, Polytechnic of Bari, Italy.

Hager, W.H. 1983. Hydraulics of Plane Free Overfall. *Jl of Hyd. Engrg.*, ASCE, Vl. 109, No. 12: 1683-1697.

Hager, W.H., Bremen, R., & Kawagoshi, N. 1990. Classical Hydraulic Jump: Length of Roller. *Jl of Hyd. Res.*, IAHR, Vol. 28, No. 5: 591-608.

Hakoishi, N., & Sumi, T. 2000. Hydraulic Design of Nakasujigawa Dam Stepped Spillway. In H.E. Minor & W.H. Hager (ed.) *International Workshop on Hydraulics of Stepped Spillways, Zürich, Switzerland*: 27-34. Balkema.

Hansen, K.D., & McLean, G.G. 1995. Roller Compacted Concrete for Overtopping Protection: an Overview. *Proc. 1st Intl Conf. on Water Resources Eng.*, ASCE, San Antonio (TX). Vol. 2: 1016-1020.

Hartung, F., & Kuros, G.R. 1987. Historische Talsperren im Iran. (Historical Dams in Iran) In *Historische Talsperren*, Vol. 1, K. Wittwert, Stuttgart, Germany: 221-274 (in German).

Hartung, F., & Scheuerlein, H. 1970. Design of Overflow Rockfill Dams. *Proc. 10th ICOLD Congress*, Montréal, Q. 36, R. 35: 587-598.

Haugen, H.L., & Dhanak, A.M. 1966. Momentum Transfer in Turbulent Separated Flow past a Rectangular Cavity. *Jl of Applied Mech.*, Trans. ASME, Sept.: 641-464.

Hauser, G.E., & Morris, D.I. 1995. High-Performance Aerating Weirs for Dissolved Oxygen Improvement. *Proc. Intl Conf. on Hydropower*, Waterpower'95, ASCE, San Francisco, July, Vol. 2: 1695-1705.

Hauser, G.E. et al. 1992. Innovative Reregulation Weirs. *Civil Engineering*, ASCE, Vol. 62, No. 5: 64-66.

He, G., & Zeng, X. 1995. The Integral RCC Dam Design Characteristics and Optimization Design of its Energy Dissipator in Shuidong Hydropower Station. *Proc. Intl Symp on RCC Dams*, Santander, Spain, IECA-CNEGP, Vol. 1: 405-412.

Helot-Lécroart, D. 1985. *Le Domaine de Richelieu à Rueil de 1600 à 1800.* (The Domain of Richelieu at Rueil from 1600 to 1800. Millau: Maury Imprimeur (in French).

Henderson, F.M. 1966. *Open Channel Flow.* New York: MacMillan.

Hewlett, H.W.M., Boorman, L.A., & Bramley, M.E. 1987. Design of Reinforced Grass Waterways. *CIRIA Report No. 116*, London.

Hibbert, M., Kwade, M., & Scharf, R. 1982. Influence of Drag Reducing Additives on the Structure of Turbulence in a Mixing Layer. *Rheology Acta*, Vol. 21: 582-586.

Hinds, J. 1932. 200-Year-Old Masonry Dams in Use in Mexico. *Engineering News-Record*, Vol. 109, Sept. 1: 251-253.

Hinds, J. 1953. Continuous Development of Dams since 1850. *Transactions*, ASCE, Vol. CT: 489-520.

Hodge, A.T. 1992. *Roman Aqueducts & Water Supply*. London: Duckworth.

Hodge, F.W. 1893. Prehistoric Irrigation in Arizona. *The American Anthropologist*, Vol. VI: 323-330.

Hoek, J.P. van der, Kappelhof, J.W.N.M., & Hijnen, W.A.M. 1992. Biological Nitrate Removal from Ground Water by Sulphur/Limestone Denitrification. *Jl Chem. Tech. Biotechnol.*, Vol. 54: 197-200.

Holler, A.G. 1971. The Mechanism Describing Oxygen Transfer from the Atmosphere to Discharge through Hydraulic Structures. *Proc. 14th IAHR Congress*, Vol. 1, Paper A45, Paris, France: 372-382.

Hollingworth, F., & Druyts, F.H.W. 1986. Rollcrete: Some Applications to Dams in South Africa. *Intl Water Power and Dam Construction*, Vol. 38, No. 1, Jan.: 13-16.

Horner, M.W. 1969. *An Analysis of Flow on Cascades of Steps*. Ph.D. thesis, University of Birmingham, UK.

Houston, K.L. 1987. Hydraulic Model Studies of Upper Stillwater Dam Stepped Spillway and Outlet Works. *Report No. REC-ERC-87-6*, US Bureau of Reclamation, Denver (CO).

Howe, J.W., Shieh, G.C., & Obadia, A.O. 1955. Aeration Demand of a Weir Calculated. *Civil Engineering*, ASCE, Vol. 25, No. 5: 289.

Huanxiong, X., & Caiyan, H. 1991. Stability of Protection Gabions on the Downstream Slope of Overflow Rockfill Cofferdams. *Proc. 17th ICOLD Congress*, Vienna, Austria, Q. 67, R. 31: 507-523.

Humber, W. 1876. *Comprehensive Treatise on the Water Supply of Cities and Towns with Numerous Specifications of Existing Waterworks*. London: Crosby Lockwood.

Hungr, O. et al. 1987. Debris Flow Defenses in British Columbia. *Reviews in Engineering Geology*, Geological Society of America, Vol. VII: 20-222.

Hunt, B. 1988. An Asymptotic Solution for Dam Break Floods in Sloping Channels. In P.N. Cheremisinoff, N.P. Cheremisinoff and S.L. Cheng (ed.) *Civil Engineer-*

ing Practice. Vol. 2, Section 1, Chapter 1: 3-11. Lancaster (Penn.): Technomic.

International Organization for Standardization 1979. *Units of Measurements*. ISO Standards Handbook, No. 2, Switzerland.

Ippen, A.T., & Harleman, R.F. 1956. Verification of Theory for Oblique Standing Waves. *Transactions*, ASCE, Vol. 121: 678-694.

Jamme, G. 1974. Travaux Fluviaux. (Fluvial Works) *Eyrolles*, EDF-DRE collection, Paris.

Jansen, R.B. 1983. *Dams and Public Safety*. US Govt. Printing Office, US Bureau of Reclamation, Denver.

Japan Dam Association 1963. *New Horizons - Top Most Dams of the World*. Tokyo, Japan: Hakuodo Print.

Jellicoe, S., & Jellicoe, G. 1971. *Water. The Use of Water in Landscape Architecture*. London: Adam & Charles Black.

Jevdjevich, V., & Levin, L. 1953. Entrainment of Air in flowing Water and Technical Problems connected with it. *Proc. 5th IAHR Congress*, IAHR-ASCE, Minneapolis: 439-454.

Johnson, P.L. 1984. Prediction of Dissolved Gas Transfer in Spillway and Outlet Works Stilling Basin Flows. In W. Brutsaert and G.H. Jirka (ed.) *Proc. 1st Intl Symp on Gas Transfer at Water Surfaces*, Gas Transfer at Water Surfaces: 605-612.

Johnson, N.A., & Templar, J. 1974. Lowering the Warren Dam Spillway. *ANCOLD Bulletin*, No. 44, May: 14-19.

Judd, H.E., & Peterson, D.F. 1969. Hydraulics of Large Bed Element Channels. *UWRL Research Report*, PRWG 17-6, Utah State University, Logan: 115 pages.

Kawase, Y., & Moo-Young, M. 1992. Correlations for Liquid-Phase Mass Transfer Coefficients in Bubble Column Reactors with Newtonian and Non-Newtonian Fluids. *Can. Jl of Chem. Eng.*, Vol. 70, Feb.: 48-54.

Kazemipour, A.K., & Apelt, C.J. 1983. Effects of Irregularity of Form on Energy Losses in Open Channel Flow. *Aust. Civil Engrg Trans.*, I.E.Aust., Vol. CE25: 294-299.

Kelen, N. 1933. *Gewichtsstaumauern und Massive Wehre*. (Gravity Dams and Large Weirs) Berlin: Verlag von Julius Springer (in German).

Kells, J.A. 1993. Spatially Varied Flow over Rockfill Embankments. *Can. Jl of Civ. Engrg.*, Vol. 20: 820-827. Discussion: Vol. 21, No. 1: 161-166.

Kells, J.A. 1995. Comparison of Energy Dissipation between Nappe and Skimming Flow Regimes on Stepped Chutes - Discussion. *Jl of Hyd. Res.*, IAHR, Vol. 33, No. 1: 128-133.

Kistler, A.L., & Tan, F.C. 1967. Some Properties of Turbulent Separated Flows. *Physics of Fluids*, Vol. 10, No. 9, Pt II: S165-S173.

Knauss, J. 1979. Computation of Maximum Discharge at Overflow Rockfill Dams (a comparison of different model test results). *Proc. 13th ICOLD Congress*, New Delhi, Q. 50, R. 9: 143-159.

Knauss, J. 1995. ΤΗΣ ΓΡΙΑΣ ΤΟ ΠΗΔΗΜΑ, der Altweibersprung. Die Rätselhafte Alte Talsperre in der Glosses-Schlucht bei Alyzeia in Akarnanien. *Archäologischer Anzeiger*, Heft 5: 138-162 (in German).

Knight, R.G. 1938. The Subsidence of a Rockfill dam and the Remedial Measures employed at Eildon Reservoir, Australia. *Proc. Instn Civ. Engrs.*, UK, Vol. 8, 1937/38: 111-191. Discussion: Vol. 8: 192-208 & Vol. 9: 451-495.

Knight, D.W., & MacDonald, J.A. 1979. Hydraulic Resistance of Artificial Strip Roughness. *Jl of Hyd. Div.*, ASCE, Vol. 105, No. HY6, June: 675-690.

Kraijenhoff, D.A., & Dommerholt, A. 1977. Brink Depth Method in Rectangular Channel. *Jl of Irrig. and Drain.*, ASCE, Vol. 103, No. IR2: 171-177.

Krest'yaninov, A.M., & Pravdivets, Y.P. 1986. Stepped Spillways for Small Dams. *Gidrotekhnicheskoe Melioratsiya*, No. 8: 27-30 (in Russian).

Kuperman, S.C. 1995. Brazilian Roller Compacted Concrete Dams - Past, Present and Future. *Proc. Intl Symp on RCC Dams*, Santander, Spain, IECA-CNEGP, Vol. 3: 1323-1349.

Labrum, E.A. 1994. *Civil Engineering Heritage. Eastern and Central England*. London: Thomas Telford.

Lavagne, H. 1998. Fonticuli. Deux Fontaines à Escaliers d'eau en Narbonnaise. Beaurepaire (Isère) et Nissan-lez-Ensérune (Hérault). (Fonticuli. Two Water Staircase Fountains in Narbonnaise. Beaurepaire (Isère) et Nissan-lez-Ensérune (Hérault).) *Revue des Etudes Anciennes*, Vol. 100, No. 1-2: 269-287 (in French).

Lawson, J.D. 1987. Protection of Rockfill Dams and Cofferdams against Overflow and Throughflow. The Australian Experience. *Civil Engrg Trans. I.E.Aust.*, Vol. CE29, No. 3: 138-147.

LeBaron Bowen Jr, R. & Albright, F.P. 1958. *Archaelogical Discoveries in South Arabia*. Baltimore: The John Hopkins Press.

Lefranc, M. 1992. Evolution dans l'Exploitation des Evacuateurs de Crues et Rénovations Récentes sur les Barrages d'Electricité de France. (Evolution in Spillway Operations and Recent Refurbishings on the Dams of the French Electricity Commission) *Jl La Houille Blanche*, No 2/3: 163-174. Discussion: 248.

Lejeune, A., & Lejeune, M. 1994. Some Considerations on the Hydraulic Behaviour of Stepped Spillways. *Proc. Intl Conf. Modelling, Testing and Monitoring for Hydo Powerplants*, UNESCO-IAHR, Budapest, July.

Lemperiere, F. 1991. Overspill Rockfill Dams: Conventional and Unconventional Designs. Technology - Costs - Safety. *Proc. 17th ICOLD Congress*, Vienna, Austria, Q. 67, R. 7: 111-127.

Lemperiere, F. 1993. Dams that have Failed by Flooding: an Analysis of 70 Failures. *Intl Water Power and Dam Construction*, Vol. 45, No. 9/10: 19-24.

Lesieur, M. 1994. *La Turbulence*. (The Turbulence) Presses Universitaires de Grenoble, Collection Grenoble Sciences (in French).

Levin, L. 1968. *Formulaire des Conduites Forcées, Oléoducs et Conduits d'Aération*. (Handbook of Pipes, Pipelines and Ventilation Shafts) Paris: Dunod (in French).

Liggett, J.A. 1994. *Fluid Mechanics*. New York: McGraw-Hill.

Lysne, D.K. 1992. New Concepts in Spillway Design. In E. Broch and D.K. Lysne (ed.) *Proc. 2nd Intl Conf. on Hydropower*, LilleHammer, Norway, June: 415-423. Balkema.

McDonald, G., & Yadon, D.M. 1998. Over the Top Concrete. *Civil Engineering*, ASCE, Vol. 68, No. 1, Jan.: 72-75.

McLean, F.G., & Hansen, K.D. 1993. Roller Compacted Concrete for Embankment Overtopping Protection. In L.R. Anderson (ed.) *Proc. Spec. Conf. on Geotechnical Practice in Dam Rehabilitation*, ASCE, Raleigh (NC): 188-209.

Mackey, P.G. 1988. Derwent Valley and Ladybower Dams. Flood Accommodation Works and Raising to meet Probable Maximum Flood (PMF). *Proc. 16th ICOLD Congress*, San Francisco, Vol. IV, Q. 63, R. 84: 1455-1469.

Mamak, W. 1964. *River Regulation*. Warsaw: Arkady.

Manning, R. 1890. *On the Flow of Water in Open Channels and Pipes*. Instn of Civil Engineers of Ireland.

Marchi, E. 1993. On the Free-Overfall. *Jl of Hyd. Res.*, IAHR, Vol. 31, No. 6: 777-790.

Marsh, F.B. 1957. Danger of Fixed Flashboards shown by Flood at Croton Dam. *Civil Engineering*, ASCE, Vol. 64: 408-409.

Mastropietro, M.A. 1968. Effects of Dam Reaeration on Waste Assimilation Capacities of the Mohawk River. *Proc. 23rd Industrial Waste Conf.*, Engineering Extension Series, No. 132, Part 2, Purdue University: 754-765.

Matos, J. 2000. Hydraulic Design of Stepped Spillways over RCC Dams. In H.E. Minor & W.H. Hager (ed.) *International Workshop on Hydraulics of Stepped Spillways, Zürich, Switzerland*: 187-194. Balkema.

Matos, J. et al. 1999. Characteristic Depth and Pressure Profiles in Skimming Flow over Stepped Spillways. *Proc. 28th IAHR Congress*, Graz, Austria, Session B14, 6 pages (CD-ROM).

Maull, D.J., & East, L.F. 1963. Three-Dimensional Flow in Cavities. *Jl of Fluid Mech.*, Vol. 16: 620-632.

May, R.W.P. 1987. Cavitation in Hydraulic Structures: Occurrence and Prevention. *Hydraulics Research Report*, No. SR 79, Wallingford, UK.

May, R.W.P., & Willoughby, I.R. 1991. Impact Pressures in Plunge Basins Due to Vertical Falling Jets. *Hydraulics Research Report*, No. SR 242, Wallingford, UK.

Metropolitan Water 1980. Investigation into Spillway Discharge Noise at Avon Dam. *ANCOLD Bulletin*, No. 57, Aug.: 31-36.

Michels, V. 1966. Some Recent Spillway Designs. *ANCOLD Bulletin*, No. 19: 15-24.

Mickelson, H., & Moorwood, R.P. 1984. The Maranoa River Weir - Design for a Typical Inland Stream. *ANCOLD Bulletin*, No. 67: 37-42. Discussion: No. 67: 42-43.

Miller, A.P., Pravdivets, Y.P., & Salov, V.A. 1987. Earth Overflow Dam. *Gidrotekhnicheskoe Stroitel'stvo*, No. 8: 49-52 (in Russian). (Translated in Hydrotechnical Construction (1988), Plenum: 503-507).

Ministerio de Obras Públicas, Transportes y Medio Ambiante 1993. *Obras Hidráulicas en América Colonial*. Madrid: Cehopu, Cedex & Tabapress (in Spanish).

Montes, J.S. 1992. A Potential Flow Solution for the Free Overfall. *Proc. Intn. Civ. Engrs Wat. Marit. & Energy*, Vol. 96, Dec.: 259-266.

Montes, J.S. 1992b. Curvature Analysis of Spillway Profiles. *Proc. 11th Australasian Fluid Mechanics Conference AFMC*, Vol. 2, Paper 7E-7, Hobart: 941-944.

Montes, J.S. 1994. Private Communication.

Montes, J.S. 1998. *Hydraulics of Open Channel Flow*. New York: ASCE Press.

Moore, W.L. 1943. Energy Loss at the Base of a Free Overfall. *Transactions,* ASCE, Vol. 108: 1343-1360. Discussion: Vol. 108: 1361-1392.

Morelli, C. 1971. The International Gravity Standardization Net 1971 (I.G.S.N.71). *Bureau Central de l'Association Internationale de Géodésie*, Paris.

Murray, W.M. 1984. The Ancient Dam of the Mytikas Valley. *American Jl of Archaeology*, Vol. 88: 195-203 & Plates 32-33.

Nakao, T. 1993. Research and Practice of Hydraulic Engineering in Japan - Sabo (Erosion Control). *Jl of Hydroscience and Hyd. Eng. in Japan*, No. SI-4 River Engineering: 175-229.

Nakasone, H. 1987. Study of Aeration at Weirs and Cascades. *Jl of Envir. Engrg.*, ASCE, Vol. 113, No. 1: 64-81.

Needham, J. 1971. *Science and Civilisation in China*. Cambridge University Press, London, UK: Vol. 4, Part 3, Section 28.

Netzer, E. 1983. Water Channels and a Royal Estate from the Late Hellenistic Period in the Western Plains of Jericho. *Symp. on Historical Water Development Projects in the Eastern Mediterranian*, Jerusalem, Israël (Mitteilungen, Institut für Wasserbau der Technischen Universität Braunschweig, Germany, No. 82, 1984).

Nielsen, P. 1993. Turbulence Effects on the Settling of Suspended Particles. *Jl of Sedimemtary Petrology*, Vol. 63, No. 5: 835-838.

Nielsen, P., & Zeng, Q.C. 1998. Turbulence Effects on the Settling of Heavy Particles and the Rise of Bubbles. *Proc. 3rd Intl Conf. Multiphase Flow*, ICMF'98, Lyon, France.

Nikuradse, J. 1933. Strömungsgesetze in rauhen Rohren. (Laws of Turbulent Pipe Flow in Rough Pipes) *VDI-Forschungsheft*, No. 361 (in German) (Translated in NACA Tech. Memo. No. 1292, 1950).

Noetzli, F.A. 1931. High Dams - A Symposium - Discussion. *Transactions*, ASCE, Vol. 95: 171-174.

Noori, B.M.A. 1984. Form Drag Resistance of Two Dimensional Stepped Steep Open Channels. In K.V.H. Smith (ed.) *Proc. 1st Intl Conf. on Hyd. Design in Water Resources Engineering*, Channels and Channel Control Structures, Southampton, UK. Berlin: Springer-Verlag.

Ohtsu, I., & Yasuda, Y. 1997. Characteristics of Flow Conditions on Stepped Channels. *Proc. 27th IAHR Biennal Congress*, San Francisco, Theme D: 583-588.

Ohtsu, I, Yasuda, Y., & Takahashi, M. 2000. Characteristics of Skimming Flow over Stepped Spillways. Discussion. *Jl of Hyd. Engrg.*, ASCE, Vol. 126, No. 11: 869-871.

Olivier, H. 1967. Through and Overflow Rockfill Dams - New Design Techniques. *Proc. Instn. Civil Eng.*, March, 36: 433-471. Discussion, 36: 855-888.

Osaka, H., & Mochizuki, S. 1991. Turbulent Drag Reduction of a D-Type Rough Wall Boundary Layer with Longitudinal Thin Ribs Placed Within the Transverse Grooves. In K.S. Choi (ed.) *Recent Developments in Turbulence Managements*: 163-180. Dordrecht: Kluwer.

Outland, C.F. 1963. *Man-Made Disaster: the Story of the St. Francis Dam*. Glendale (Cal): Arthur H. Clark Comp.

Pariset, E. 1955. Etude sur la Vibration des lames Déversantes. (Study of the Vibration of Free-Falling Nappes) *Proc. 6th IAHR Congress*, The Hague, Vol. 3, paper C21: 1-15.

Peterka, A.J. 1953. The Effect of Entrained Air on Cavitation Pitting. *Joint Meeting Paper*, IAHR/ASCE, Minneapolis, Aug. 1953: 507-518.

Petrikat, K. 1958. Vibration Tests on Weirs and Bottom Gates. *Water Power*: 52-57, 99-104, 147-149 & 190-197.

Petrikat, K. 1978. Model Tests on Weirs, Bottom Outlet Gates, Lock Gates and Harbours Moles. *M.A.N. Technical Bulletin*, Nurnberg.

Peyras, L., Royet, P., & Degoutte, G. 1991. Ecoulement et Dissipation sur les Déversoirs en Gradins de Gabions. (Flows and Dissipation of Energy on Gabion Weirs) *Jl La Houille Blanche*, No. 1: 37-47 (in French).

Peyras, L., Royet, P., & Degoutte, G. 1992. Flow and Energy Dissipation over Stepped Gabion Weirs. *Jl of Hyd. Engrg.*, ASCE, Vol. 118, No. 5: 707-717.

Pinheiro, A.N., & Fael, C.S. 2000. Nappe Flow in Stepped Channels - Occurrence and Energy Dissipation. In H.E. Minor & W.H. Hager (ed.) *International Workshop on Hydraulics of Stepped Spillways, Zürich, Switzerland*: 119-126. Balkema.

Plumptre, G. 1993. *The Water Garden*. London: Thames & Hudson.

Post, G. et al. 1987. Spillways for Dams. *ICOLD Bulletin*, No. 58.

Prandtl, L. 1925. Über die ausgebildete Turbulenz. (On Fully Developed Turbulence) *Z.A.M.M.*, Vol. 5: 136-139 (in German).

Pravdivets, Y.P. 1992. Stepped Spillways in World and Domestic Hydraulic Engineering. *Gidrotekhnicheskoe Stroitel'stvo*, No. 10, Oct.: 28-32 (in Russian). (Translated in Hydrotechnical Construction (1993) Vol. 27, No. 10, Plenum: 589-594).

Pravdivets, Y.P., & Bramley, M.E. 1989. Stepped Protection Blocks for Dam Spillways. *Intl Water Power and Dam Construction*, Vol. 41, No. 7, July: 49-56.

Queensland Water Resources 1988. Construction of Bucca Weir. *Water Resources Commission Report*, Contract No. 2541, Post-Construction Report, April, Brisbane, Australia.

Quintela, A.C., Cardoso, J.L., & Mascarenhas, J.M. 1987. Roman Dams in Southern Portugal *Intl Water Power and Dam Construction*, Vol. 39, No. 5: 38-40 & 70.

Raikes, R.L. 1964-1965. The Ancient Garbalands of Baluchistan. *East and West*: 26-35.

Rajaratnam, N. 1990. Skimming Flow in Stepped Spillways. *Jl of Hyd. Engrg.*, ASCE, Vol. 116, No. 4: 587-591. Discussion: Vol. 118, No. 1: 111-114.

Rajaratnam, N., & Chamani, M.R. 1995. Energy Loss at Drops. *Jl of Hyd. Res.*, IAHR, Vol. 33, No. 3: 373-384. Discussion: *Jl of Hyd. Res.*, IAHR, Vol. 34, No. 2: 273-278.

Rajaratnam, N., & Muralidhar, D. 1968. Characteristics of the Rectangular Free Overfall. *Jl of Hyd. Res.*, IAHR, Vol. 6, No. 3: 233-258.

Rakob, F. 1974. Das Quellenheigtum in Zaghouan und die Römische Wasserleitung nach Karthago. *Mitt. des Deutschen Archaeologischen Instituts Roemische Abteilung*, Vol. 81: 41-89, Plates 21-76 & Maps (in German).

Rand, W. 1955. Flow Geometry at Straight Drop Spillways. *Proceedings*, ASCE, Vol. 81, No. 791, Sept.: 1-13.

Redeker, M. 2000. Wave Phenomena in the Cascade Type Spillway of the Sorpe Dam. *Interim Report*, Ruhrverband, Dept of Dams and Water Courses, Aug.

Resch, F.J., Leutheusser, H.J., & Alemu, S. 1974. Bubbly Two-Phase Flow in Hydraulic Jump. *Jl of Hyd. Div.*, ASCE, Vol. 100, No. HY1: 137-149.

Rhone, T.J. 1990. Development of Hydraulic Structures. In A.M. Alsaffar (ed.) *50th Anniversary of the Hydraulics Division 1938-1988*, ASCE: 132-147. New York.

Richter, J.P. 1939. *The Literary Works of Leonardo da Vinci*. London: Oxford University Press. 2 volumes.

Riediger, S. 1989. Influence of Drag Reducing Additives on a Plane Mixing Layer. *Proc. 4th Intl Conf. on Drag Reduction*, Davos, Switzerland No. 10.4: 303-310. Chichester: Ellis Horwood.

Rindels, A.J., & Gulliver, J.S. 1986. Air-Water Oxygen Transfer at Spillways and Hydraulic Jumps. *Proc. Water Forum '86: World Water Issues in Evolution*, ASCE, New York: 1041-1048.

Rindels, A.J., & Gulliver, J.S. 1989. Measurements of Oxygen Transfer at Spillways and Overfalls. *St Anthony Falls Hydraulic Laboratory Project Report*, No. 266, University of Minnesota, Minneapolis (MN).

Rizk, T.A., & Hauser, G.E. 1993. Chatuge Aerating Infuser Physical Model Study. *Report No. WR28-1-17-101*, Tennessee Valley Authority, Norris (TN) June.

Robinson, K.M. 1989. Hydraulic Stresses on an Overfall Boundary. *Transactions*, ASAE, Vol. 32, No. 4: 1269-1274.

Robison, R. 1994. Chicago's Waterfalls. *Civil Engineering*, ASCE, Vol. 64, No. 7: 36-39.

Rolt, L.T.C. 1973. *From Sea to Sea - The Canal du Midi*. London: Allen Lane.

Rouse, H. 1936. Discharge Characteristics of the Free Overfall. *Civil Engineering*, Vol. 6, April: 257.

Rouse, H. 1938. *Fluid Mechanics for Hydraulic Engineers*. New York: McGraw-Hill.

Rouse, H. 1943. Energy Loss at the Base of a Free Overfall - Discussion. *Transactions*, ASCE, Vol. 108: 1383-1387.

Ru, S.X. et al. 1994. Stepped Dissipator on Spillway Face. *Proc. 9th APD-IAHR Congress*, Singapore, Vol. 2: 193-200.

Ruff, J.F., & Frizell, K.H. 1994. Air Concentration Measurements in Highly-Turbulent Flow on a Steeply-Sloping Chute. *Proc. Hydraulic Engineering Conf.*, ASCE, Buffalo, USA, Vol. 2: 999-1003.

Russell, S.O., & Sheehan, G.J. 1974. Effect of Entrained Air on Cavitation Damage. *Can. Jl of Civil Engrg.*, Vol. 1: 97-107.

Saladin, H. 1886. Rapport sur la Mission Faite en Tunisie de Novembre 1882 à Avril 1883. *Archives des Missions Scientifiques et Littéraires*, Ministère de l'Instruction Publique, France, Serties 2, Vol. 13: 1-225 (in French).

Sams, E.W. 1952. Experimental Investigation of Average Heat-Transfer and Friction

Coefficients for Air Flowing in Circular Tubes Having Square-Thread-Type Roughness. *NACA report*, RM E52D17, USA.
Sanchez-Bribiesca, J.L., & Gonzalez-Villareal, F. 1996. Spilling Floods Cost effectively. *Intl Water Power and Dam Construction*, Vol. 48, No. 5: 16-20.
Sankey, R.H. 1867. Irrigation in Mysore. *Professional Papers on Indian Engineering*, Vol. IV, No. CL: 212.
Sankey, R.E. 1871. Report on the Coliban and Geelong Water Schemes of Water Supply. *Report*, presented to both Houses of Parliament, Victoria (Australia), 11 November.
Schleicher, F. 1942. *Taschenbuch für Bauingenieure*. (Handbook for Civil Engineers). Berlin: publisher unknown (in German).
Schlichting, H. 1979. *Boundary Layer Theory*. New York: McGraw-Hill, 7th edition.
Schnitter, N.J. 1991. Roman Dams and Weirs in the Iberian Peninsula. *8th Intl Symp. on History of Hydraulics*, Merida, Spain (Mitteilungen, Inst. für Wasserbau der Technische Univ. Braunschweig, Germany, Col. 117, 1992: 161-177).
Schnitter, N.J. 1994. *A History of Dams: the Useful Pyramids*. Rotterdam: Balkema.
Schodek, D.L. 1987. *Landmarks in American Civil Engineering*. Cambridge (MA): MIT Press.
Schoklitsch, A. 1937. Hydraulic structures. *ASME*, New York, USA, 2 volumes.
Schrader, E., & Stefanakos, J. 1995. Roller Compacted Concrete Cavitation and Erosion Resistance. *Proc. Intl Symp on RCC Dams*, Santander, Spain, IECA-CNEGP, Vol. 2: 1175-1188.
Schuyler, J.D. 1909. Reservoirs for Irrigation, Water-Power and Domestic Water Supply. New York: John Wiley & Sons, 2nd edition.
Schwartz, H.I. 1964a. Projected Nappes Subject ot Harmonic Pressures. *Proc. Instn. Civ. Engrs. (London)*, Vol. 28: 313-36.
Schwartz, H.I. 1964b. Nappe Oscillation. *Jl of Hyd. Div.*, ASCE, Vol. 90, No. HY6: 129-143. Discussion: Vol. 91, No. HY3: 389-392.
Schwartz, H.I., & Nutt, L.P. 1963. Projected Nappes Subject to Transverse Pressure. *Jl of Hyd. Div.*, ASCE, Vol. 89, July: 97-104.
Shvainshtein, A.M. 1999. Stepped Spillways and Energy Dissipation. *Gidrotekhnicheskoe Stroitel'stvo*, No. 5: 15-21 (in Russian). (Also Hydrotechnical Construction, Vol. 3, No. 5, 1999: 275-282.)
Simons, D.B., Chen, Y.H., & Swenson, L.J. 1984. Hydraulic Test to Develop Design Criteria for the Use of Reno Mattresses. *Report*, Civil Engineering Dept., Colorado State University, Fort Collins.
Smith, H.A. Jr. 1973. A Detrimental Effect of Dams on Environment: Nitrogen Supersaturation. *Proc. 11th ICOLD Congress*, Madrid, Q. 40, R. 17: 237-253.
Smith, N. 1971. *A History of Dams*. London: The Chaucer Press.
Soil and Water 1992. *Soil and Water Newsletter*, Council of Agriculture, Executive Yuan, Taiwan, No. 13, Sept..
Sorensen, R.M. 1985. Stepped Spillway Hydraulic Model Investigation. *Jl of Hyd. Engrg.*, ASCE, Vol. 111, No. 12: 1461-1472. Discussion: Vol. 113, No. 8: 1095-1097.
Soyer, G. 1992. Evacuateur de Crues du Barrage de Pinet sur le Tarn. (Spillway of the Pinet Dam on the Tarn River) *Jl La Houille Blanche*, No 2/3: 195-196. Discussion: 248.
Stein, A. 1940. Surveys on the Roman Frontier in Iraq and Trans-Jordan. *The Geographical Journal*, Vol. XCV: 428-438.
Stephenson, D. 1979a. Gabion Energy Dissipators. *Proc. 13th ICOLD Congress*, New Delhi, Q. 50, R. 3: 33-43.
Stephenson, D. 1979b. *Rockfill in Hydraulic Engineering*. Amsterdam: Elsevier.
Stephenson, D. 1980. The Stability of Gabion Weirs. *Intl Water Power and Dam Construction*, Vol. 31, No. 4: 24-28.

Stephenson, D. 1988. Stepped Energy Dissipators. *Proc. Intl Symp. on Hydraulics for High Dams*, IAHR, Beijing, China: 1228-12-35.

Stephenson, D. 1991. Energy Dissipation down Stepped Spillways. *Intl Water Power and Dam Construction*, Vol. 43, No. 9, Sept.: 27-30.

Straub, L.G., & Anderson, A.G. 1958. Experiments on Self-Aerated Flow in Open Channels. *Jl of Hyd. Div.*, Proc. ASCE, Vol. 84, No. HY7, paper 1890: 1890-1 to 1890-35.

Streeter, V.L., & Wylie, E.B. 1981. *Fluid Mechanics*. Singapore: McGraw-Hill. 1st SI Metric edition.

Tantirige, S.C., Itibarne, A.P., Ojhas, M., & Trass, O. 1994. The Turbulent Boundary Layer over Single V-Shaped Grooves. *Intl Jl of Heat and Mass Transfer*, Vol. 37, No. 5: 2261-2271.

Thomas, H.H. 1976. *The Engineering of Large Dams*. London: John Wiley & Sons. 2 volumes.

Thompson, R.C., & Hutchinson, R.W. 1929. The Excavations of the Temple of Nabû at Nineveh. *Archaeologia (London)*, Vol. 79: 103-148.

Thorne, C.R., & Zevenbergen, L.W. 1985. Estimating Velocity in Mountain Rivers. *Jl of Hyd. Engrg.*, ASCE, Vol. 111, No. 4: 612-624.

Toms, E.A., & Gauthiere, J. 1997. Improved method for Upgrading Old Spillways. *Public Works*, Vol. 128, No. 1: 38-40.

Toombes, L., & Chanson, H. 2000. Air-Water Flow and Gas Transfer at Aeration Cascades: a Comparative Study of Smooth and Stepped Chutes. In H.E. Minor & W.H. Hager (ed.) *International Workshop on Hydraulics of Stepped Spillways, Zürich, Switzerland*: 77-84. Balkema.

Toso, J.W., & Bowers, C.E. 1988. Extreme Pressures in Hydraulic-Jump Stilling Basins. *Jl of Hyd. Engrg.*, ASCE, Vol. 114, No. 8: 829-843.

Townes, H.W., & Sabersky, R.H. 1966. Experiments on the Flow over a Rough Surface. *Intl Jl of Heat and Mass Transfer*, Vol. 9: 729-738.

Tozzi, M.J. 1992. *Caracterização/Comportamento de Escoamentos em Vertedouros com Paramento em Degraus*. (Hydraulics of Stepped Spillways) Ph.D. thesis, University of Sao Paulo, Brazil (in Portuguese).

Tozzi, M., Taniguchi, E., & Ota, J. 1998. Air Concentration in Flows over Stepped Spillways. *Proc. 1998 ASME Fluids Eng. Conf.*, FEDSM'98, Washington DC, USA, June 21-25, Paper FEDSM98-5053 (CD-ROM).

US Department of the Interior 1965. *Design of Small Dams*. Bureau of Reclamation, Denver (CO), USA, 1st edition, 3rd printing.

Utrillas, J.L., Gamo, A., & Soriano, A. 1992. Reconstruction of the Tous Dam. *Intl Water Power and Dam Construction*, Vol. 44, No. 9, Sept.: 55-65.

Vandine, D.F. 1985. Debris Flows and Debris Torrents in Southern Canada Cordillera. *Can Jl of Geotech.*, Vol. 22: 44-68.

Vasiliev, O.F., & Bukreyev, V.I. 1967. Statistical Characteristics of Pressure Fluctuations in the region of Hydraulic jump. *Proc. 12th IAHR Congress*, Fort-Collins, USA, Vol. 2, paper B1: 1-8.

Villanueva, A.V. 1993. El Abastecimiento de Agua a la Cordoba Romana. I: El Acueducto de Valdepuentes. (The Water Supply of the Roman Cordoba. I: Aqueduct of Valdepuentes) *Monografias No. 197*, Universidad de Cordoba, Servicio de Publicaciones, Cordoba (in Spanish).

Vischer, D., & Hager, W.H. 1998. *Dam Hydraulics*. Chichester: John Wiley.

Vita-Finzi, C. 1961. Roman Dams in Tripolitania. *Antiquity*, Vol. 35: 14-20.

Vittal, N., & Porey, P.D. 1987. Design of Cascade Stilling Basins for High Dam Spillways. *Jl of Hyd. Engrg.*, ASCE, Vol. 113, No. 2: 225-237.

Volkart, P. 1980. The Mechanism of Air Bubble Entrainment in Self-Aerated Flow. *Intl Jl of Multiphase Flow*, Vol. 6: 411-423.

Wahrheit-Lensing, A. 1996. *Selbstbeluftung und Energieumwandlung beim Anfluss uber Treppenformige Entlastungsanlagen*. (Self-Aeration and Energy Dissipation in Flow over Stepped Spillways) Ph.D. thesis, Karlsruhe University, Germany (in German).

Wark, R.J. et al. 1998. Aspects of the Design and Construction of North Dandalup Dam. *ANCOLD Bulletin*, No. 108, April: 39-56.

Wark, R.J., Kerby, N.E., & Mann, G.B. 1991. New Victoria Dam Project. *ANCOLD Bulletin*, No. 88, Aug.: 14-32.

Wark, R.J., & Mann, G.B. 1992. Design and Construction Aspects of New Victoria Dam. *Intl Water Power and Dam Construction*, Vol. 44, No. 2, Feb.: 24-29.

Water Power 1992. Five TBMs Break Records at Svartisen. *Intl Water Power and Dam Construction*, Vol. 44, No. 6, June: 23-27.

Wegmann, E. 1907. The Design of the New Croton Dam. *Transactions*, ASCE, Vol. LXVIII, No. 1047: 398-457.

Wegmann, E. 1911. *The Design and Construction of Dams*. New York: John Wiley & Sons, 6th edition.

Wegmann, E. 1922. *The Design and Construction of Dams*. New York: John Wiley & Sons, 7th edition.

Wei, C.Y., & de Fazio, F.G. 1982. Simulation of Free Jet Trajectories for Design of Aeration Devices on Hydraulic Structures. *Proc. 4th Intl. Conf. on Finite Elements in Water Resources*, Hannover, June: 17/45-17/55.

Weiss, R.F. 1970. The Solubility of Nitrogen, Oxygen and Argon in Water and Seawater. *Deep-Sea Research*, Vol. 17: 721-735.

White, M.P. 1943. Energy Loss at the Base of a Free Overfall - Discussion. *Transactions*, ASCE, Vol. 108: 1361-1364.

Wiley, A.J. 1931. High Dams - A Symposium - Closure. *Transactions*, ASCE, Vol. 95: 212-217.

Wilhelms, S.C. et al. 1981. Gas Transfer in Hydraulic Jumps. *Technical Report E-81-10*, US Army Engineer Waterways Experiment Station, CE, Vicksburg (Miss).

Wilhelms, S.C., & Gulliver, J.S. 1989. Self-Aerating Spillway Flow. In M.A. Ports (ed.) *National Conference on Hydraulic Engineering*, ASCE, New Orleans: 881-533.

Witham, G.B. 1955. The Effects of Hydraulic Resistance in the Dam-Break Problem. *Proc. Roy. Soc. of London*, Ser. A, Vol. 227: 399-407.

Wood, J.N. 1933. Features of Construction of the Ryburn Dam. *Trans. Instn. Water Engrs*, UK, Vol. 38: 28-67.

Wood, I.R. 1983. Uniform Region of Self-Aerated Flow. *Jl Hyd. Eng.*, ASCE, Vol. 109, No. 3: 447-461.

Wood, I.R. 1985. Air Water Flows. *Proc. 21st IAHR Congress*, Melbourne, Australia, Keynote address: 18-29.

Wood, I.R. 1991. Air Entrainment in Free-Surface Flows. *IAHR Hydraulic Structures Design Manual No. 4*, Hydraulic Design Considerations. Rotterdam: Balkema.

Wyganski, I., & Fiedler, H.E. 1970. The Two-Dimensional Mixing Region. *Jl of Fluid Mech.*, Vol. 41, Part 2: 327-361.

Yasuda, Y., & Ohtsu, I. 1999. Flow Resistance of Skimming Flow in Stepped Channels. *Proc. 28th IAHR Congress*, Graz, Austria, Session B14 (CD-ROM).

Yuan, Mingshun 1990. On Numerical Simulations of Aerated Jet Flows about aerators on Spillway. *Proc. 7th Congress of APD-IAHR*, Vol. II, Beijing, Nov.: 166-170.

Zhou, Hui 1996. Hydraulic Performances of Skimming Flow over Stepped Spillway. *Proc. 10th APD-IAHR Congress*, Langkawi Island, Malaysia, Aug.: 1-8.

Abbreviations of journals and institutions

AFMC	Australasian Fluid Mechanics Conference
AGU	American Geophysical Union (USA)
AIAA Jl	Journal of the American Institute of Aeronautics and Astronautics (USA)
ANCOLD	Australian Committee on Large Dams
Ann. Chim. Phys.	Annales de Chimie et Physique, Paris (France)
ANSSR	Academy of Sciences of the USSR, Moscow
APHA	American Public Health Association
ARC	Aeronautical Research Council (UK) Australian Research Council
ARC RM	Aeronautical Research Council Reports and Memoranda
ARC CP	Aeronautical Research Council Current Papers
ASAE	American Society of Agricultural Engineers
ASCE	American Society of Civil Engineers
ASME	American Society of Mechanical Engineers
AVA	Aerodynamische Versuchanstalt, Göttingen (Germany)
BHRA	British Hydromechanics Research Association (BHRA Fluid Engineering)
BSI	British Standards Institution, London
CIRIA	Construction Industry Research and Information Association, UK
EDF	Electricité de France
Ergeb. AVA Göttingen	Ergebnisse Aerodynamische Versuchanstalt, Göttingen (Germany)
Forsch. Ing. Wes.	Forschung auf dem Gebiete des Ingenieur-Wesens (Germany)
Forschunsheft	Research supplement to Forsch. Ing. Wes. (Germany)
Gid. Stroit.	Gidrotekhnicheskoe Stroitel'stvo (Russia) (translated in *Hydrotechnical Construction*)
IAHR	International Association for Hydraulic Research
IAWQ	International Association for Water Quality
ICOLD	International Committee on Large Dams
IEAust.	Institution of Engineers, Australia
IIHR	Iowa Institute of Hydraulic Research, Iowa City (USA)
Ing. Arch.	Ingenieur-Archiv (Germany)
JAS	Journal of Aeronautical Sciences (USA) (replaced by JASS in 1959)
JASS	Journal of AeroSpace Sciences (USA) (replaced by AIAA Jl in 1963)
Jl Fluid Mech.	Journal of Fluid Mechanics (Cambridge, UK)
Jl Roy. Aero. Soc.	Journal of the Royal Aeronautical Society, London (UK)
JSCE	Japanese Society of Civil Engineers
JSME	Japanese Society of Mechanical Engineers
Luftfahrt-Forsch.	Luftfahrt-Forschung (Germany)
NACA	National Advisory Committee on Aeronautics (USA)

NACA Rep.	NACA Reports (USA)
NACA TM	NACA Technical Memoranda (USA)
NACA TN	NACA Technical Notes (USA)
NASA	National Aeronautics and Space Administration (USA)
NBS	National Bureau of Standards (USA)
ONERA	Office National d'Etudes et de Recherches Aérospatiales (France)
Phil. Mag.	Philosophical Magazine
Phil. Trans. R. Soc. Lond.	Philosophical Transactions of the Royal Society of London (UK)
Proc. Cambridge Phil. Soc.	Proceedings of the Cambridge Philosophical Society (UK)
Proc. Instn. Civ. Engrs.	Proceedings of the Institution of Civil Engineers (UK)
Proc. Roy. Soc.	Proceedings of the Royal Society, London (UK)
Prog. Aero. Sci.	Progress in Aerospace Sciences
Proc. Cambridge Phil. Soc.	Transactions of the Cambridge Philosophical Society (UK)
SAF	St Anthony Falls Hydraulic Laboratory, Minneapolis (USA)
SHF	Société Hydrotechnique de France
SIA	Société des Ingénieurs et Architectes (Switzerland)
Trans. Soc. Nav. Arch. Mar. Eng.	Transactions of the Society of Naval Architects and Marine Engineers
USBR	United States Bureau of Reclamation, Department of the Interior
VDI Forsch.	Verein Deutsche Ingenieure Forschungsheft (Germany)
Wat. Res. Res.	Water Resources Research Journal
WES	US Army Engineer Waterways Experiment Station
Z.A.M.M.	Zeitschrift für angewandete Mathematik und Mechanik (Germany)
Z.A.M.P.	Zeitschrift für angewandete Mathematik und Physik (Germany)
Z. Ver. Deut. Ingr.	Zeitschrift Verein Deutsche Ingenieure (Germany)

Common abbreviations

Conf.	Conference
Cong.	Congress
Intl.	International
Jl	Journal
Mitt.	Mitteilungen
Ph.D.	Doctor of Philosophy
Proc.	Proceedings
Symp.	Symposium
Trans.	Transactions

INDEX

Prototype spillways are listed per country. For example, Dartmouth dam (Australia) is listed under : Australia, Dartmouth dam.

Note : (*f*) figure or photograph

Bold major section

T